I0146428

Pacific Northwest Ferns and Their Allies

Published in association with the
University of British Columbia by
University of Toronto Press

Pacific Northwest Ferns and Their Allies

Thomas M. C. Taylor

© University of Toronto Press 1970
Reprinted in 2018
Printed in Canada by
University of Toronto Press,
Toronto and Buffalo
ISBN 978-1-4875-8256-2 (paper)
ISBN 0-8020-5227-4

To B.D.T. with love and great admiration

Contents

Preface

The present work is the outcome of many years of interest in the ferns
and their allies of the Pacific Northwest. Since the publication of Frye's
Ferns of the Northwest in 1934, nothing comprehensive for this area
has appeared. It is true that Frye's book was reprinted in 1956, but as no
revisions were made we still lack a modern summary of the state of our
knowledge of this group of plants. *Pacific Northwest Ferns and Their
Allies* is an attempt to fill a need expressed by both professional botanists
and interested amateurs.

The maps showing the distribution of the various species are based on
herbarium specimens mostly examined by the writer, but supplemented
in some cases by verification of records by the curators of herbaria other
than those visited. Specimens have been studied in the following herbaria
and the writer's sincere thanks are extended to their curators for their
co-operation and assistance: Dudley Herbarium, Stanford University;
University of California, Berkeley (general collection as well as the
McCabe collection from British Columbia); Jepson Herbarium, Berkeley;
California Academy of Sciences Herbarium, San Francisco; University of
Oregon; Oregon State University; the Peck Herbarium, Willamette University; University of Washington; Provincial Museum, Victoria, British
Columbia, and the University of British Columbia. A number of persons
have been kind enough to give freely of advice and information; among
these it is a pleasure to record the names of Mr. J. A. Calder, Mr. C. V.
Morton, Dr. W. H. Wagner, Jr., Dr. R. M. Tryon, and Dr. J. R. Slater. In
particular the writer is indebted to Dr. R. Holm, Director, Division of
Systematic Biology, Stanford University, for providing working space
and other facilities during a period of five months, and to Dr. J. H. Thomas,
curator of the Dudley Herbarium, for his co-operation and innumerable
kindnesses.

The illustrations were drawn by Katherine Jones except for that of
Asplenium septentrionale which was kindly prepared by Dr. T. C.
Brayshaw.

In presenting this book to the public the author is keenly aware of a
number of inadequacies; without doubt there are also errors of which he
is ignorant. He hopes that such will be brought to his attention so that
they may be corrected subsequently. It is his hope that professional botanists will find some value in the work and that amateur readers may become
interested in learning more about this special segment of the plant
kingdom.

T. M. C. TAYLOR

Pacific Northwest Ferns and Their Allies

Introduction

The ferns and their allies are usually grouped together by botanists as pteridophytes, vascular plants which reproduce by spores rather than seeds. This and the general inability to increase the diameter of their stems each year are about the only features that hold them together as a division of the plant kingdom. They are a small group of about 10,000 species in contrast to about 275,000 species of seed plants; what they lack in numbers, however, they make up for by the interest found in their antiquity. Their earliest known fossil ancestor was thriving in Siberia about 500 million years ago; by the Devonian period of approximately 400 million years ago, natural selection had established several types of organisms. Some of these apparently could not keep pace with changing conditions and have now been extinct for millions of years; others, though, have persisted in one form or another to our own times. It is the modern representatives of these ancient lines that today we classify as pteridophytes. We recognize them as being of noble lineage for their ancestors in the remote Carboniferous period of 300 million years ago were the dominant type of plant found in the swamp-forests of that period. Although contemporary pteridophytes may have relatively small economic significance, we are indebted to their Carboniferous ancestors for most of our coal deposits and all the economic benefits that have come from them.

It is quite possible that there are actually more species of pteridophytes alive at the present time than when they were at their zenith. Nevertheless, they have been surpassed by the younger and more vigorous seed plants. The pteridophyte lines that have survived are well adapted to special ecological niches, but their very specialization makes them vulnerable to any marked changes in their environment. They may be considered as one of Nature's reasonably successful experiments, more successful than some, less so than others. What have survived are a number of distinct lines without common ancestry for scores of millions of years. As a result, the pteridophytes show great variety in shape and size, in basic organization, and in the conditions under which they thrive.

The grouping of these various lines of descent is primarily into orders but, as many of the orders contain only a single family, it is simplest for us in the present work to concern ourselves only with families, the first subdivision of an order. Families are divided into genera and species. The plan of the book is to provide keys to the families and to the genera within each family, and finally to provide each genus with a key to help in the determination of its species. A key is a device for quickly placing a plant in its proper taxonomic category. It will be noticed that the keys take the form of a number of indented lines, each preceded by a numeral. At each step in the key are two statements, one of which is true and the other false for the specimen in question. The first decision to be made is whether the plant has the characters mentioned following the first "1," or those after the second "1." One then goes on to the next numeral under the true statement and so on until finally a statement is reached which ends in a name, of a family, genus, or species as the case may be. It is wise at this point to check the description given on the appropriate page to make sure that there has been no error in going through the key. If the specimen does not agree with the description and illustration it is necessary to go back through the key to find where the faulty decision was made and then proceed on a new track. To help the beginner understand new technical words both in the keys and the descriptions, a glossary is provided.

The arrangement of families, genera, and species in a book of this sort is always a matter of concern as any sequence will raise objections in the minds of some. It is doubtful that any modern taxonomist interested in matters of evolutionary relationship would be satisfied with the time-honoured sequences of families and genera. In fact certain outstanding botanical philosophers have indicated their views by suggesting radical re-arrangements and new groupings within certain families and genera. At the present time, however, the measure of disagreement is so great that no useful purpose would be served by following one of these authors rather than another. Furthermore, such arrangements are mystifying and frustrating to the layman and even to the non-specialist professional. As this book is frankly intended as a catalogue in which anyone can quickly find what he wishes, it is arranged alphabetically throughout by family, genus, and species. In general only passing reference is made to subspecific elements present in a number of species. It should be noted that for ease of reference all the "true ferns" are treated as members of the family Polypodiaceae, although the author is well aware that modern pteridologists divide this large family into a number of smaller ones.

In recent years there has been considerable interest in the numbers of chromosomes in plant species, including the ferns and their allies. After appropriate chemical treatments, chromosomes are evident in the nuclei of dividing cells and so may be counted under a microscope. If counts are made in nuclei of the sexually reproductive (haploid) generation they are expressed as n; if, on the other hand, they are made in the nuclei of the substantive spore-producing (diploid) plant they are given as $2n$. Chromosome numbers are included at the end of the description of each species for which they are known. In Appendix I will be found a summary of the chromosome situation together with references to the authors reporting the various numbers. The reader must always bear in mind that these counts are usually based on a relatively few individuals in a single population. Each time the same number of chromosomes is counted in another population the greater is the likelihood that this figure is characteristic for the species as a whole.

This book is an attempt to bring together in convenient form the judgments of monographers and others who have written about the pteridophytes of the Northwest. Plants are no respecters of political divisions, so that in this, as in all floristic works, certain arbitrary boundary lines had to be drawn. Examination of the distribution of our West Coast species demonstrates that the Siskiyou Mountains of southern Oregon and northern California apparently constitute a barrier to the southward movement of a number of species and, incidentally, to the northward migration of others. South of these mountains there is a new and different pteridophyte flora. The inclusion of California would add considerably to the number of species treated and would also add many taxonomic problems at present poorly understood. Northward, of course, the number of species gradually decreases, so the inclusion of the pteridophytes of Alaska adds few complications. The area covered in the present work, therefore, comprises Oregon, Washington, British Columbia, the Yukon Territory, and Alaska to the Arctic Ocean.

This great area extends from about 42° to beyond 70° north latitude and from 115° to 170° west longitude. The climate varies from mild, virtually frost-free winters to arctic cold; there are coastal forests with well over 200 inches of annual precipitation and inland semi-deserts with less than 10 inches. The altitude varies from sea-level to over 20,000 feet. It is an area of diverse and often very rugged terrain corrugated lengthwise by three main sets of mountains that run in a northwesterly direction. On the western side of each range there is an area of higher precipitation while the eastern side lies in a "rain-shadow." Mountains of granite, limestone, and pumice fill the needs of certain specialized species and account in part for the somewhat erratic distribution patterns shown in some cases.

The present work includes descriptions, illustrations, and distribution maps for 97 species, about a quarter of the total number found in North America north of the Mexican border. Of the species in our area 44 are circumboreal in their distribution; 13 are endemic to North America but widely distributed; 27 are endemic to western North America; 7 are found in western North America and Eurasia, and 6 are species with peculiar ranges. Lists of the species in each of these groups are given in Appendix II.

Under the provisions of the International Code of Botanical Nomenclature each taxon can have only one legitimate name and this in general is the one first given to it. Other names applied to the species in the past are said to be synonyms. Here no exhaustive attempt has been made to list all synonyms for each species described. Only those that are necessary to relate the current name to that which has been used in one or other of the general floras of the region have been included.

Literature referred to can be found starting on p. 242. It is arranged alphabetically by author and chronologically under each author. Reference is made to this list by giving the name of the author with the year of publication in brackets after his name.

In addition to numerous specific references made in the text to published papers, the following floras and works dealing with the Pacific Northwest have been freely consulted:

MACOUN, *Catalogue of Canadian Plants*, pt. 5. Montreal: Wm. Foster, Brown and Co., 1890.

HENRY, *Flora of Southern British Columbia*. Toronto: Gage and Co., 1915.

PIPER and BEATTIE, *Flora of the Northwest Coast*. Lancaster, Penn.: New Era Printing Co., 1915.

ABRAMS, *Illustrated Flora of the Pacific States*, vol. 1. Stanford: Stanford University Press, 1923, reprinted in 1940.

FRYE, *Ferns of the Northwest*. Portland: Metropolitan Press, 1934.

PECK, *Manual of the Higher Plants of Oregon*. Corvallis: Oregon State University Press, 1961.

ANDERSON, *Flora of Alaska*. Ames: Iowa State University Press, 1959.

WIGGINS and THOMAS, *Flora of the Alaskan Arctic Slope*. Toronto: University of Toronto Press, 1962.

TAYLOR, *Ferns and Fern Allies of British Columbia*. B.C. Provincial Museum Handbook No. 12. Victoria: Queen's Printer, 1956, revised 1963.

HULTÉN, *Flora of Alaska and Neighboring Territories*. Stanford: Stanford University Press, 1968.

CALDER and TAYLOR, *Flora of the Queen Charlotte Islands*, Pt. 1. Ottawa: Queen's Printer, 1968.

ST. JOHN, *Flora of Southeastern Washington and Adjacent Idaho*. Escondido, Calif.: Outdoor Pictures, 1963.

Key to Families

1   Leaves small, narrow, scalelike or linear-subulate.

  2   Stems conspicuously jointed and ridged lengthwise.

EQUISETACEAE

  2   Stems not obviously jointed and ridged lengthwise.

    3   Leaves quill-like and more or less terete, particularly above.

      4 Leaves in rosettes, enlarged at base to contain spores; stem short and thick, cormlike.   ISOETACEAE

      4 Leaves in rows along a spreading rhizome; spores in sporocarps.

MARSILIACEAE

    3   Leaves short with flat expanded blade.

      5 Plants very small, mosslike, floating on water; if rooted in mud, leaves in a single plane, each leaf divided into 2 lobes.

SALVINIACEAE

      5 Plants larger, more than 3 cm long, terrestrial; leaves in 2 or more planes, not lobed.

        6 Stems extensively creeping with erect branches or, if otherwise, stems tufted and dichotomously branched, or a single branch from a short rhizome; strobilus terete, conspicuous, homosporous.

LYCOPODIACEAE

        6 Stem short-creeping, tending to form mats; branches prostrate, not dichotomously branched; strobili 4-sided and inconspicuous, heterosporous.   SELAGINELLACEAE

1   Leaves large, more than 2 cm long; part, at least, expanded into a blade.

  7   Leaves 4-foliate, petioles long; sporangia in oval sporocarps borne in pairs on short stalks at base of vegetative leaves. MARSILEACEAE

  7   Plants otherwise.

    8   Fronds very delicate, blade only 1 cell thick; rhizome threadlike, much branched, and spreading.   HYMENOPHYLLACEAE

    8   Fronds much coarser, blade several cells thick; rhizomes stouter, vertical, or creeping; if creeping, not threadlike.

      9 Sporangia borne on specialized branch arising from base of lateral leafy blade.   OPHIOGLOSSACEAE

      9 Sporangia borne on backs of but slightly modified vegetative frond or inside tubelike or berrylike segments.   POLYPODIACEAE

Equisetaceae/Horsetail Family

Perennial herbs. Rhizomes deeply subterranean, branched, widely spreading, dark-coloured, jointed with the nodes marked by toothed sheaths, aerial stems arising at intervals. Stems, either all alike and green or of two kinds with the fertile stems lacking chlorophyll; usually erect, simple or branched, cylindrical, grooved, prominently ridged between the grooves with siliceous tubercles or bands; internodes with a large central cavity (except in E. *scirpoides*) and a ring of smaller vallecular cavities in the cortex; stomata conspicuous, sunken in bands or in regular rows. Leaves very small, scalelike, in whorls of 3 or more at a node, fused laterally to form a sheath, usually dark-coloured and not green, with free, sometimes deciduous tips, usually of the same number as the grooves of the stem. Branches, if present, in whorls, growing out through the nodal sheath; branch sheaths much smaller and with fewer teeth. Cones terminal, apiculate, or obtuse, made up of whorled, peltate sporophylls with flat polygonal tips, in some species at the tips of vegetative growth and in others on specialized reproductive stems that lack chlorophyll; spores greenish, each provided with four hygroscopic bands with widened tips (elaters).

The family contains a single, almost cosmopolitan, genus of about 20 species. (Named from the Latin *equus*, horse, and *seta*, bristle.)

EQUISETUM L., Sp. Pl. 1061/1753

Characters are those of the family.

1   Cones long-peduncled, rounded at the top; stems annual; sterile stems, as a rule, with regularly whorled branches.

   2   Fertile and sterile stems not alike; first internode of the primary branches considerably longer than the stem sheath; coning in spring.

      3   Coning stem somewhat fleshy, whitish, or in shades of pink and brown, soon withering, usually unbranched; branches on sterile stems usually not again branched; teeth of branch sheaths triangular-lanceolate or narrower.

         4   Fertile stems 5–25 cm tall, stem sheaths with 8–12 teeth; cones 2–3 cm long; sterile stems up to 60 cm tall, 6–14 furrows; branches 3- or 4-angled.   *E. arvense*

         4   Fertile stems 30–60 cm tall; stem sheaths with 20–30 teeth; cones 4–8 cm long; sterile stems 0.5–3.0 m tall, 20–40 furrows; branches 4–6-angled.   *E. telmateia*

3   Coning stems becoming green and branched, persistent although cones and tips of stems soon wither; branches on sterile stems re-branched, but if not so, teeth of branch sheaths deltoid.

    5   Stem sheaths chestnut brown, flaring upwards, teeth cohering in several broad lobes; branches usually rebranched.   *E. sylvaticum*

    5   Stem sheaths green, rather tight, teeth white-margined, free or nearly so; branches usually unbranched.   *E. pratense*

2   Fertile and sterile stems similar, green; first internode of the primary branches equalling or mostly shorter than the stem sheath, or the stem unbranched; coning in summer.

    6   Central cavity of main stem small, about 1/6 the diameter of the stem; vallecular cavities nearly as large; stem sheaths mostly 10–14 mm long, loose, expanding upwards with 10, usually white-margined, teeth.   *E. palustre*

    6   Central cavity larger, 1/2 to 4/5 the diameter of the main stem; vallecular cavities much smaller or wanting; stem sheaths 5–10 mm long, usually appressed, teeth scarcely white-margined.   *E. fluviatile*

1   Cones short-peduncled, the peduncle not, or only slightly, exceeding the subtending sheath, apiculate; stems evergreen (except *E. laevigatum*), usually unbranched or at least without regular whorls of branches.

    7   Stems lacking a central cavity; vallecular cavities usually 3; stems low and flexuous.   *E. scirpoides*

    7   Stems with a central cavity; vallecular cavities 5 or more; stems other than low and flexuous.

    8   Stems slender, 5- to 10- (rarely 12-) ridged, the ridges sulcate and bearing 2 rows of tubercles; teeth of stem sheaths persistent.
                E. variegatum

    8   Stems stout, normally 14- to 40-ridged, the ridges rounded with a single row of tubercles or crossbands.

      9   Sheaths longer than broad, somewhat flaring upwards with a single dark band above; stems smoothish, mostly annual.   *E. laevigatum*

      9   Sheaths nearly or quite as broad as long, nearly cylindrical, tight, mostly ashy at maturity with 2 dark bands; stems rough, evergreen.
                E. hyemale

Equisetum arvense L.

Sp. Pl. 1061, 1753; Macoun, Cat. Can. Pl. pt. 5, 249, 1890; Henry, Fl. S. British Columbia 8, 1915; Piper & Beattie, Fl. Northwest Coast 11, 1915; Maxon [in] Abrams, Ill. Fl. Pac. States 1:39, *Fig.* 81, 1923; Victorin, Contr. Lab. Bot. Univ. Montreal No. 9:114, 1927; Frye, Ferns Northwest 59, *Fig.* VIII, 1934; Peck, Man. Higher Pl. Ore. 54, 1941; Morton [in] Gleason, Ill. Fl. 1:13, *Fig.*, 1952; Anderson, Fl. Alaska 18, 1959; Wiggins & Thomas, Fl. Alaskan Arctic Slope 37, 1962; St. John, Fl. Southeast. Wash. 11, 1963; Hultén, Fl. Alaska 38, 1968; Calder & Taylor, Fl. Queen Charlotte Islands pt. 1, 128, 1968.

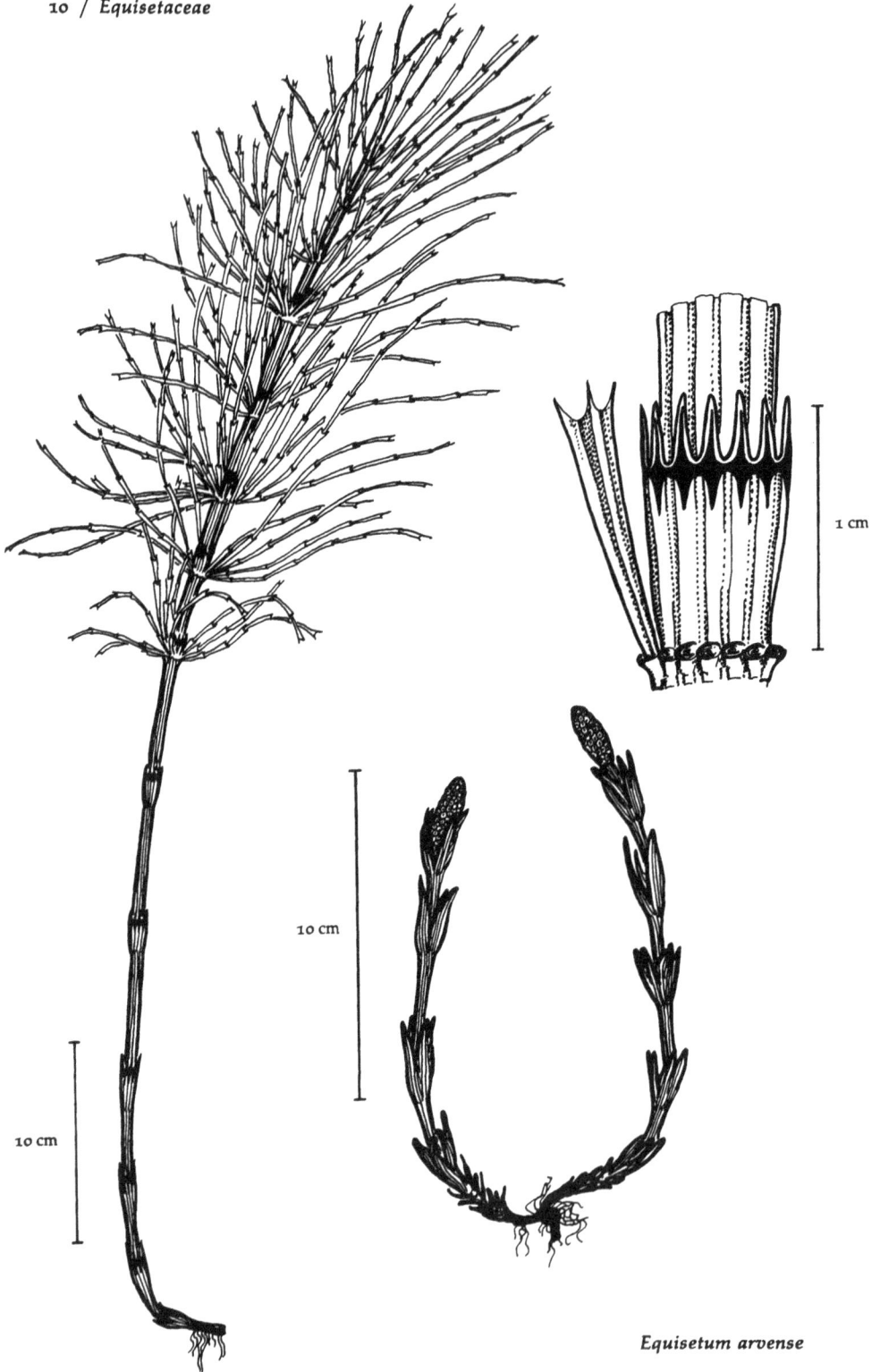

Equisetum arvense

Rhizome felted, bearing ovoid tubers, stems annual, up to 75 cm tall, of two kinds. Sterile stems solitary or clustered, erect or decumbent, the central cavity about 1/2 the diameter of the stem, vallecular cavities large. Sheaths gradually widened upwards, green; teeth as many as the grooves, triangular-lanceolate to lance-attenuate, persistent, scarious-margined, less than 1/4 the length of the tube. Branches solid, numerous, spreading, regularly whorled, 3- or 4-angled, branch sheath 3- or 4-toothed; first internode longer than the subtending stem sheath. Fertile stems flesh-coloured to brown, stout, up to 25 cm tall and 8 mm thick, unbranched, appearing before the sterile stems but soon withering. Sheaths pale brown, teeth 6–12, darker, lanceolate, partly connate. Cone obtuse, long-peduncled. ($n = 108$)

HABITAT Very tolerant of soil conditions, usually growing in full sun or part shade; often invades railway embankments or roadsides, a common garden weed in some parts.

RANGE Cosmopolitan throughout the northern parts of the northern hemisphere.

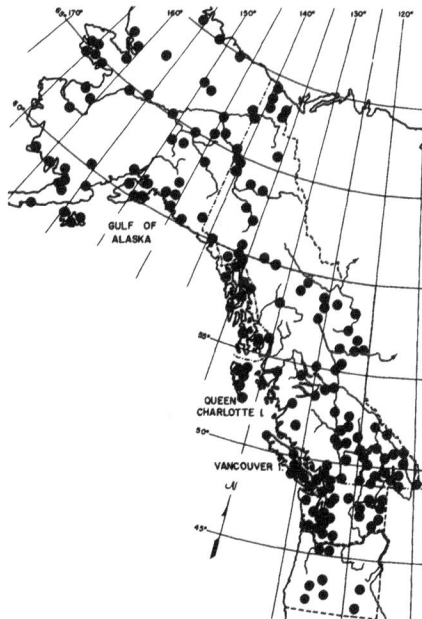

COMMENTS Generally common everywhere. A variable species with numerous forms that apparently result from differences in the environment. Many have been given names as formae or in some cases as varieties. One of these, var. *boreale* (Bong.) Rupr., may have some taxonomic merit. It is characterized by 3-angled, rather than 4-angled branchlets and is said to be most common in calcareous regions in the northern part of the species range. Sterile shoots are likely to be confused with those of *E. pratense* but usually can be distinguished by the teeth of the branch sheath; in

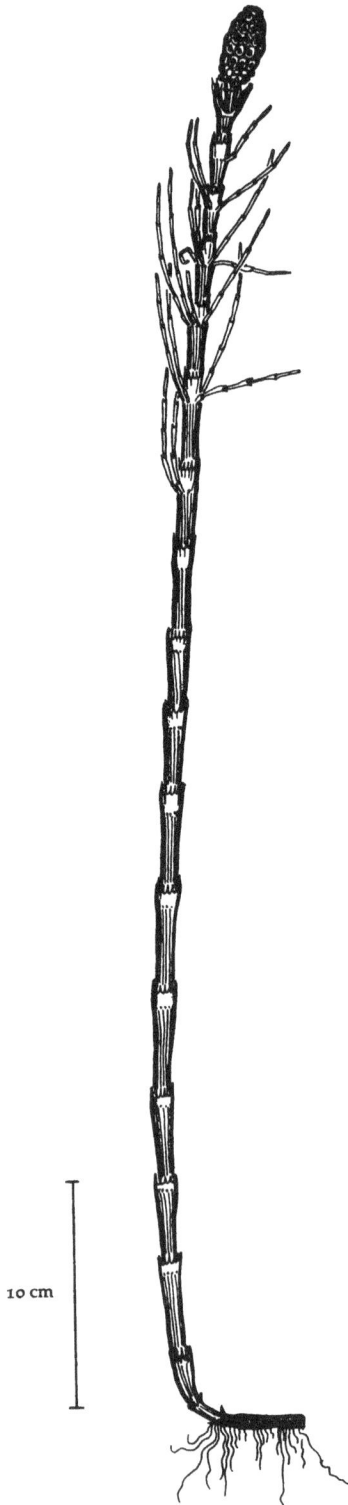

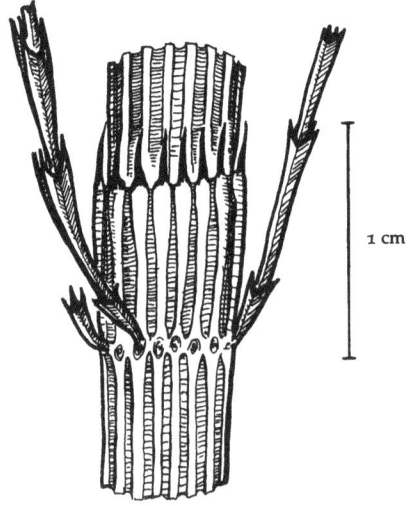

Equisetum fluviatile

10 cm

1 cm

E. arvense these are narrowly triangular-lanceolate, gradually tapering to a long point.

This species forms sterile hybrids with *E. fluviatile* (*E* × *litorale* Kuhl.). The hybrid origin of *E* × *litorale* was presumed for many years but was finally established on the basis of cytological evidence by Manton (1950).

Equisetum fluviatile L.

Sp. Pl. 1062, 1753; Macoun, Cat. Can. Pl. pt. 5, 251, 1890; Henry, Fl. S. British Columbia 9, 1915; Piper & Beattie, Fl. Northwest Coast 12, 1915; Maxon [in] Abrams, Ill. Fl. Pac. States 1:40, *Fig.* 84, 1923; Victorin, Contr. Lab. Bot. Univ. Montreal No. 9:125, 1927; Frye, Ferns Northwest 53, 1934; Peck, Man. Higher Pl. Ore. 54, 1941; Morton [in] Gleason, Ill. Fl. 1:14, *Fig.*, 1952; Anderson, Fl. Alaska 19, 1959; Wiggins & Thomas, Fl. Alaskan Arctic Slope 38, 1962; St. John, Fl. Southeast. Wash. 11, 1963; Hultén, Fl. Alaska 36, 1968; Calder & Taylor, Fl. Queen Charlotte Islands pt. 1, 128, 1968.
E. limosum L., Sp. Pl. 1062, 1753.

Rhizomes glabrous, shining, reddish, widely creeping, rarely tuber-bearing. Stems annual, all alike, erect, mostly solitary, up to 1 m tall, 3–8 mm thick, shallowly ridged; ridges smooth not spinulose; grooves 10–30, fine. Central cavity at least 4/5 the diameter of the stem; vallecular cavities absent except near the base. Sheaths subcylindrical, mostly tight-appressed or the upper somewhat widened upwards; teeth persistent, free, not ribbed, lanceolate, acuminate, dark brown, scarcely scarious-margined. Branching varies from none to many, mostly near the middle of the stem, branches themselves unbranched; branch sheaths with 4–6 teeth, the basal dark brown, the first internode slightly shorter than the stem sheath. Cones borne at the tip of the main stem, obtuse, long-stalked, deciduous. ($n = 108$)

HABITAT Shallow water at the edge of lakes, ponds, and ditches, or marshy places.

COMMENTS This species, which is common throughout, often occurs in numbers sufficient to form conspicuous communities in shallow water. The large diameter of the central cavity, the tendency to branch mostly above the middle, and the reddish, shining, basal part of the stem are useful distinguishing features. Hybrids between *E. fluviatile* and *E. arvense* have been named *E.* × *litorale* Kuhl. They combine, in various degrees, characters of both species; the central cavity of the hybrid is 1/2 to 2/3 the diameter of the stem, the vallecular cavities are often well developed, the spores are mostly aborted, and meiosis is irregular.

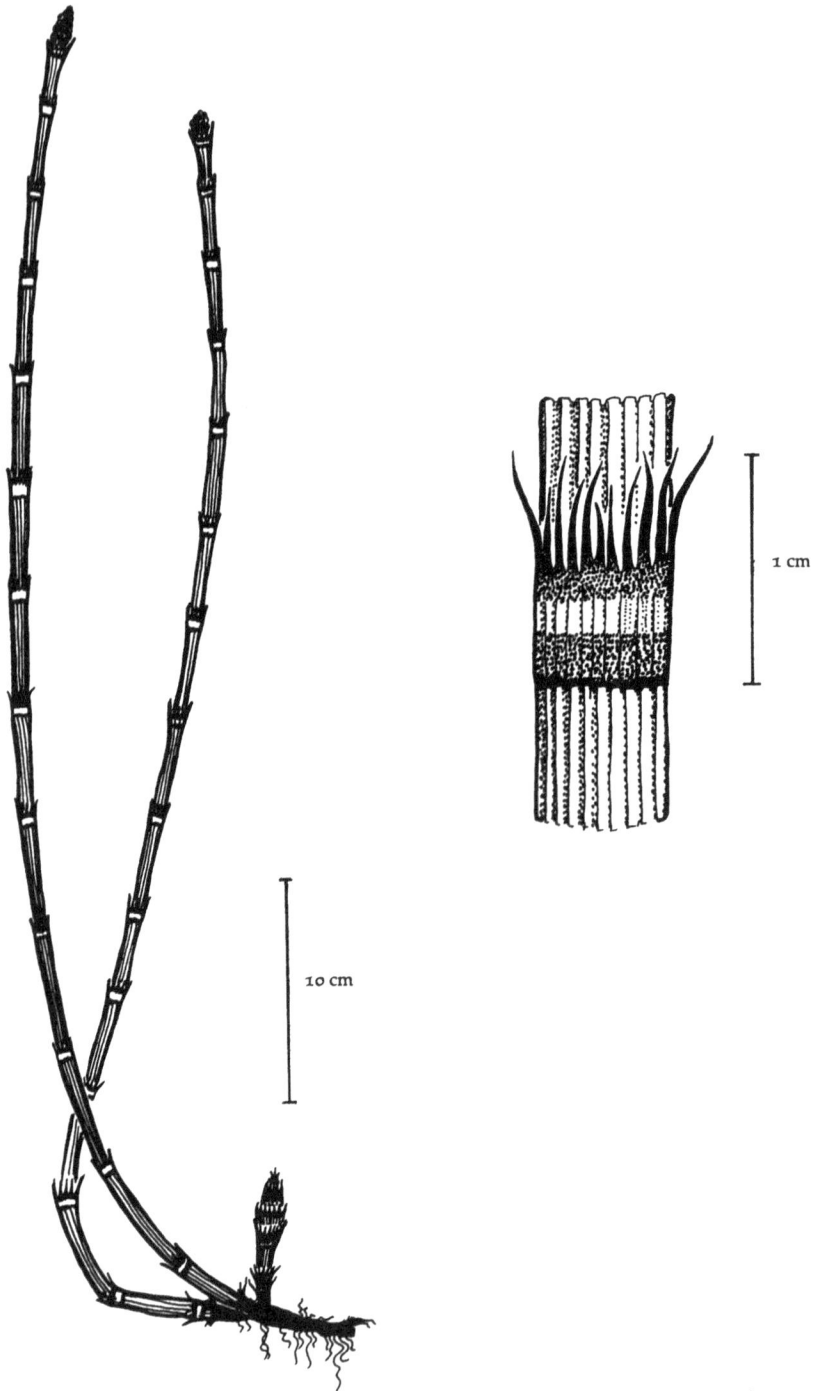

10 cm

1 cm

Equisetum hyemale

RANGE Alaska to Newfoundland, south to Oregon, Wyoming, Nebraska, Illinois, Ohio, and Virginia; Eurasia.

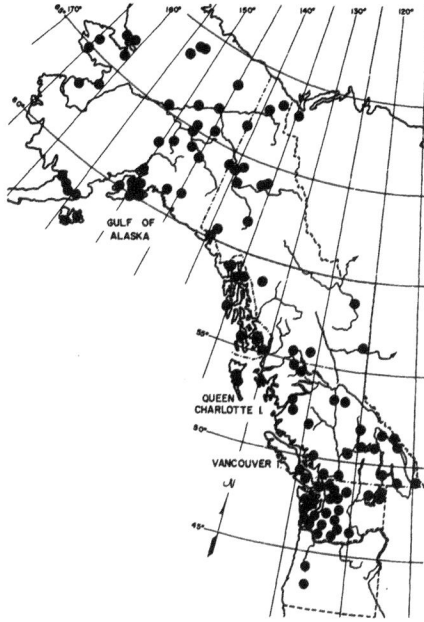

Equisetum hyemale L.

Sp. Pl. 1062, 1753; Macoun, Cat. Can. Pl. pt. 5, 251–2, 1890; Henry, Fl. S. British Columbia 9, 1915; Piper & Beattie, Fl. Northwest Coast 12, 1915; Maxon [in] Abrams, Ill. Fl. Pac. States 1:42, *Figs. 90, 91,* 1923; Victorin, Contr. Lab. Bot. Univ. Montreal No. 9:128, 1927; Frye, Ferns Northwest 48, *Fig.* VIII, 1934; Peck, Man. Higher Pl. Ore. 54, 1941; Morton [in] Gleason, Ill. Fl. 1:14, *Fig.,* 1952; Anderson, Fl. Alaska 20, 1959; Hauke, Am. Fern J. **52**:58, 1962; St. John, Fl. Southeast. Wash. 12, 1963; Hultén, Fl. Alaska 34, 1968; Calder & Taylor, Fl. Queen Charlotte Islands pt. 1, 128, 1968.

E. prealtum Raf., Fl. Ludovic. 13, 1817.
E. robustum A. Br., Am. J. Sci. **46**:88, 1844.
E. robustum A. Br. γ *affine* Engelm., Am. J. Sci. **46**:88, 1844.
E. hiemale var. *californicum* Milde, Verh. Zool.-Bot. Ges. Wien **12**:1264, 1862.

Rhizome felted, widely creeping, branched, growing deep in the ground. Stems glaucous-green, all alike, erect, unbranched except sometimes following injury, evergreen, up to 15 dm tall, 4–6 mm in diameter, mostly 18- to 40-ridged; ridges broad, rounded, roughened with prominent crossbands of silex or 1 or 2 rows of fine tubercles. Central cavity 2/3 the diameter of the stem; vallecular cavities small. Sheaths cylindrical, appressed, about as long as broad, the ridges 2-furrowed, at first green, later with a black band above and below. Teeth as many as the grooves, lanceolate, fused at the base, dark brown with broad scarious margins, subulate-tipped. Cones stoutly apiculate, short-stalked, almost sessile in the uppermost sheath. ($n = 108$)

HABITAT Damp alluvial situations, often in partial shade.

COMMENTS A widely ranging, highly variable species found generally throughout. Many forms have been named, but there is so much inter-gradation that the value of this practice is doubtful. Most North American floras divide the western hemisphere *E. hyemale* into three varieties. These are var. *pseudohiemale* (Farw.) Morton, with teeth of the sheaths promptly and uniformly deciduous; var. *elatum* (Engelm.) Morton, with

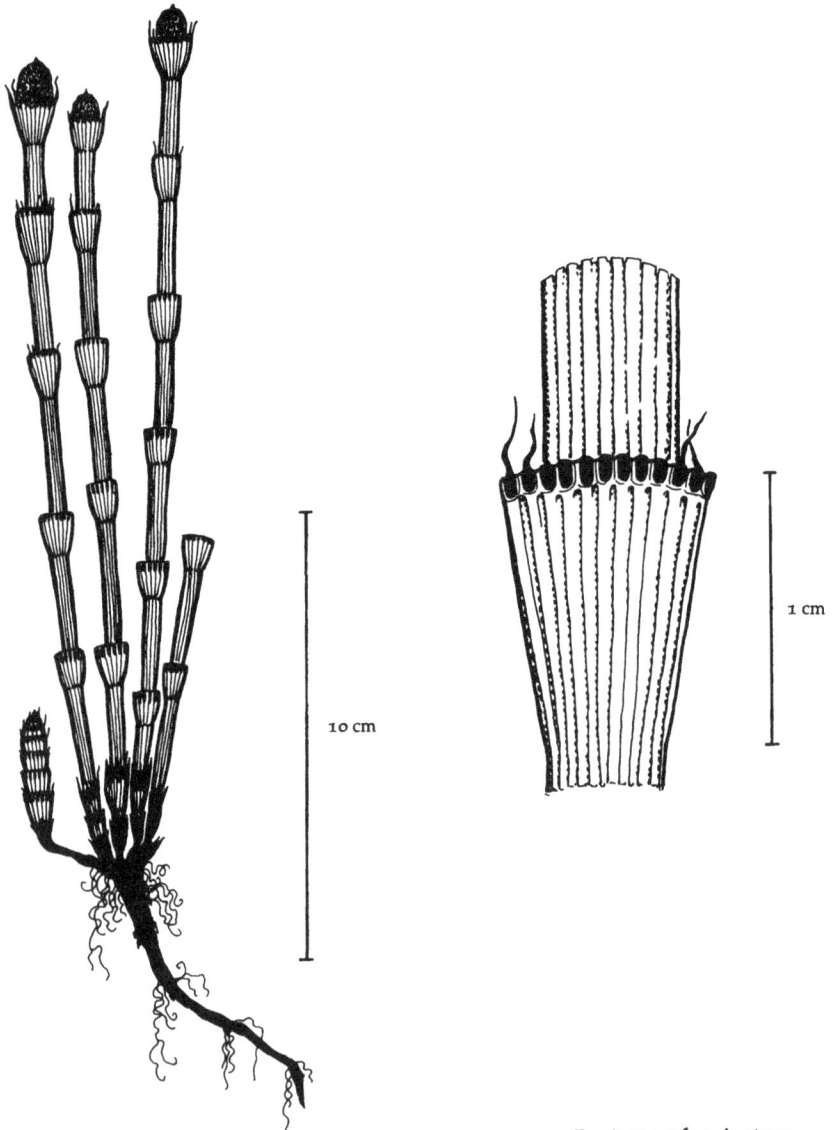

10 cm

1 cm

Equisetum laevigatum

teeth persistent and a single row of tubercles on the ridges; and var. *californicum* Milde, similar to the last but with 2 rows of tubercles. Hauke (1963) has shown that there is an independent and clinal relationship between the characters that differentiate these varieties. Furthermore, the variation in any one character is so gradual that no real discontinuities can be recognized. For these reasons he considers *E. hyemale* present as var. *affine* (Engelm.) A. A. Eaton from Guatemala to Quebec westward to the Pacific ocean and across the Aleutian Islands to eastern Asia to be a single variable variety. In his opinion even the differences between the eastern and western hemisphere populations are not sufficiently marked to warrant subspecific ranks. This species hybridizes with *E. variegatum* to produce *E.* × *trachyodon* A. Br.

R A N G E Alaska to Nova Scotia, south to Guatemala, Texas, Georgia, and Florida; Eurasia.

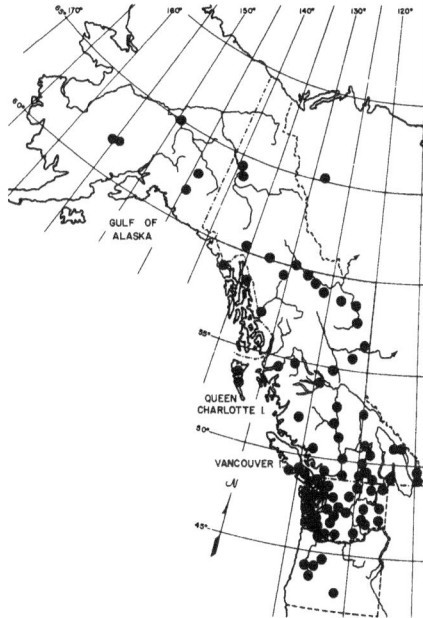

Equisetum laevigatum A. Br.

Am. J. Sci. **46**:87, 1844; Macoun, Cat. Can. Pl. pt. 5, 251, 1890; Henry, Fl. S. British Columbia 10, 1915; Maxon [in] Abrams, Ill. Fl. Pac. States **1**:40–1, *Figs. 85, 86*, 1923; Frye, Ferns Northwest 46, *Fig.* VIII, 1934; Peck, Man. Higher Pl. Ore. 54, 1941; Morton [in] Gleason, Ill. Fl. **1**:14, *Fig.*, 1952; Hauke, Am. Fern J. **52**:34, 1962; St. John, Fl. Southeast. Wash. 12, 1963.
E. funstonii A. A. Eaton, Fern Bull. **11**:10, 1903.
E. kansanum Schaffn., Ohio Nat. **13**:21, 1912.
E. fontinale Copel., Madroño **3**:367, 1936.

Rhizome glabrous, felted only at the nodes, branched, creeping. Stems usually annual, all alike, pale yellow-green or glaucous, erect, normally unbranched, up to 10 dm tall, mostly 14- to 26-ridged; ridges roundish, smoothish, tubercles scarcely, if at all, developed. Central cavity 3/4 the diameter of the stem or less; vallecular cavities small; stomata borne in 2 rows in the grooves. Sheaths subcylindric-funnel form, contracted at the base, flaring towards the apex, black-banded at the apex only, or the lowest black at the base as well or throughout; ridges with 2 furrows. Teeth lanceolate-subulate, dark with broad scarious margins, quickly and uniformly deciduous. Cones blunt or acute, rarely apiculate, short-stalked. (According to Hauke (1958) meiosis is irregular suggesting a hybrid origin for the species.)

HABITAT Dry sandy or clay soil in the open.

RANGE British Columbia to Ontario, south to California, Texas, Louisiana, Missouri, North Dakota, Minnesota, Illinois, Georgia, and New Jersey; Mexico and south to Guatemala.

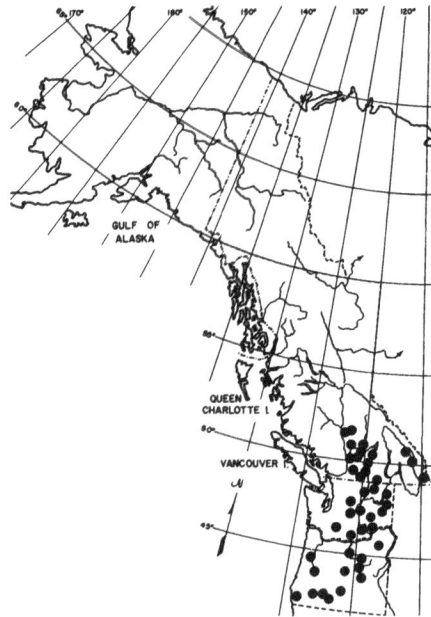

COMMENTS This species has been the cause of a good deal of taxonomic confusion in the past, now resolved by Hauke's (1963) monographic study. In the southwestern part of its range, perhaps due to the absence of winter-killing, it tends to be coarser-stemmed and the cones to be slightly apiculate. This is the form sometimes named *E. funstonii*. Although there is a general similarity to *E. fluviatile* owing to the smooth stems and blunt cones, *E. laevigatum* can be readily distinguished by its deciduous sheath teeth and stomata in single lines. This is the only species of *Equisetum* endemic to North America. It is similar to *E. ramosissimum* Desf., its Eurasian counterpart. *Equisetum laevigatum* forms natural hybrids with *E. hyemale* (*E.* × *ferrissii* Clute) and *E. variegatum* (*E.* × *nelsonii* (A. A. Eaton) Schaffner). These are essentially intermediate in character between the parents and have a very high proportion of aborted spores.

Equisetum palustre L.

Sp. Pl. 1061, 1753; Macoun, Cat. Can. Pl. pt. 5, 250, 1890; Henry, Fl. S. British Columbia 9, 1915; Maxon [in] Abrams, Ill. Fl. Pac. States 1:40, *Fig.83*, 1923; Victorin, Cont. Lab. Bot. Univ. Montreal No. 9, 120, 1927; Frye, Ferns Northwest 55, *Fig.* VIII, 1934; Peck, Man. Higher Pl. Ore. 54, 1941; Morton [in] Gleason, Ill. Fl. 1, 13, *Fig.*, 1952; Anderson, Fl. Alaska 19, 1959; Wiggins & Thomas, Fl. Alaskan Arctic Slope 38, 1962; Hultén, Fl. Alaska 37, 1968; Calder & Taylor, Fl. Queen Charlotte Islands pt. 1, 129, 1968.

Rhizome glabrous, shining black, creeping and somewhat branched, lacks both felt and tubers, deeply subterranean. Stems annual, all alike, erect or somewhat decumbent, solitary or clustered, up to 6 dm tall, deeply 7- to 10-ridged; ridges smooth or slightly rough but lacking spinules. Central cavity mostly about 1/6 the diameter of the stem, vallecular cavities nearly as large. Sheaths gradually widened upwards, teeth persistent, free, lanceolate, acuminate, 1-ribbed, with a blackish-brown central portion and a broad white hyaline margin. Branches few and irregular or numerous and regularly whorled, slender, nearly smooth; the sheaths 5- or 6-toothed, the basal ones blackish, the first internode much shorter than the subtending stem sheath. Cones borne at the tip of the main stem, long-stalked, obtuse, deciduous; occasionally smaller ones form at the tip of branches. ($n = 108$)

H A B I T A T Shallow water and wet marshy places.

R A N G E Alaska to Labrador, south to Oregon, Montana, Wyoming, Nebraska, Illinois, northern Pennsylvania, Vermont, Maine, and Newfoundland; Eurasia.

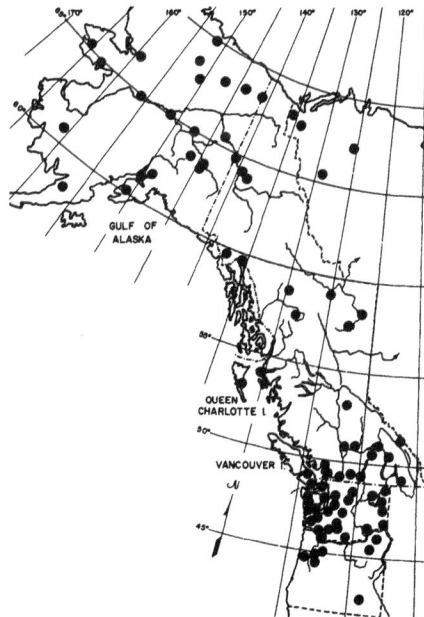

10 cm

1 cm

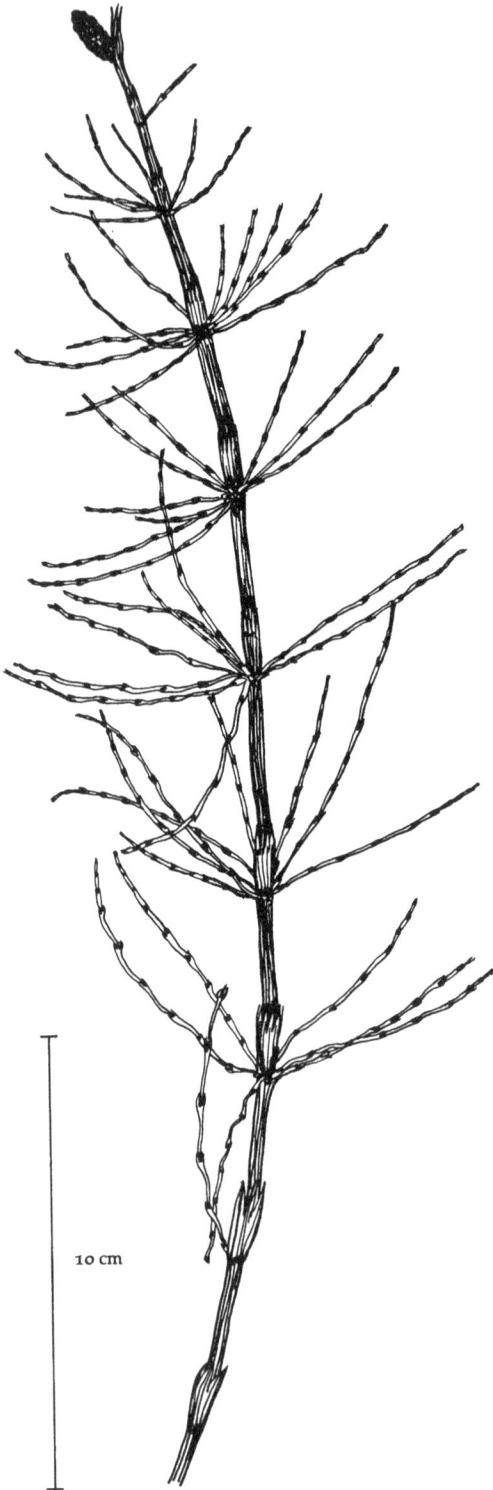

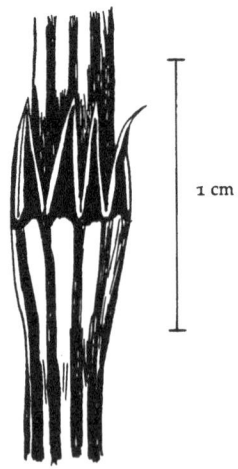

Equisetum palustre

COMMENTS To be expected in suitable habitats throughout. Sterile stems may be confused with those of *E. arvense* but can be recognized by the small size of the central cavity, by the lowermost internode on the branches being shorter than the adjacent stem sheath, and by the 5- or 6-toothed branch sheaths. The eastern North American form has been named var. *americanum* Vict. It differs from the European in the teeth of the sheaths which are as long or longer than the sheath itself, acute, triangular with straight sides and a narrow, uniformly broad scarious margin. The European plant has shorter, blunt teeth with a broad scarious margin, broadest at the middle point of the teeth.

Equisetum pratense Ehrh.

Hannov. Mag. **9**, 138, 1784; Macoun, Cat. Can. Pl. pt. 5, 249, 1890; Henry, Fl. S. British Columbia 8, 1915; Piper & Beattie, Fl. Northwest Coast 11, 1915; Victorin, Contr. Lab. Bot. Univ. Montreal No. **9**, 118, 1927; Frye, Ferns Northwest 57, *Fig.* XIII, 1934; Morton [in] Gleason, Ill. Fl. **1**:13, *Fig.*, 1952; Anderson, Fl. Alaska 18, 1959; Wiggins & Thomas, Fl. Alaskan Arctic Slope 37, 1962; Hultén, Fl. Alaska 38, 1968.

Rhizome black, creeping. Stems annual, erect, up to 50 cm tall, of two kinds. Sterile stems pale green, mostly solitary; central cavity 1/3 to 1/2 the diameter of the stem; vallecular cavities 10–18 with as many ridges; ridges much broader than the deep grooves, slightly roughened with fine spinules. Sheaths subcylindric, green, teeth persistent, lanceolate, acuminate, coherent at the base, each with a dark brown central stripe and a scarious margin. Branches solid, numerous, regularly whorled, slender, mostly 3-angled, smooth, unbranched. Branch sheaths rather loose and flaring upwards, 3-toothed; teeth deltoid, acute, white-margined, about 1/3 the length of the tube of the sheath; the lowermost are pale brown. First internode mostly equalling or shorter than the subtending stem sheath. Fertile stems normally unbranched at first; later, developing whorls of branches similar to those on the sterile stems; sheaths somewhat longer and more conspicuous. Cone obtuse, long-peduncled, soon withering and deciduous. ($n = 108$)

HABITAT Grassy banks of streams and edges of damp woods.

COMMENTS A boreal, circumpolar species found only east of the coastal mountains in British Columbia but coming to the coast in Alaska. Sterile shoots may be mistaken for those of *E. arvense* but can be identified by the flaring branch sheaths with short, white-margined, deltoid teeth.

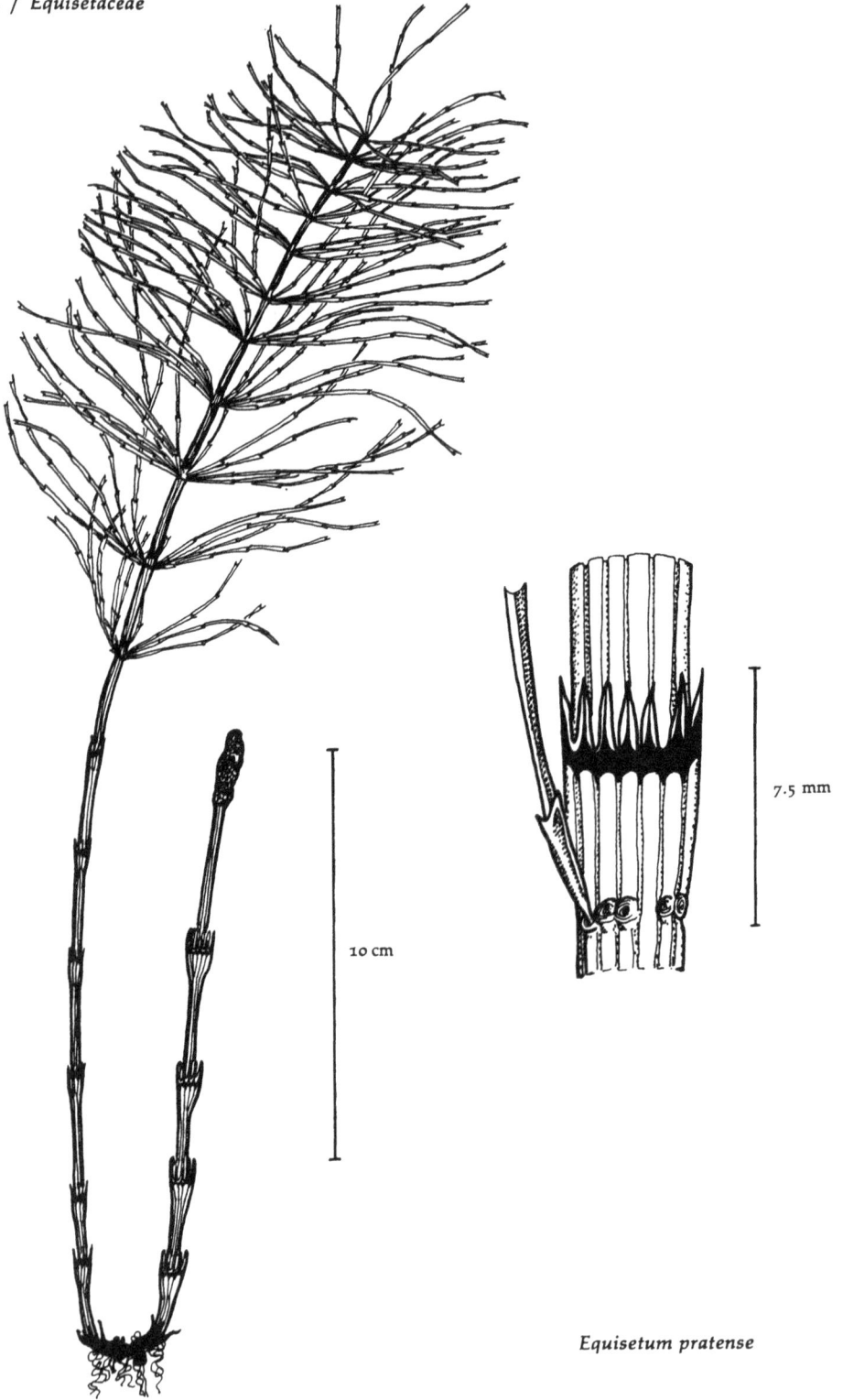

Equisetum pratense

10 cm

7.5 mm

RANGE Alaska to Nova
Scotia, south to British Co-
lumbia, Wyoming, Colorado,
Iowa, and New Jersey;
Eurasia.

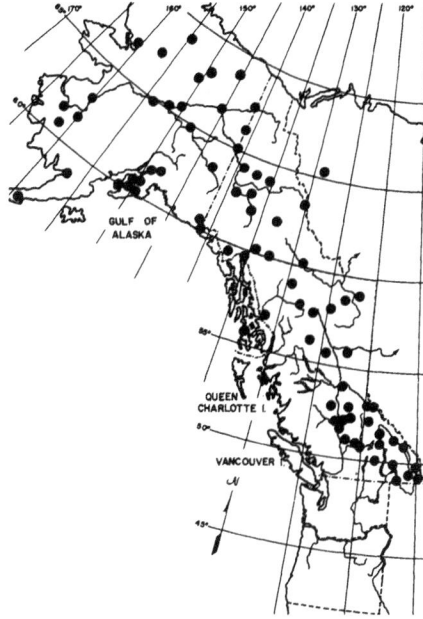

Equisetum scirpoides Michx.

Fl. Bor. Am. 281, 1803; Macoun, Cat. Can. Pl. pt. 5, 252, 1890; Henry, Fl. S.
British Columbia 9, 1915; Piper & Beattie, Fl. Northwest Coast 12, 1915; Maxon
[in] Abrams, Ill. Fl. Pac. States 1:41, *Fig.* 88, 1923; Victorin, Contr. Lab. Bot.
Univ. Montreal No. 9:133, 1927; Frye, Ferns Northwest 51, 1934; Morton [in]
Gleason, Ill. Fl. 1:14, *Fig.*, 1952; Anderson, Fl. Alaska 19, 1959; Wiggins &
Thomas, Fl. Alaskan Arctic Slope 39, 1962; Hauke, Am. Fern J. 52:63, 1962;
Hultén, Fl. Alaska 36, 1968.

Rhizome creeping, shallow, much branched, filiform. Stems evergreen, all
alike, numerous, densely tufted, prostrate or ascending, arched-recurving
and flexuous or zigzag, very slender, up to 16 cm long, unbranched or with
a few irregular, elongated branches, 3-ridged. Ridges broad and deeply
grooved so that the stem appears 6-ridged. Central cavity lacking; vallecular cavities conspicuous, 3 (or 4). Sheaths enlarged and darkened upwards, teeth 3, triangular, acuminate, scarious-margined, the basal part
persistent. Cones black, small, subsessile, apiculate. ($2n = 216$)

HABITAT Open springy banks and damp coniferous woods.

COMMENTS This curious little species is often overlooked because it
grows among mosses or is partially buried by humus in coniferous woods.
The thin, flexuous, zigzag stems and its small size are characteristic. It is
largely confined to mountain areas in the interior·of British Columbia and
northward.

5 mm

5 mm

5 cm

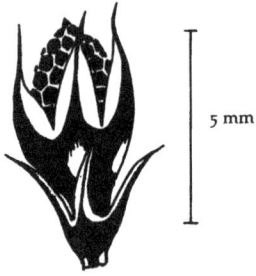

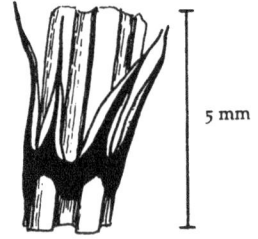

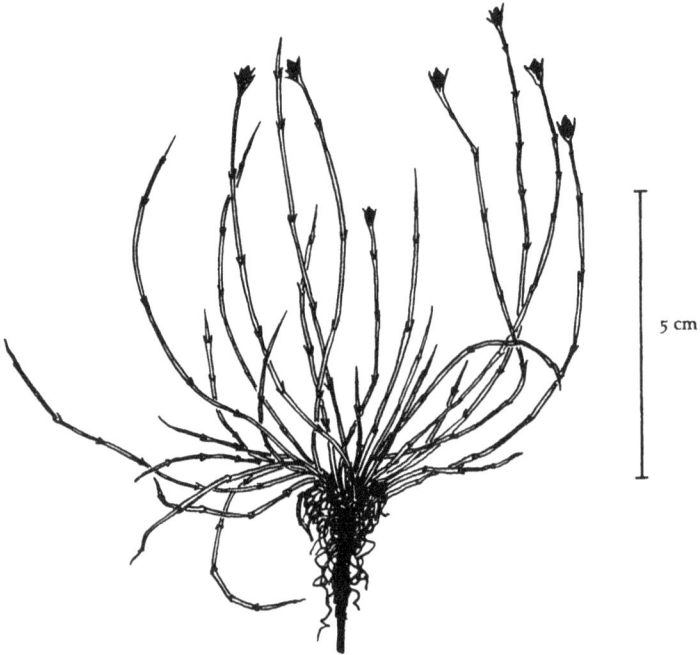

Equisetum scirpoides

RANGE Circumboreal;
Alaska to Labrador, south to
Washington, Montana,
Minnesota, northern Illinois,
Ontario, New York, and
Connecticut; Greenland;
Eurasia.

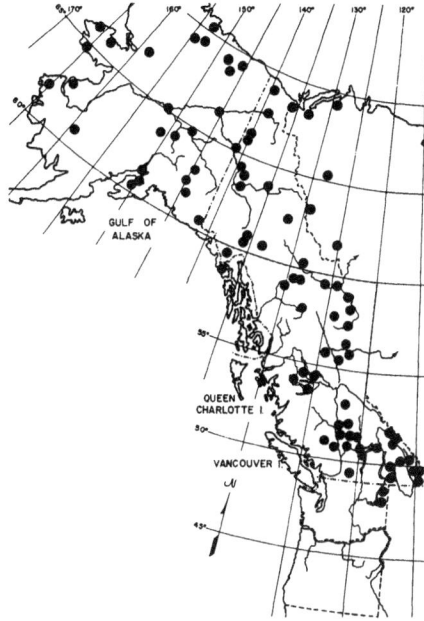

Equisetum sylvaticum L.

Sp. Pl. 1061, 1753; Macoun, Cat. Can. Pl. pt. 5, 250, 1890; Henry, Fl. S. British
Columbia 9, 1915; Victorin, Contr. Lab. Bot. Univ. Montreal No. 9, 119, 1927;
Frye, Ferns Northwest 55, *Fig.* XII, 1934; Morton [in] Gleason, Ill. Fl. 1:13, *Fig.*,
1952; Anderson, Fl. Alaska 19, 1959; St. John, Fl. Southeast. Wash. 12, 1963;
Hultén, Fl. Alaska 37, 1968.

Rhizome creeping. Stems annual, erect, up to 70 cm tall, of two kinds.
Sterile stems green, mostly solitary, smooth or nearly so, the central cavity
more than 1/2 the diameter of the stem, 10–18 conspicuous vallecular
cavities and the same number of ridges, each ridge with two rows of
delicate, sharply hooked spinules. Sheaths subcylindric, green at the base,
chestnut-brown above; teeth persistent, irregularly coherent into 3–5
groups. Branches numerous, tending to droop at the ends, irregularly
whorled; first internode longer than the subtending stem sheath, deeply
4- or 5-angled, simple or mostly branched again. Ultimate branches delicate, filiform, 3-angled, their sheaths deeply 3-toothed. Fertile stems
shorter, green to flesh-coloured, at first unbranched but later producing
whorls of mostly compound branches. Stem sheaths larger, loosely inflated, and flaring upwards. Cone obtuse, long peduncled, soon withering.
($2n = 216$)

HABITAT Cool, moist, somewhat open woods in mildly acid soils.

COMMENTS Another markedly inland and variable species in which
many forms and varieties have been named. None of these seems to have

10 cm

2.5 cm

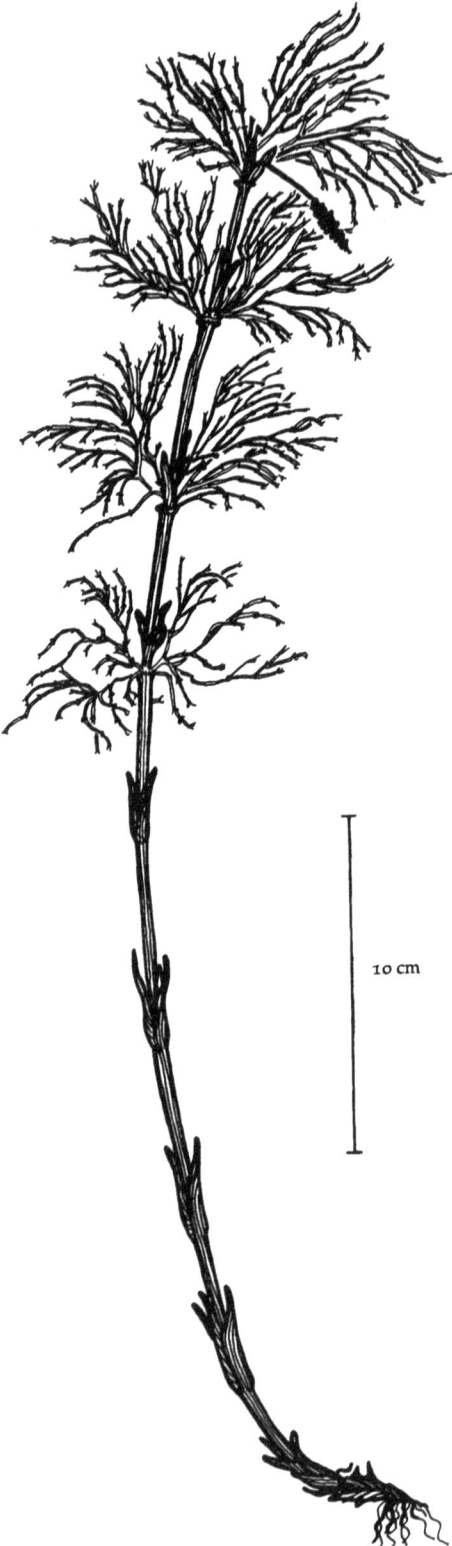

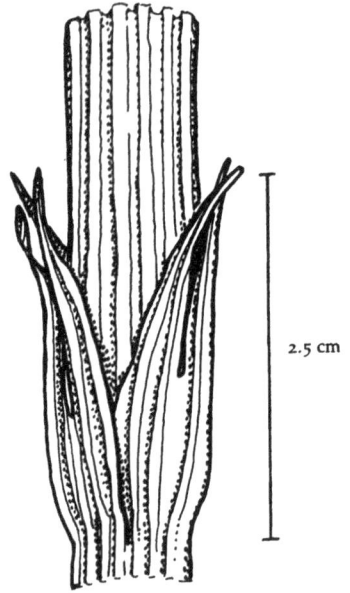

Equisetum sylvaticum

taxonomic significance except possibly the Eurasian var. *sylvaticum* and the North American var. *pauciramosum* Milde. The former has branches that are scabrous, particularly at their bases, while in the latter they are smooth right to the base. These distinctions have a general geographic significance, but on both continents individuals with the opposing characteristic can be found.

RANGE Alaska to Labrador, south to Washington, Montana, South Dakota, Iowa, Ohio, West Virginia, North Carolina, and Maryland; Eurasia.

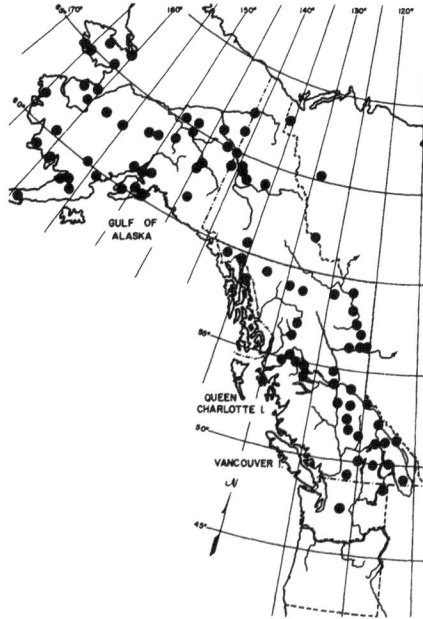

Equisetum telmateia Ehrh.

Hannov. Mag. **1783**: 287, 1783; Macoun, Cat. Can. Pl. pt. 5, 249, 1890; Henry, Fl. S. British Columbia 8, 1915; Piper & Beattie, Fl. Northwest Coast 12, 1915; Maxon [in] Abrams, Ill. Fl. Pac. States **1**:39, *Fig. 82*, 1923; Frye, Ferns Northwest 57, *Figs.* VIII, XI, 1934; Peck, Man. Higher Pl. Ore. 54, 1941; Calder & Taylor, Fl. Queen Charlotte Islands pt. 1, 130, 1968.
E. maximum Lam., Fl. Françoise, 2nd ed., **1**:1245, 1795.

Rhizome felted, often bearing pear-shaped tubers. Stems annual, up to 2 m tall, of two kinds. Sterile stems coarse, erect, mostly solitary, dirty white or pale green, grooves fine, 20–40. Ridges of internodes somewhat scabrous, central cavity nearly 5/6 the diameter of the stem, vallecular cavities large. Stem sheaths cylindrical, appressed, pale, blackish above; teeth 1/4 to 1/2 as long as the tube, persistent, united in groups of 2 or 3, elongated, with broad hyaline margins. Branches solid, numerous, in dense spreading whorls, 4- to 6-angled, their sheaths with 2-ribbed teeth. First internode much longer than the subtending stem sheath. Fertile stems unbranched, very stout, succulent, whitish or pale brownish, up to 60 cm tall, appearing before the sterile stems but soon withering. Sheaths

10 cm

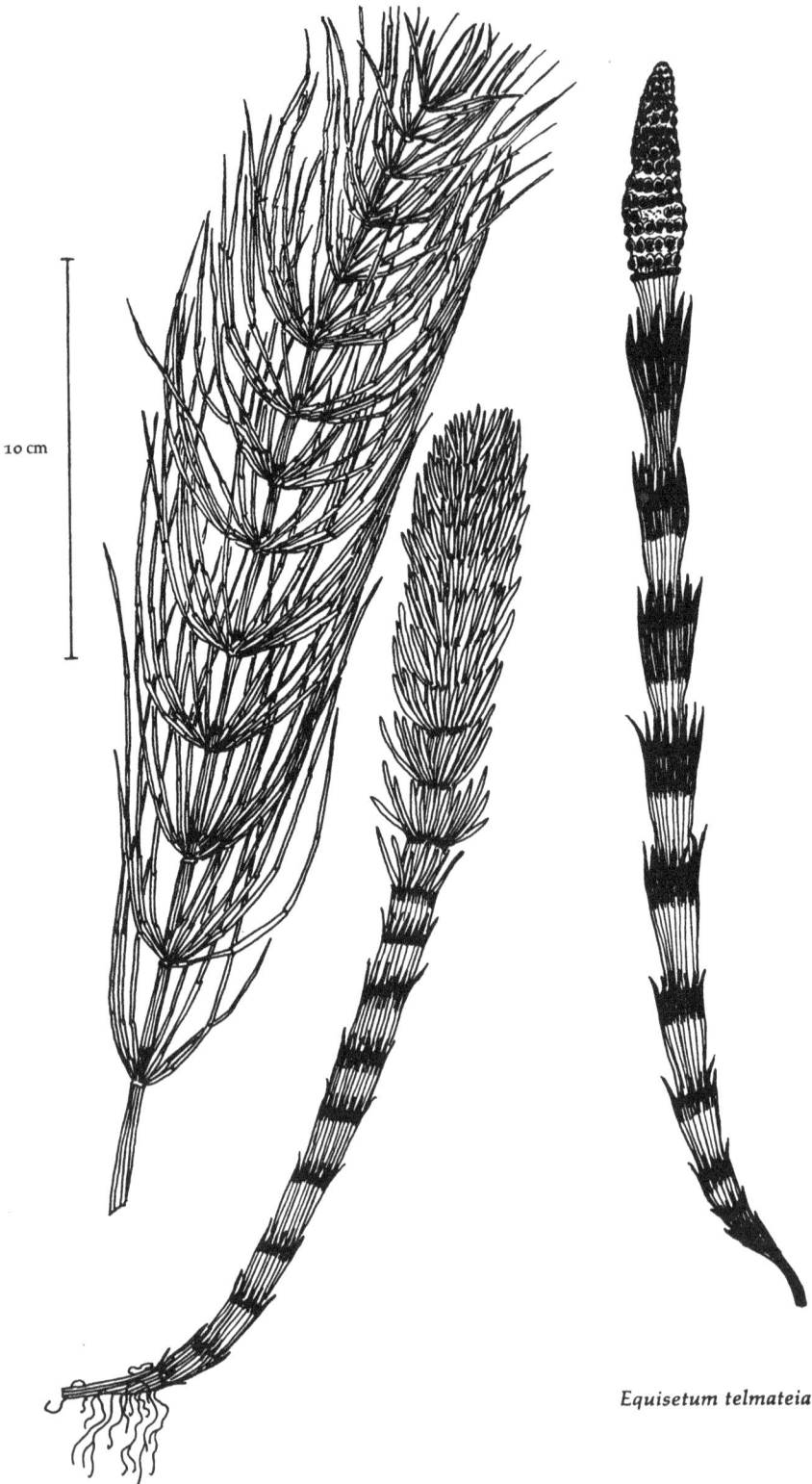

Equisetum telmateia

loose, membranous, whitish below, brown above, the lower one often longer than the internodes; teeth brown, lance-attenuate. Cone obtuse, 4–8 cm long, very thick, hollow, long-peduncled. ($2n = 216$)

HABITAT Low, wet ground, swamps, borders of streams.

RANGE British Columbia to California; Eurasia; North Africa.

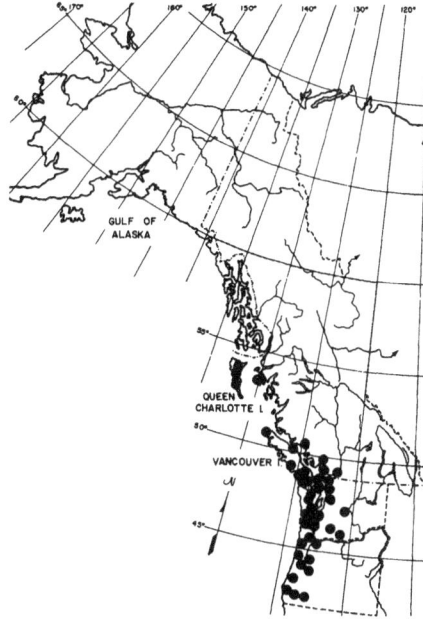

COMMENTS Limited to the coastal regions, except for some extension up the Columbia river; not reported from east of the coastal mountains. Small individuals can be distinguished from large specimens of *E. arvense* by the larger, looser sheath with 2-ribbed teeth; the greater number of teeth in both branch and stem sheaths also distinguishes this species from *E. arvense*. The North American specimens, which may actually constitute a distinct species, are usually referred to as var. *braunii* (Milde) Milde which differs from the Eurasian and North African var. *telmateia* in the coherence of the sheath teeth, longer cone, rough internodes, and more robust growth.

The reported occurrence of this species in the Keeweenaw Peninsula of Michigan is based on a record of more than 70 years ago. It has apparently not been re-collected since then. Its presence there should be regarded as doubtful and needing verification.

Equisetum variegatum Schleich.

[In] Usteri, Neue Ann. Bot. **21**:120–35, 1797; Macoun, Cat. Can. Pl. pt. 5, 252, 1890; Henry, Fl. S. British Columbia 9, 1915; Piper & Beattie, Fl. Northwest Coast 12, 1915; Maxon [in] Abrams, Ill. Fl. Pac. States **1**:41, *Fig. 87*, 1923;

Victorin, Contr. Lab. Bot. Montreal No. **9**, 131, 1927; Frye, Ferns Northwest 50, *Figs.* VII, VIII, 1934; Peck, Man. Higher Pl. Ore. 54, 1941; Morton [in] Gleason, Ill. Fl. **1**:14, *Fig.*, 1952; Anderson, Fl. Alaska 19, 1959; Hauke, Am. Fern J. **52**:61, 1962; Wiggins & Thomas, Fl. Alaskan Arctic Slope 39, 1962; St. John, Fl. Southeast. Wash. 12, 1963; Hultén, Fl. Alaska 35, 1968; Calder & Taylor, Fl. Queen Charlotte Islands pt. 1, 130, 1968.

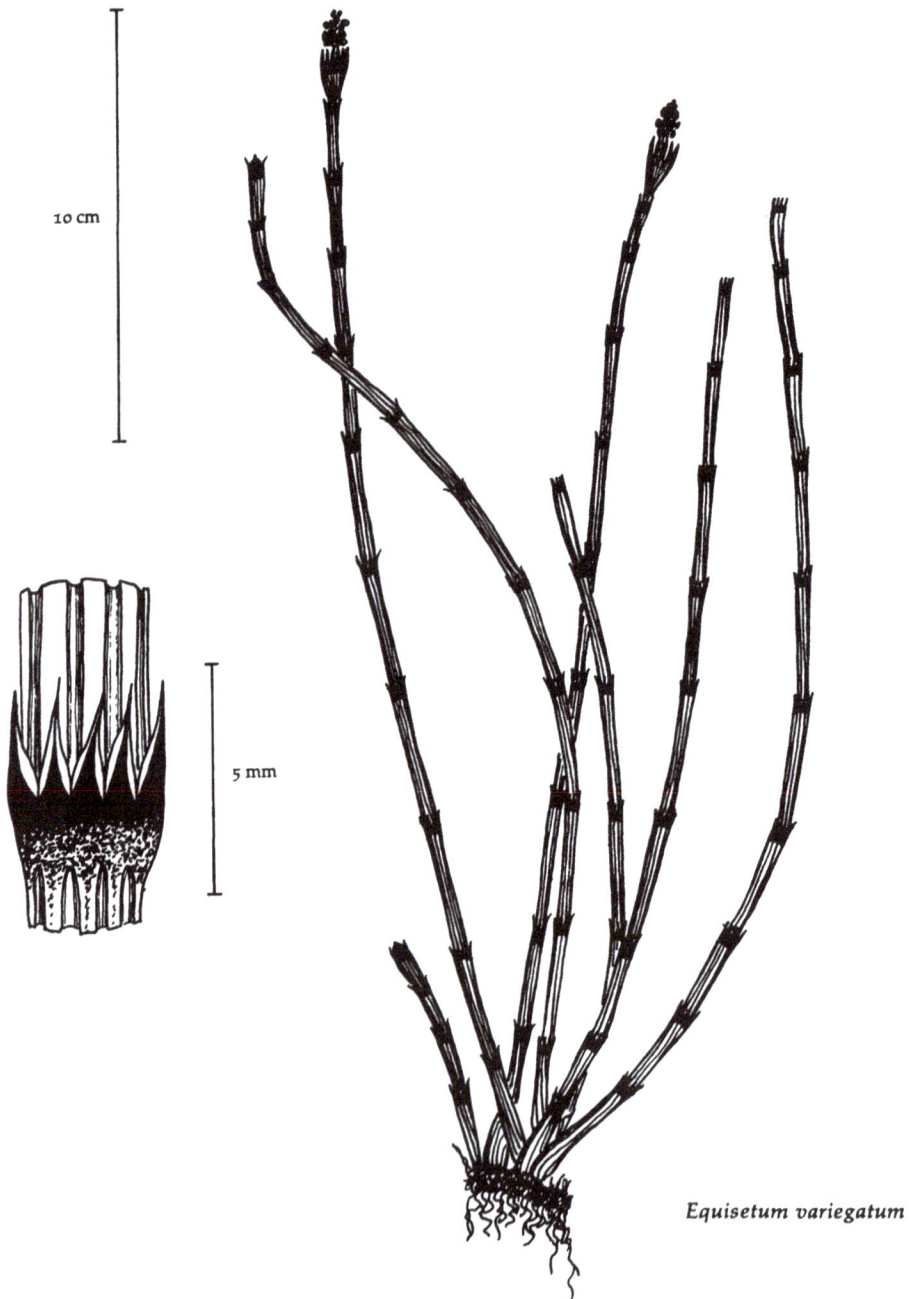

Equisetum variegatum

Rhizome black and shining, widely creeping near the surface of the ground, much branched. Stems slender, evergreen, all alike, clustered, ascending to erect, up to 4 dm tall, simple or branched from near the base but without whorls, 5- to 10-ridged; the ridges finely 1-grooved with 2 rows of minute tubercles. Central cavity about 1/3 the diameter of the stem; vallecular cavities conspicuous. Sheaths green to ashy with a broad black band around the top, tight below, rather loose above, the ridges 3-furrowed. Teeth persistent, triangular-ovate to broadly lanceolate, white scarious-margined with a blackish centre, abruptly acuminate, tips deciduous. Cones subsessile, sharply apiculate. ($2n = 216$)

HABITAT Sandy shores, wet thickets, banks of mountain streams.

RANGE Arctic regions; Alaska to Labrador, south to Oregon, Utah, Colorado, Wisconsin, Illinois, New York, and Connecticut; Greenland; Eurasia.

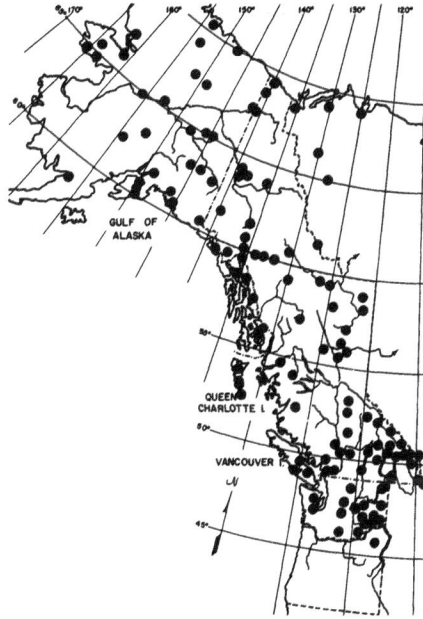

COMMENTS Common in suitable sites throughout. This is a variable species and many names have been applied to different forms of it. According to Hauke (1963) only two deserve varietal status, viz., the widely ranging var. *variegatum* and var. *alaskanum* A. A. Eaton; the latter is found along the Pacific Coast from the Aleutians to Vancouver Island. In Washington and the mainland of Alaska it intergrades with the typical variety. It is characterized by its greater diameter of 4 mm and tight, not flaring, sheaths with black incurved teeth. Hybrids between this species and *E. hyemale* are known in North America as *E. variegatum* var. *jesupii* A. A. Eaton and in Europe as *E.* × *trachyodon* A. Br. Its hybrids with *E. laevigatum* are *E.* × *nelsonii* (A. A. Eaton) Schaffner (*E. variegatum* var. *nelsonii* A. A. Eaton).

Hymenophyllaceae - Filmy Fern Family

Rhizomes slender, branching, creeping, bearing fronds in two rows. Fronds circinate, small and delicate with blades only one cell thick; veins free, sori marginal at the ends of veins; indusium tubular or bivalvate, usually sessile or sunken in the frond tissue. Sporangia stalked or sessile; annulus complete, oblique; spores numerous and all alike.

A predominantly southern hemisphere family of several hundred species, many of which are epiphytes. Only two genera and seven species are known to occur in continental United States and Canada: one genus (*Trichomanes*) with six species in eastern North America, and the other with a single species in the West.

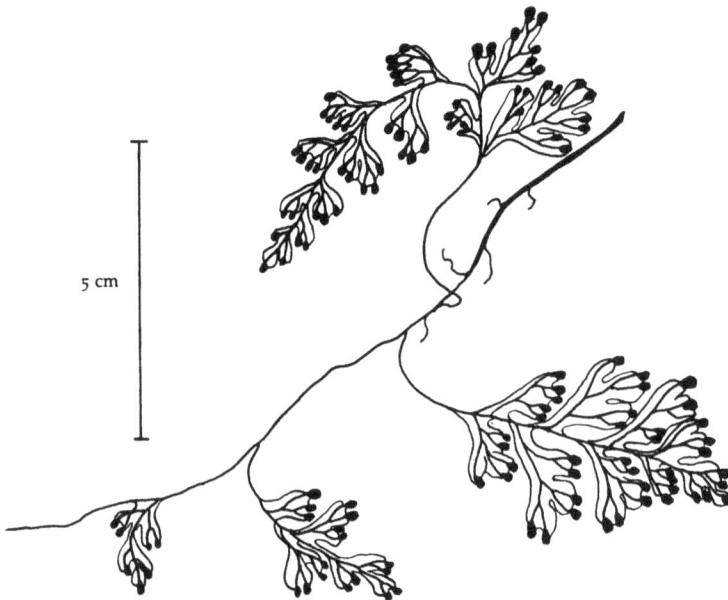

5 cm

1 cm

Mecodium wrightii

MECODIUM Copeland, Philip. J. Sci. **67**:17, 1938

Epiphytic or on wet mossy rocks. Rhizome very slender; fronds remote, pinnately compound with entire margins, glabrous. Sori terminal on frond-segments; indusium with two deeply cleft valves, receptacle usually included.

A genus of about 100 species, chiefly in the southern hemisphere and tropics. A few range into the north temperate zone. Only one is known from North America.

Mecodium wrightii (van den Bosch) Copel.

Philip. J. Sci. **67**:23, 1938; Iwatsuki, Am. Fern J. **51**:141–4, 1961.
Hymenophyllum wrightii van den Bosch, Synopsis 51, 1859.

Rhizomes threadlike, extensively creeping and branched. Fronds 3–5 cm long, delicate and lax, glabrous, pinnate. Stipes threadlike, up to 15 mm long, blackish, glabrous except for a small tuft of hairlike scales at the base, very narrowly winged through the decurrence of the basal pinnae. Pinnae pale green, pinnatifid through 3 or 4 dichotomous branchings into a few, linear, blunt, almost filamentous lobes, glabrous. Veins conspicuous, dark, dichotomously branched. Sori at tips of veinlets; indusia bivalvate to the base, receptacle with sporangia included.

HABITAT Wet, shady rock faces, or epiphytic at base of trees.

COMMENTS First collected in our area in 1957 at Dawson Inlet, Graham Island, more recently the gametophyte phase alone has been found near Prince Rupert on Moresby Island, British Columbia, on Biorka Island, Alaska and on the west coast of Vancouver Island. These are the only known stations for this genus in North America. This species is the only filmy fern so far discovered in western North America.

RANGE South Japan to Korea; Queen Charlotte Islands and adjacent mainland, Vancouver Island, Biorka Island, Alaska.

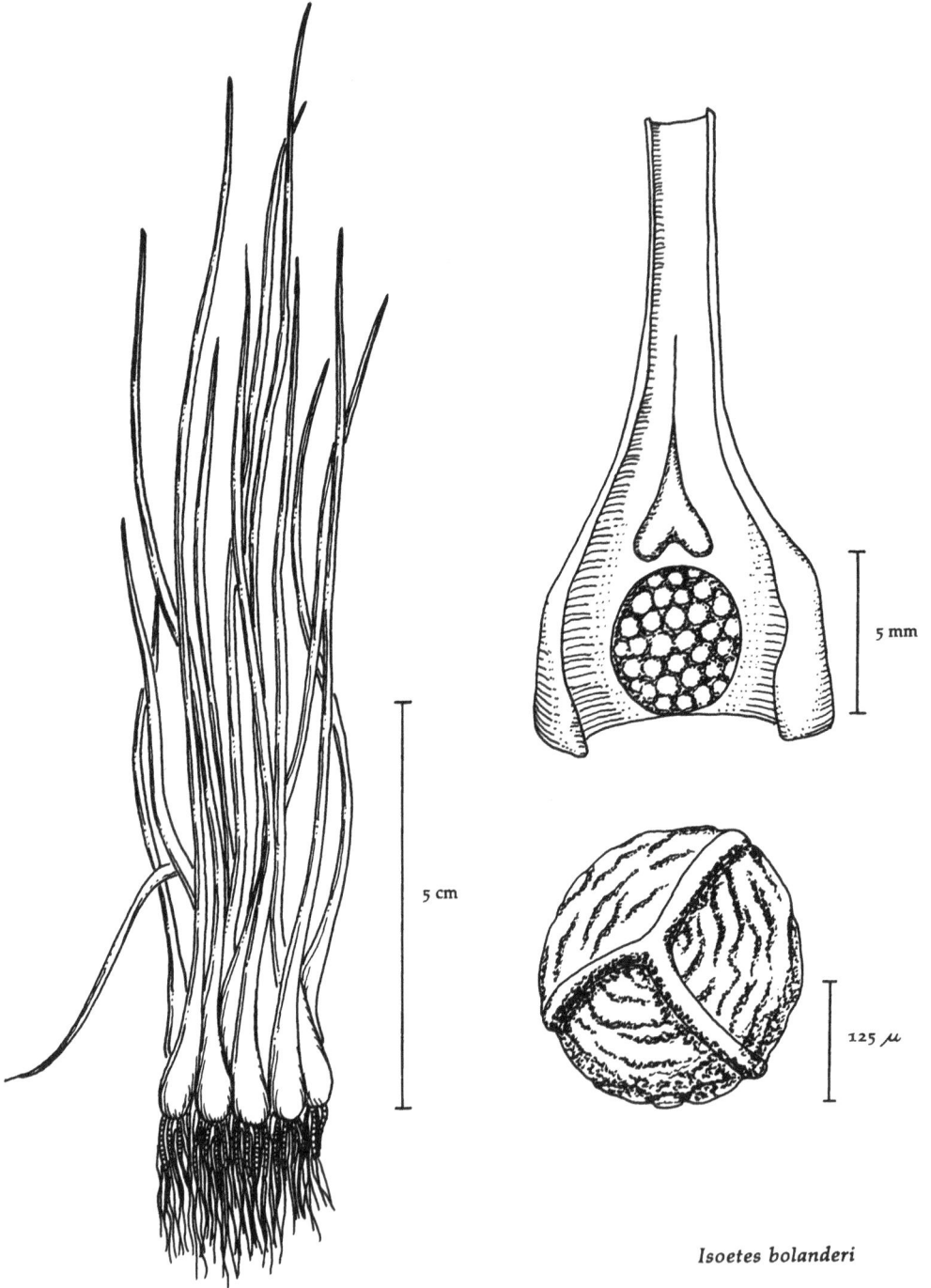

Isoetes bolanderi

5 cm

5 mm

125 μ

Isoetaceae - Quillwort Family

Submerged, amphibious, or terrestrial perennials. Stem (a fleshy corn-like structure) flattened on top, 2–3-lobed, up to 2.5 cm in diameter, crowned by leaf-bases, and giving rise to numerous roots. Leaves few to many, spirally arranged, simple, elongate-subulate, up to 50 cm long, dilate and imbricate at base, rounded, quadrate, or triangular in cross-section above, hollow, divided into 4 transverse, septate cavities. Sporangia of two sorts but solitary, large, globose to oblong, immersed in a cavity in the ventral side of the leaf base, more or less covered by a membrane (the velum); ligule, a small triangular or elongate, delicate tissue borne just above the sporangium. Spores of two kinds; microspores very small and numerous, powdery, essentially smooth; megaspores few and up to 0.25 mm in diameter, chalky white when dry, hemispherical below the equatorial ridge and with three triradiate lines joining at the apex; the walls variously sculptured by spines, tubercles, crests, or wrinkles, rarely smooth.

A family consisting of a single genus of world-wide distribution. (Named from Greek, *iso*, equal and *etos*, year, perhaps having reference to the evergreen nature of the leaves.)

ISOETES L., Sp. Pl. 1100, 1753

Characters are those of the family.

1   Surface of megaspores chiefly papillate, tuberculate, spiny, or with low ridges.

  2   Megaspores papillate or tuberculate.

    3   Corms 3-lobed; velum completely covering the sporangium; terrestrial.   *I. nuttallii*

    3   Corms 2-lobed; velum narrow, usually covering not more than 1/3 of the sporangium; amphibious or submerged.

      4   Amphibious; stomata and peripheral strands usually evident.   *I. howellii*

      4   Submerged; stomata few, peripheral strands lacking.   *I. bolanderi*

  2   Megaspores with distinct spines.   *I. echinospora*

1   Surface of megaspores irregularly crested.   *I. occidentalis*

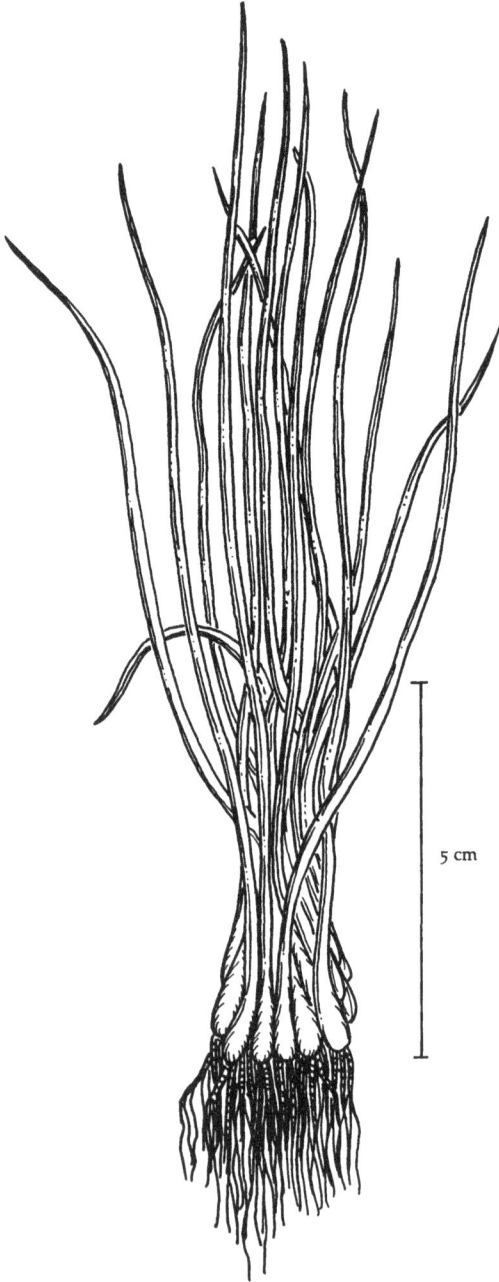

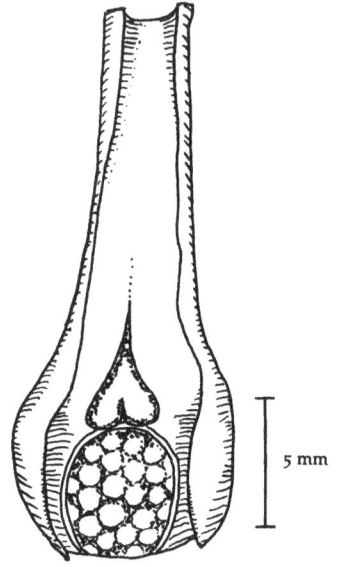

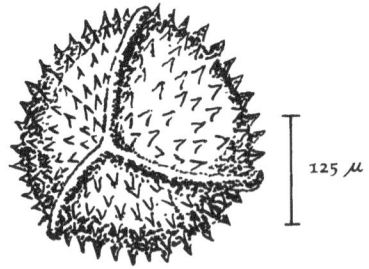

Isoetes echinospora

Isoetes bolanderi Engelm.

Am. Nat. 8:214, 1864; Macoun, Cat. Can. Pl. pt. 5, 293, 1890; Piper, Fl. Wash. 89, 1906; Henry, Fl. S. British Columbia 11, 1915; Pfeiffer [in] Abrams, Ill. Fl. Pac. States 1:37, *Fig. 76*, 1923; Frye, Ferns Northwest 40, 1934; Peck, Man. Higher Pl. Ore. 55, 1941.

Submersed, sometimes in quite deep water. Stem 2-lobed. Leaves varying in number up to about 25, slender, up to 15 cm long, rather soft; ligule small, cordate; sporangia about 4 mm long, 1/3 covered by the velum. Megaspores white, about 400 μ in diameter, more or less marked with low tuberculate ridges or wrinkles. Microspores about 27 μ long, obscurely spinulose.

H A B I T A T Lakes and ponds, often alpine.

R A N G E Southern British Columbia to California, east to Wyoming and Arizona.

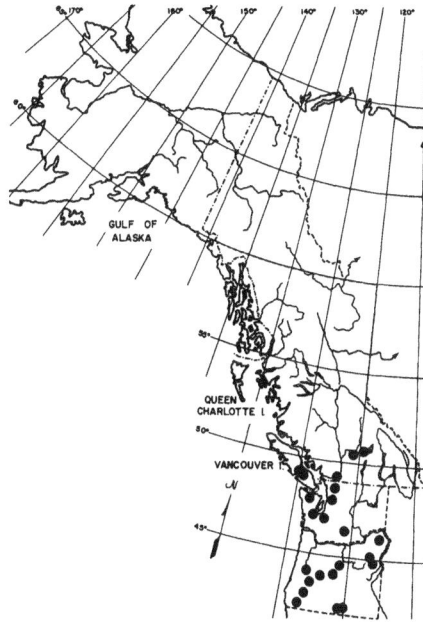

C O M M E N T S The smaller size of the megaspores and their low tubercles distinguish this species from *L. occidentalis* Henderson.

Isoetes echinospora Durieu

Bull. Soc. Bot. Fr. 8:164; 1861; Macoun, Cat. Can. Pl. pt. 5, 292, 1890; Henry, Fl. S. British Columbia 11, 1915; Piper & Beattie, Fl. Northwest Coast 15–16, 1915; Pfeiffer [in] Abrams, Ill. Fl. Pac. States 1:36, 37 *Figs. 75, 77*, 1923; Frye, Ferns Northwest 38–40, 1934; Morton [in] Gleason, Ill. Fl. 1:10, *Fig.* 1952; Löve & Löve, Bot. Not. 114:49, 1961; Hultén, Fl. Alaska 32, 1968; Calder & Taylor, Fl. Queen Charlotte Islands pt. 1, 139, 1968.
I. muricata Dur., Bull. Soc. Bot. Fr. 11:100, 1864.
I. braunii Dur., Bull. Soc. Bot. Fr. 11:101, 1864, not Unger.

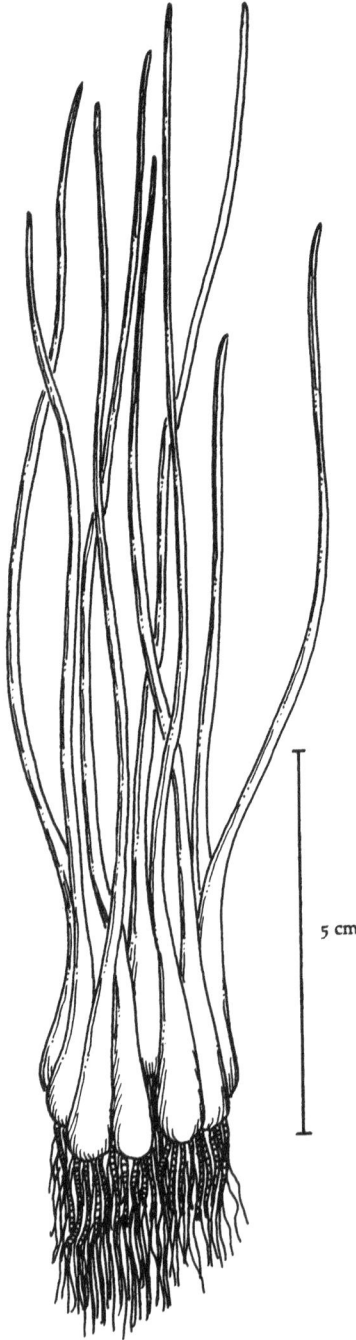

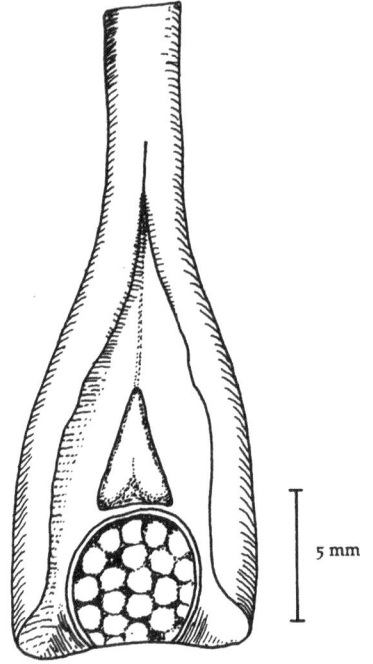

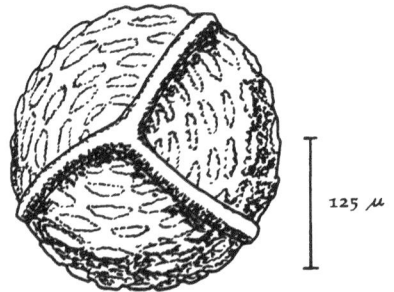

Isoetes howellii

I. maritima Underw., Bot. Gaz. **13**:94, 1888.
I. macounii A. A. Eaton, Fern Bull. **8**:12, 1900.
I. asiatica (Makino) Makino, Bot. Mag. Tokyo **28**:184, 1914.
I. flettii (A. A. Eaton) Pfeiffer, Ann. Mo. Bot. Gard. **9**:186,1923.

Usually submersed, sometimes in several feet of water, occasionally emersed for a time through drop in water level. Stem 2-lobed. Leaves 10–30, sometimes more, straight but more often recurved, up to 20 cm long; ligule deltoid; sporangia oblong about 5 mm long, often spotted, 1/2–3/4 covered by the velum. Megaspores white, about 500 μ in diameter, copiously covered by spines which may be single or even confluent. Microspores light fawn-coloured, generally smooth, about 30 μ long. ($2n = 22$)

HABITAT Lakes and ponds.

COMMENTS The relationships of this species are confused; the North American representative is probably best referred to ssp. *muricata* (Dur.) Löve & Löve (*I. braunii* Dur.). A coastal form described from the salt marshes at Alberni, British Columbia, has been named ssp. *maritima* (Underw.) Löve & Löve (*I. maritima* Underw.). This form which is said to range from coastal Washington to Alaska and the Aleutian and Commander Islands is very doubtfully distinct from ssp. *muricata*. The east Asiatic form (*I. asiatica* Makino) is intermediate between the American and European representatives.

RANGE Throughout boreal North America, south to Pennsylvania, Ohio, Michigan, Minnesota, Colorado, Utah, and northern California; Eurasia.

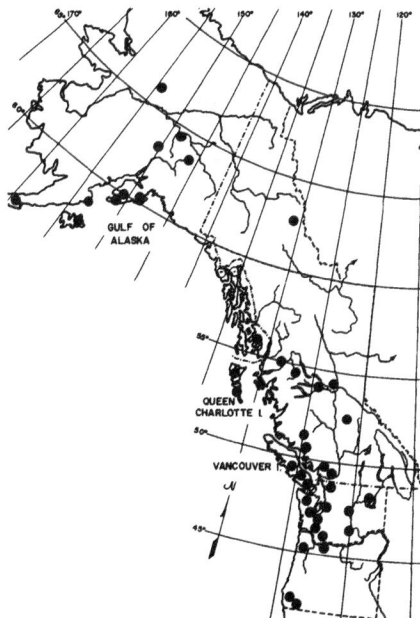

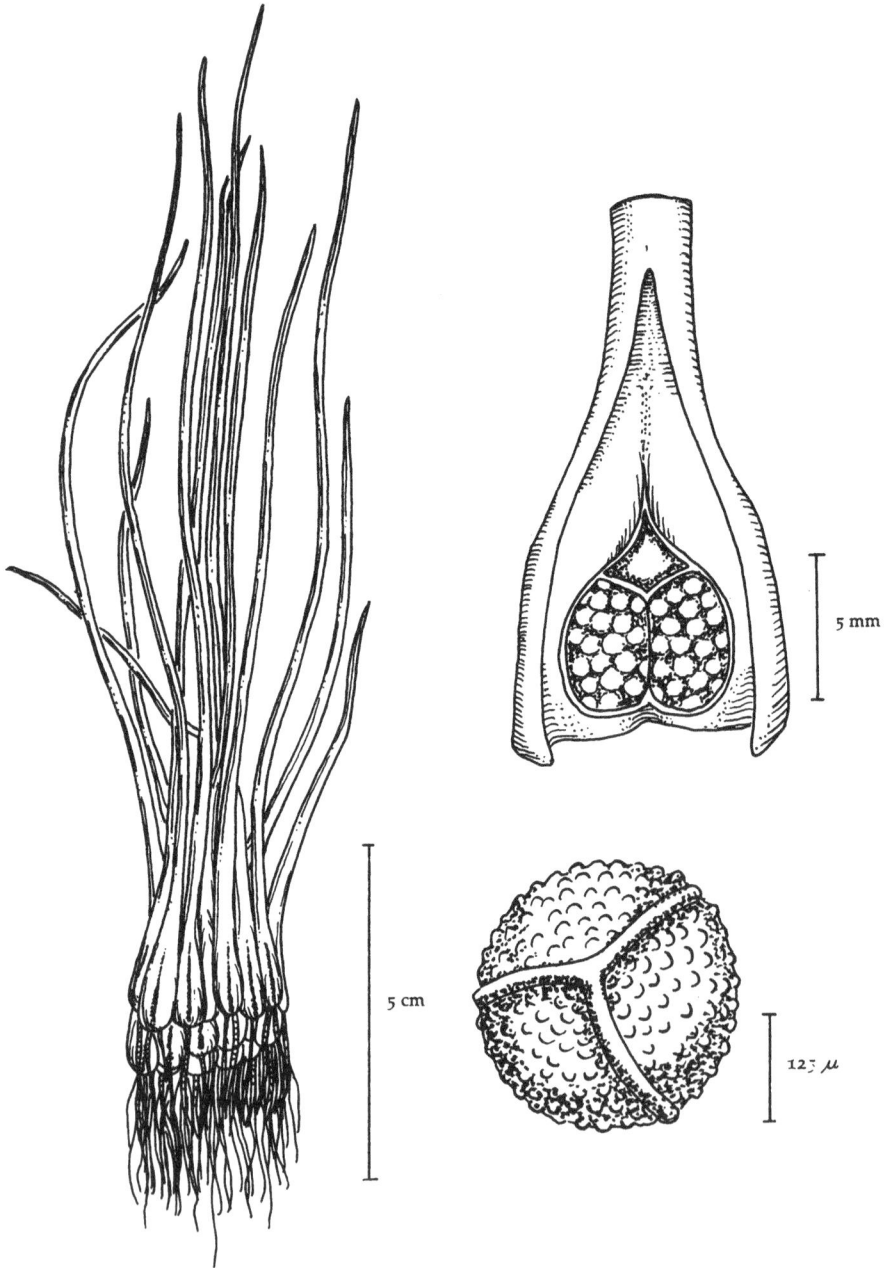

5 mm

5 cm

123 μ

Isoetes nuttallii

Isoetes howellii Engelm.

Trans. St. Louis Acad. Sci. **4**:385, 1882; Piper, Fl. Wash. 88, 1906; Pfeiffer [in] Abrams, Ill. Fl. Pac. States **1**:37, *Fig. 78*, 1923; Frye, Ferns Northwest 38, 1934; Peck, Man. Higher Pl. Ore. 55, 1941; St. John, Fl. Southeast. Wash. 3, 1963.
I. nuda Engelm., Trans. St. Louis Acad. Sci. **4**:385, 1882.
I. underwoodii Henderson, Bot. Gaz. **23**:124, 1897.
I. minima A. A. Eaton, Fern Bull. **6**:30, 1898.
I. melanopoda var. *californica* A. A. Eaton [in] Gilbert, List N. Am. Pterid. 10, 27, 1901.

Amphibious, stem 2-lobed. Leaves 5–28, up to 30 cm long, bright green, essentially erect, slender, with wide membranous margins extending well above level of the sporangia, then often abruptly narrowed; ligule narrow, elongated triangular; sporangia more or less oblong-orbicular, about 6 mm long, about 1/3 or less covered by velum. Megaspores about 475 μ in diameter, inconspicuously marked by anastomosing wrinkles or slightly tuberculate ridges; microspores about 30 μ long, more or less finely spinulose.

HABITAT Muddy shores and wet depressions.

COMMENTS If growing in shallow water this species is difficult to distinguish with certainty from *I. bolanderi*. The cordate ligule of the latter seems to be constant, and its megaspores average slightly less in diameter.

RANGE Oregon to California, east to Montana and Idaho.

5 cm

5 mm

125 μ

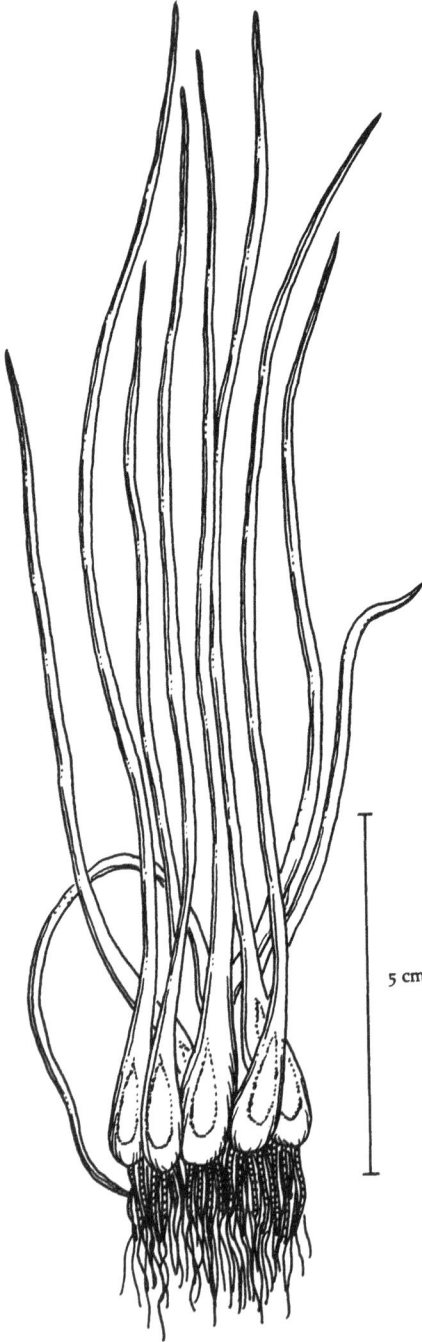

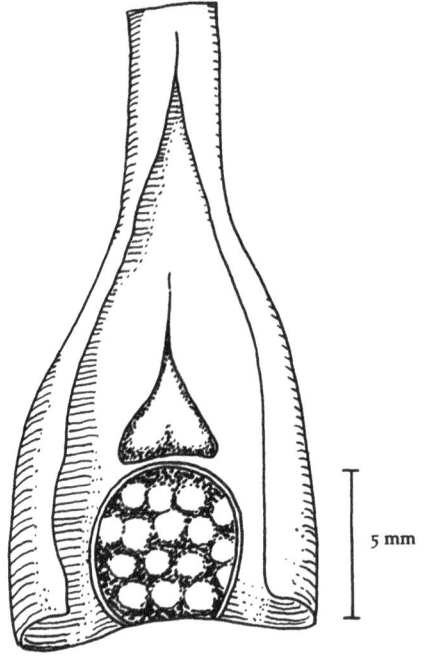

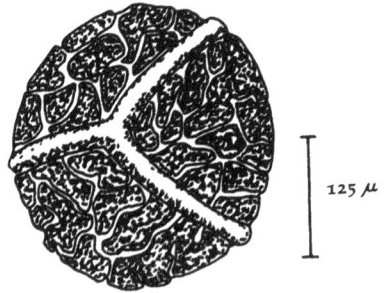

Isoetes occidentalis

Isoetes nuttallii A. Br.

[In] Engelm., Am. Nat. **8**:215, 1874; Macoun, Cat. Can. Pl. pt. 5, 293, 1890; Henry, Fl. S. British Columbia 11, 1915; Piper & Beattie, Fl. Northwest Coast 15, 1915; Pfeiffer [in] Abrams, Ill. Fl. Pac. States **1**:38, *Fig. 79*, 1923; Frye, Ferns Northwest 37, 1934; Peck, Man. Higher Pl. Ore. 55, 1941.
I. suksdorfii Baker, Fern Allies 132, 1887.

Usually terrestrial. Corm somewhat 3-lobed. Leaves setaceous, up to 60 in number, 7–17 cm long, trigonous, very slender, bright green, hyaline, basal margins very evident; ligule small, triangular; mature sporangia very conspicuous, straw-coloured, oblong, about 5 mm long, 1.5 mm wide, completely covered by the velum. Megaspores mostly 400–500 μ in diameter, surface usually rather densely short papillate, rarely smooth, triradiate lines conspicuous. Microspores white to pale brownish, 25–30 μ long, somewhat papillose.

HABITAT Springy ground, not regularly inundated.

COMMENTS Good field characters are the slender, 3-angled leaves with a conspicuous hyaline margin towards the base and the prominent, somewhat sausage-shaped sporangia completely covered by the velum.

RANGE Southern Vancouver Island to California.

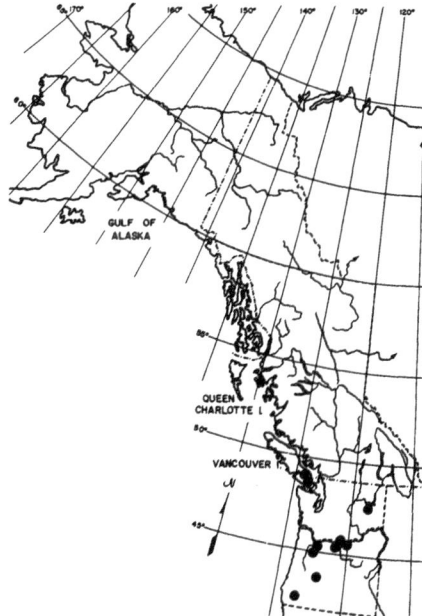

Isoetes occidentalis Henderson

Bull. Torrey Bot. Club **27**:358, 1900; Piper & Beattie, Fl. Northwest Coast 15, 1915; Pfeiffer [in] Abrams, Ill. Fl. Pac. States **1**:36, *Fig. 73*, 1923; Frye, Ferns Northwest 42–3, 1934; Peck, Man. Higher Pl. Ore. 55, 1941; St. John, Fl. Southeast. Wash. 3, 1963.

I. lacustris var. *paupercula* Engelm., Trans. St. Louis Acad. Sci. **4**:377, 1882.
I. paupercula (Engelm.) A. A. Eaton, Proc. U.S. Nat. Mus. **23**:649, 1901.
I. piperi A. A. Eaton, Fern Bull. **13**: 51, 1905.

Submersed. Stem 2-lobed. Leaves ranging from 10–30 in number, sometimes more, 5–20 cm long, dark green, often somewhat rigid, ligule short-
triangular, sporangia almost orbicular, 5–6 mm in diameter, 1/4–1/3
covered by the velum. Megaspores cream coloured, 400–500 μ in diameter, surface marked with sharp ridges or crests, occasionally tuberculate,
sometimes almost smooth. Microspores 25–40 μ long, finely spinulose or
papillose.

HABITAT Lakes and ponds.

RANGE Central British
Columbia to Colorado and
California.

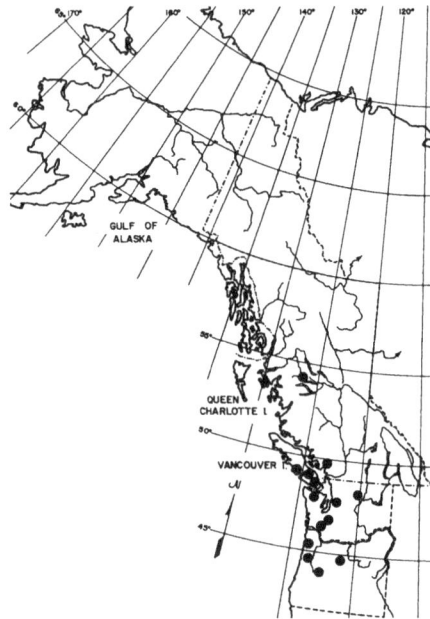

COMMENTS Can best be distinguished by the ridges or crests of the
megaspores. It seems that *I. piperi* A. A. Eaton is only a local form of this
species and is scarcely worth recognition.

Lycopodiaceae / Club-Moss Family

Perennial mosslike plants, terrestrial (our species) or epiphytic. Stems elongate, simple to much branched, upright or trailing, roots arising from the underside; leafy nearly throughout. Leaves small, simple, 1-nerved, numerous, mostly 4- to 10-ranked, reflexed or ascending, free or partly adnate to the branchlets, evergreen, entirely or slightly toothed, mostly lanceolate or linear. Sporophylls either like or unlike the vegetative leaves, in the latter case usually aggregated into a terminal cone or strobilus. Sporangia solitary, axillary or on the upper surface of the sporophylls, large, 1 mm or more in diameter, kidney-shaped or spherical. Spores tetrahedral, all alike, smoothish or variously sculptured.

The family is usually regarded as consisting very largely of a single, almost cosmopolitan, genus of about 200 species, the majority of which are found in tropical and subtropical regions. (Name derived from the Greek *lycos*, wolf, and *pous*, foot, from a fancied resemblance.)

LYCOPODIUM L., Sp. Pl. 1100, 1753

Characters are those of the family.

1	Sporangia in the axils of ordinary green leaves, not forming definite cones; broad and flat gemmae (reproductive buds) often borne in the upper axils. *L. selago*
1	Sporangia in clearly differentiated terminal cones; gemmae lacking.
2	Sterile branches creeping, never ascending; cones solitary at the end of a slender branch growing directly from the creeping base; sporophylls green, essentially like foliage leaves. *L. inundatum*
2	Sterile branches erect or strongly ascending from the horizontal, superficial or subterranean stems; cones with yellowish, scalelike sporophylls borne on erect leafy branches.
3	Ascending branches simple or few-forked; leaves uniform, spirally arranged in several ranks; slender, subulate to hairlike tips; lowermost sporophylls erose to fimbriate with subulate-aristate to hairlike tips.
4	One cone, sessile at the ends of densely leafy branches; sporophylls merely crenulate-erose, the lower one (as well as the leaves) with subulate-aristate teeth. *L. annotinum*
4	One to six cones, stalked, on leafy-bracted peduncles; sporophylls fimbriate-erose, the lower one (as well as the leaves) long attenuate to a soft tip. *L. clavatum*

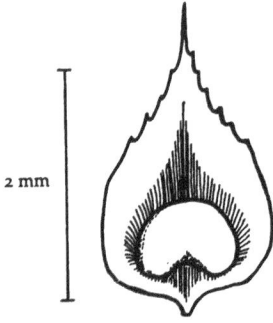

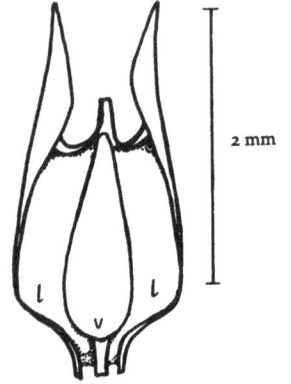

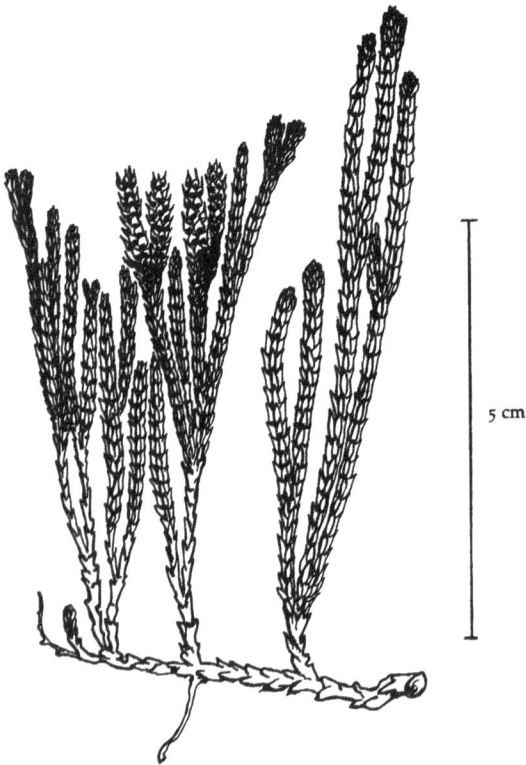

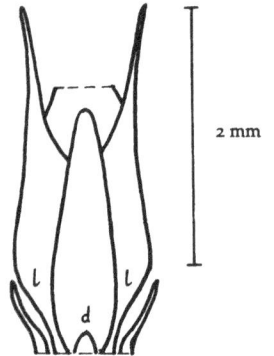

Lycopodium alpinum

3   Ascending branches freely forking, treelike, bushy-branched, or fan-like; sporophylls entire or merely erose, acuminate to subulate-tipped.

5   Ascending branches erect and tree-like with very numerous, crowded, ascending or recurved branchlets; leaves arranged spirally to somewhat dorsiventrally in 6 or 8 ranks, the free portion 3–5 mm long; cones sessile at tips of unmodified branches.   *L. obscurum*

5   Ascending branches tufted, bushy, or fanlike; branchlets flattened with minute leaves in 4 ranks, or, if terete, the leaves in 5 ranks; cones on pedunclelike branches.

   6   Sterile branchlets terete; leaves mostly in 5 ranks, all alike, the free, slender subulate tips no longer than the adnate base.   *L. sitchense*

   6   Sterile branchlets flattened; leaves in 4 ranks, those of the upper and lower faces unlike the marginal ones, which are clearly decurrent.

      7   Cones sessile, usually less than 2 cm long; marginal leaves deltoid-ovate to deltoid-lanceolate, falcate, free tips about equalling the decurrent base; leaves of the lower surface trowel-shaped.   *L. alpinum*

      7   Cones usually peduncled and more than 2 cm long; marginal leaves lance- or deltoid-subulate, free tips much shorter than the decurrent base; leaves of the lower surface subulate and appressed.
                                                                        L. complanatum

Lycopodium alpinum L.

Sp. Pl. 1104, 1753; Macoun, Cat. Can. Pl. pt. 5, 291, 1890; Henry, Fl. S. British Columbia 12, 1915; Victorin, Contr. Lab. Bot. Univ. Montreal No. 3:105, 1925; Frye, Ferns Northwest 25, *Fig.* v, 1934; Morton [in] Gleason, Ill. Fl. 1, 6, *Fig.*, 1952; Anderson, Fl. Alaska 22, 1959; Hultén, Fl. Alaska 30, 1968. *Diphasium alpinum* (L.) Rothm. Feddes Rep. Spec. Nov. 54:65, 1944.

Creeping stem nearly superficial, elongate, rooting throughout, up to 5 dm or more long. Erect branches glaucous, tufted, several times dichotomously branched, the branches short. Leaves of the somewhat flattened, sterile branchlets glaucous, in 4 ranks, acuminate, entire, more or less of 3 forms; the dorsal leaves lanceolate-subulate, appressed, adnate about half their length; the two marginal ones 4–5 mm long, the free tips deltoid-falcate about equalling the decurrent bases; the ventral shorter, trowel-shaped, scarcely adnate at the concave base. Cones essentially sessile at the end of branched, leafy, pedunclelike stems. Sporophylls yellow, broadly triangular, abruptly contracted at the base, gradually acuminate above, margins scarious, slightly erose. ($2n = 46$)

HABITAT Alpine and subalpine meadows, cool woods at high altitudes.

COMMENT This species of high elevations is very similar in general appearance and habit of growth to *L. sitchense*. It can be recognized by its more or less flattened branches with the types of leaves in 4 ranks; in particular, the difference between the lower leaf and the two lateral ones is very marked.

2 mm

5 mm

10 cm

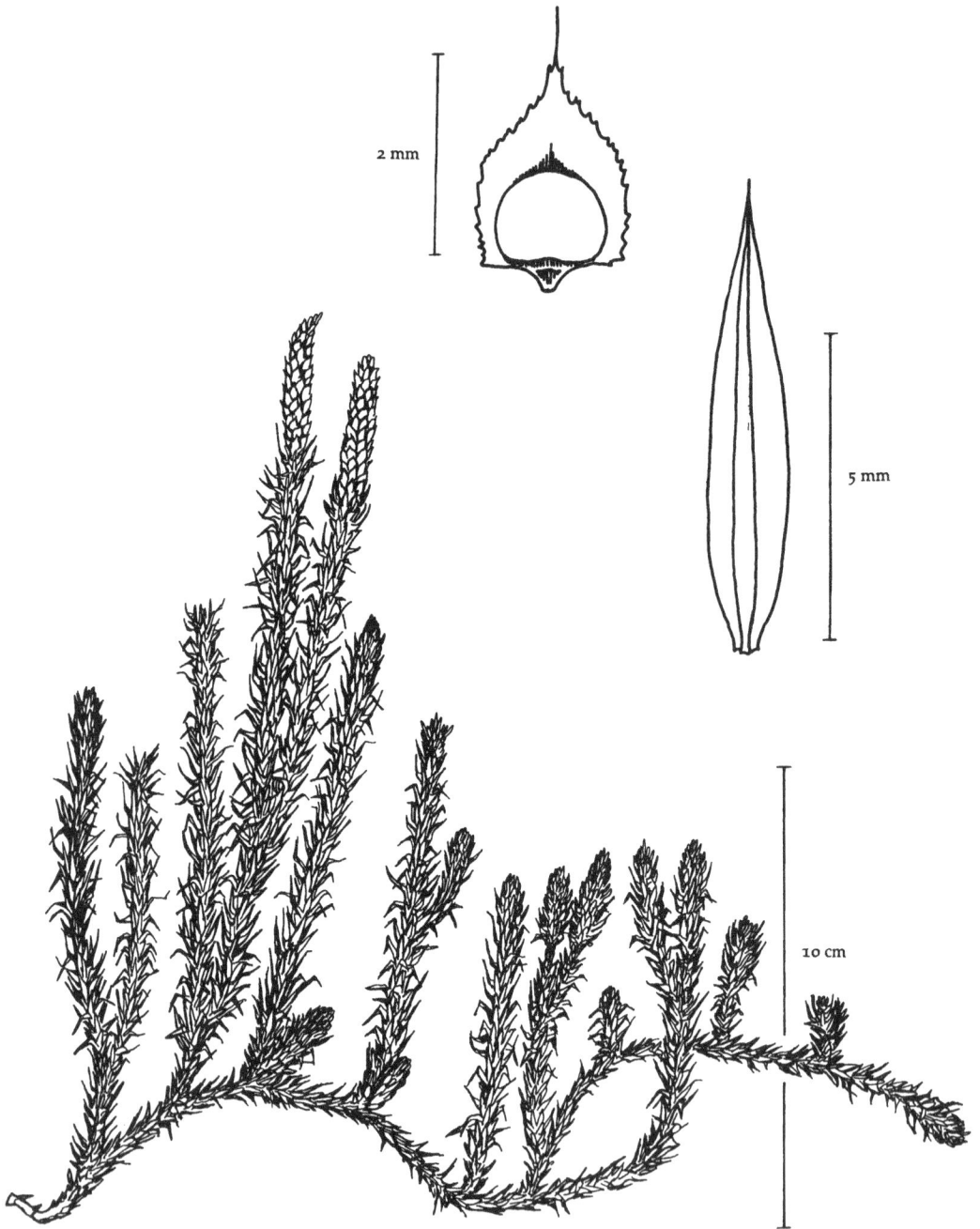

Lycopodium annotinum

RANGE Alaska to southern British Columbia and Montana; Keweenaw County, Michigan; Gaspé region, Quebec; Greenland; Eurasia.

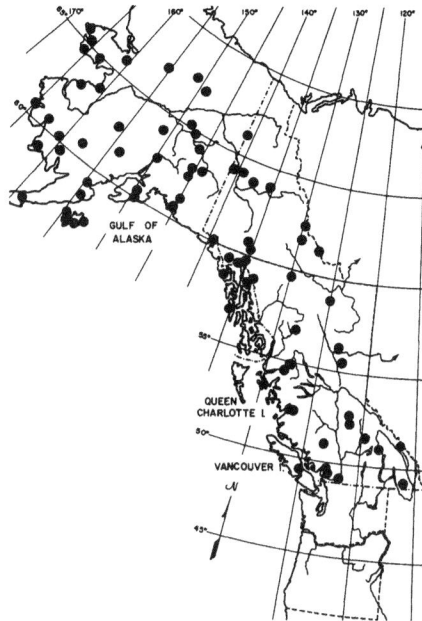

Lycopodium annotinum L.

Sp. Pl. 1103, 1753; Macoun, Cat. Can. Pl. pt. 5, 289, 1890; Henry, Fl. S. British Columbia 12, 1915; Piper & Beattie, Fl. Northwest Coast 14, 1915; Maxon [in] Abrams, Ill. Fl. Pac. States 1:45, *Fig. 97*, 1923; Victorin, Contr. Lab. Bot. Univ. Montreal No. 3, 100, 1925; Frye, Ferns Northwest 20, *Fig.* III, 1934; Peck, Man. Higher Pls. Ore. 56, 1941; Morton [in] Gleason, Ill. Fl. 1:4, *Fig.*, 1952; Anderson, Fl. Alaska 22, 1959; Wiggins & Thomas, Fl. Alaskan Arctic Slope, 35, 1962; St. John, Fl. Southeast. Wash. 1, 1963; Hultén, Fl. Alaska 27, 1968; Calder & Taylor, Fl. Queen Charlotte Islands, pt. 1, 132, 1968.

Stems trailing, elongate, up to 1 m long, superficial or shallowy subterranean, mostly unbranched, sparsely leafy. Erect branches not flattened, simple or once or twice forked, annual increase in length conspicuous. Leaves 8-ranked with 4 leaves in each whorl, uniform, firm or stiff, hard, attenuate or linear-oblanceolate, acuminate, tipped by a short, stiff point, minutely serrulate or entire, reflexed or divergent to appressed-ascending. Cones sessile and solitary at the ends of leafy branches. Sporophylls ovate, acuminate, with broad, scarious, denticulate margins, straw-coloured. Sporangia kidney-shaped, about 1.5 mm wide. ($n = 34$)

HABITAT Acid soil of wet woods or edge of bogs.

COMMENTS A widespread and common species. It shows considerable variation in different habitats, particularly those related to altitude. At low altitudes the growth is lax with relatively soft, reflexed leaves on the erect branches. At high altitudes growth is short and stiff with very firm, ascending or appressed leaves. Several of these intergrading forms have

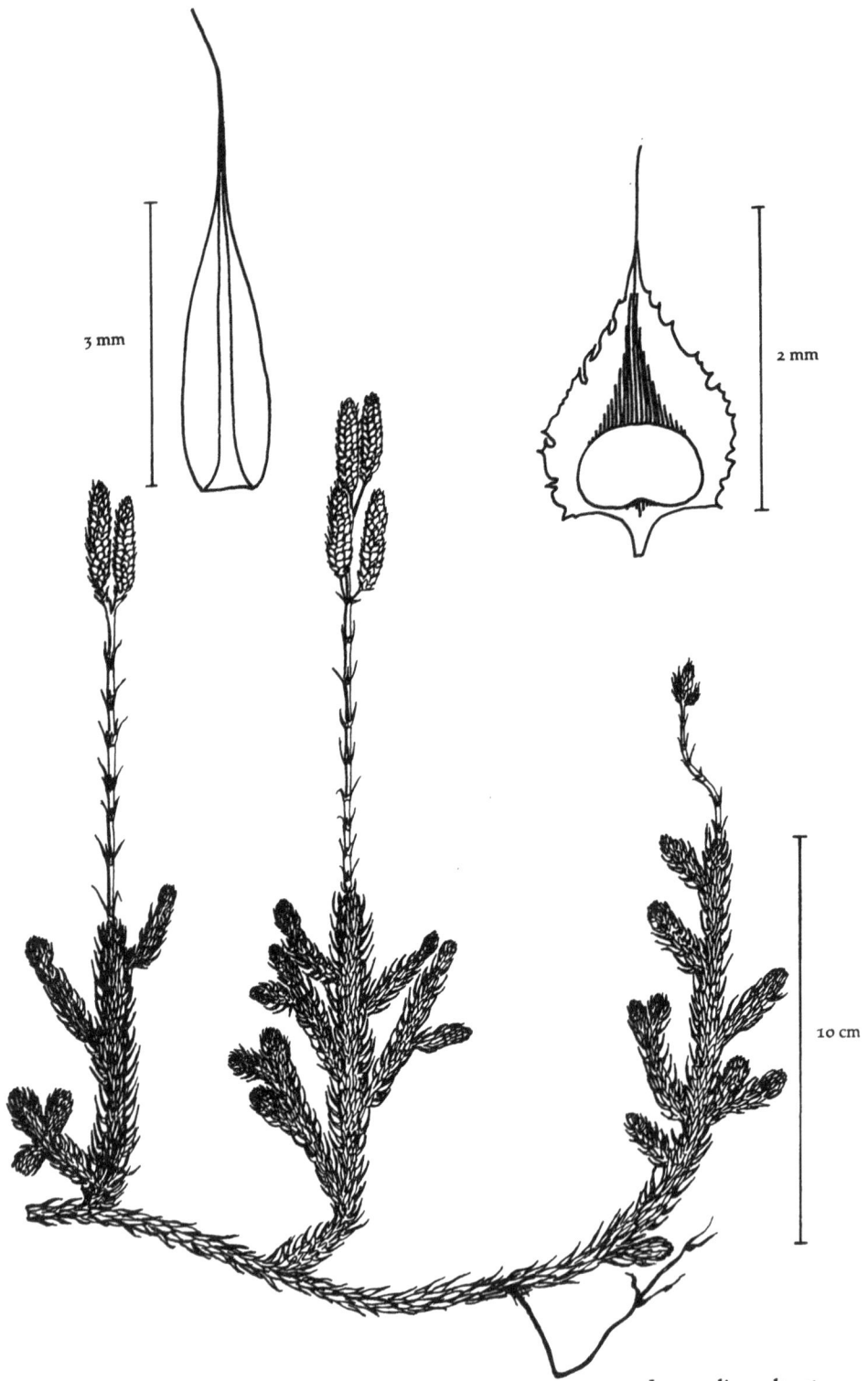

3 mm

2 mm

10 cm

Lycopodium clavatum

been named but their significance seems to be more edaphic than geographic. This species is easily recognized by its erect, leafy, fertile branches each bearing a single, sessile cone.

R A N G E Boreal regions; Alaska to Labrador and Newfoundland, south to northern Oregon, Colorado, Minnesota, Wisconsin, Michigan, mountains of Virginia and West Virginia, Pennsylvania, and New England; Eurasia.

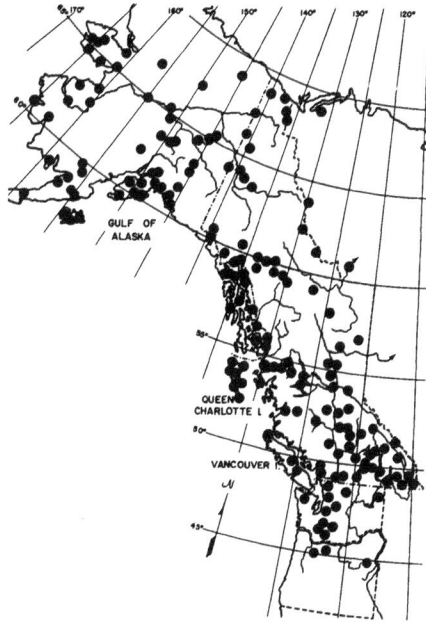

Lycopodium clavatum L.

Sp. Pl. 1101, 1753; Macoun, Cat. Can. Pl. pt. 5, 290, 1890; Henry, Fl. S. British Columbia 12, 1915; Piper & Beattie, Fl. Northwest Coast 14, 1915; Maxon [in] Abrams, Ill. Fl. Pac. States 1:45, *Fig. 98*, 1923; Victorin, Contr. Lab. Bot. Montreal No. 3:102, 1925; Frye, Ferns Northwest 20, *Fig. IV*, 1934; Peck, Man. Higher Pl. Ore. 56, 1941; Morton [in] Gleason, Ill. Fl. 1:4, *Fig.*, 1952; Anderson, Fl. Alaska 21, 1959; Hultén, Fl. Alaska 28, 1968; Calder & Taylor, Fl. Queen Charlotte Islands pt. 1, 132, 1968.

Horizontal stems on the surface of the ground, greatly elongated, up to 1 m or more, rooting at intervals, rounded. Erect stems, at first simple then more or less dichotomously branched, the branches mostly once forked. Leaves bright green, mostly in 10 ranks, uniform, incurved-spreading to appressed-ascending, linear-subulate, attenuate, usually tipped by a soft, white, hairlike bristle subentire to remotely serrulate. Peduncles with distant, yellowish, linear-subulate, scalelike leaves. Cones cylindric, 1–4 or 5 to each peduncle; sporophylls ovate with white, filiform tips, margins broadly scarious-denticulate. Sporangia reniform about 1.5 mm broad. ($n = 34$)

H A B I T A T Dry woods and rocky places with acid soil, more or less in the open.

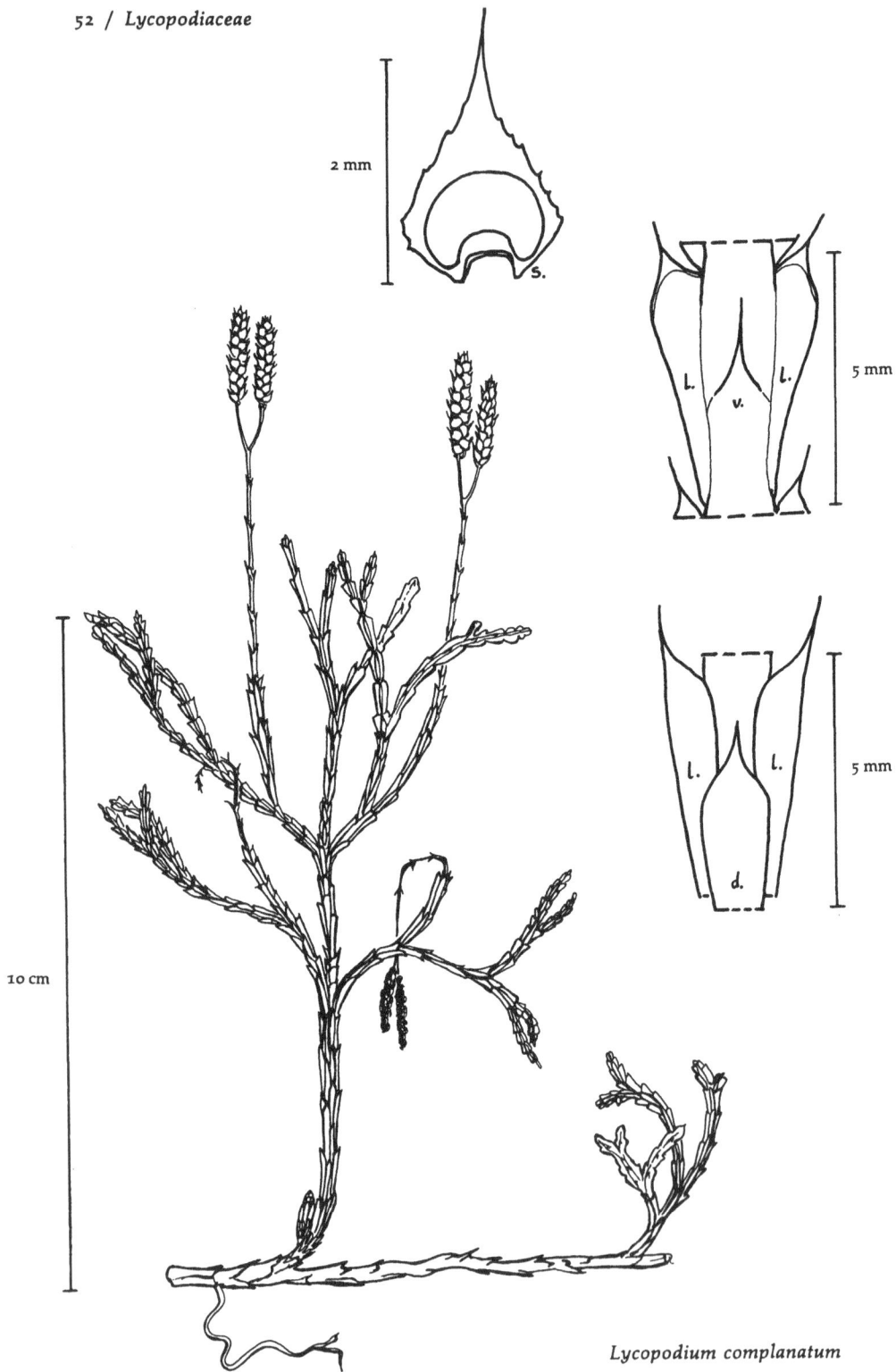

Lycopodium complanatum

RANGE Alaska to New-
foundland, south to Oregon,
Montana, Minnesota,
Michigan, southern North
Carolina, and West Virginia;
Eurasia.

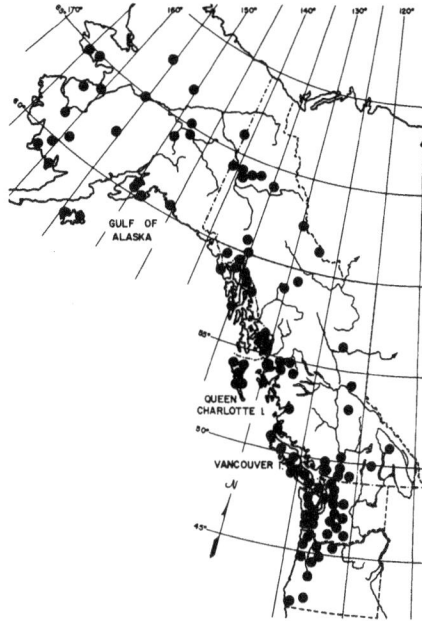

COMMENTS Very common west of the coastal mountains, much less so
in the interior. Quite variable as to the number of cones to each peduncle;
var. *monostachyon* Hook. and Grev., with a single cone on a very short
peduncle, seems to be the form of the species on exposed, rocky situations
at high altitudes and latitudes. Hultén reports this as the only variety in
the interior of Alaska and the Yukon. Also of apparent geographic significance is var. *integerrimum* Spring, characterized by leaves lacking the soft
terminal bristle. The great majority of British Columbian specimens can
best be referred to this variety, which seems to be limited largely to this
Province and Washington. In the field, good recognition features are the
extensive trailing growth, the long, narrow, soft leaves, and the erect
fertile branches with linear, awl-shaped leaves.

Lycopodium complanatum L.

Sp. Pl. 1104, 1753; Macoun, Cat. Can. Pl. pt. 5, 290, 1890; Henry, Fl. S. British
Columbia 12, 1915; Piper & Beattie, Fl. Northwest Coast 14, 1915; Maxon [in]
Abrams, Ill. Fl. Pac. States **1**:44, *Fig.* 94, 1923; Victorin, Contr. Lab. Bot. Univ.
Montreal No. **3**:107, 1925; Frye, Ferns Northwest 24, *Fig.* IV, 1934; Peck, Man.
Higher Pl. Ore. 56, 1941; Anderson, Fl. Alaska 21, 1959; Morton [in] Gleason,
Ill. Fl. **1**:6, *Fig.*, 1952; Hultén, Fl. Alaska 29, 1968; Calder & Taylor, Fl. Queen
Charlotte Islands pt. 1, 133, 1968.
L. anceps Wallr., Linnaea **12**:676, 1840.
Diphasium complanatum (L.) Rothm., Feddes Rep. Spec. Nov. **54**:64, 1944.

Horizontal stem, wide-creeping on or near the surface of the ground, 1 m

or more long. Erect stems much branched with crowded to loosely and remotely forked branchlets; branchlets obviously flattened, strongly constricted at the end of a season's growth. Leaves often somewhat glaucous,
adnate over half their length, lanceolate-subulate, sharply acuminate, entire, 4-ranked, free tips of the lateral leaves lance- or deltoid-subulate,
much shorter than the adnate portion; ventral leaves much smaller. Peduncles forked, with remote, reduced, nearly free, scalelike leaves. Cones
cylindric. Sporophylls yellowish, orbicular-deltoid, abruptly contracted at
the base and abruptly cuspidate-acuminate above, margins scarious, erose.
Sporangia reniform. ($2n = 22, 44$)

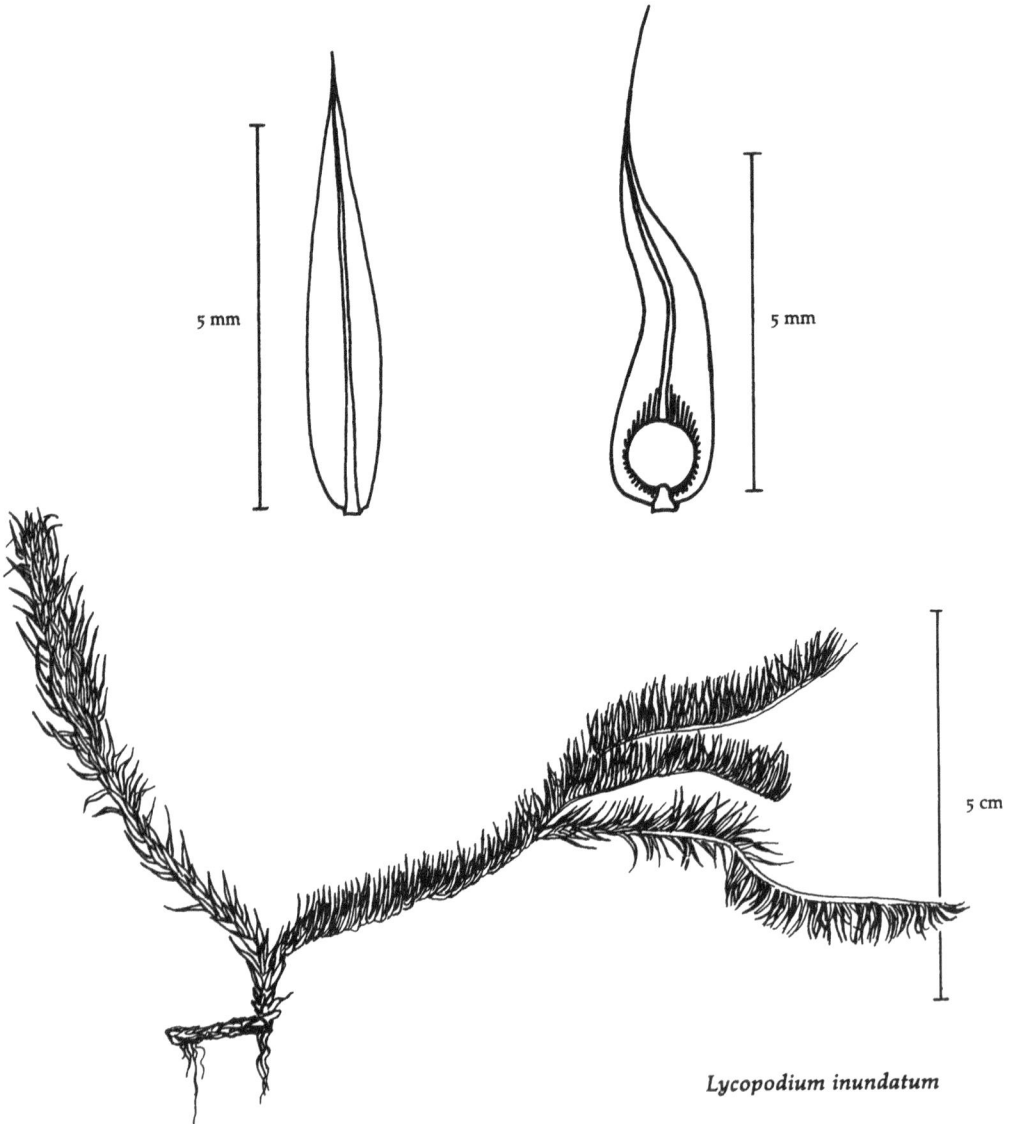

5 mm

5 mm

5 cm

Lycopodium inundatum

HABITAT Dry woods and rocky hillsides in acid soil.

RANGE Interior of Alaska to Newfoundland, south to Washington, Idaho, Montana, Minnesota, Wisconsin, northern Michigan, Ontario, and northern New England; Eurasia.

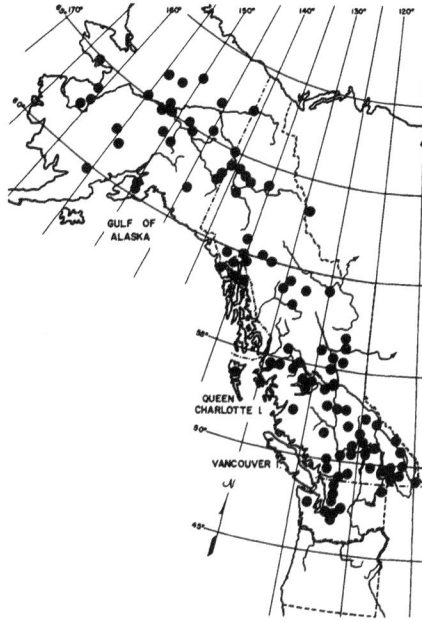

COMMENTS Widely spread and common throughout. Our form in the south is var. *complanatum*, characterized by its rather loose, irregular branching with indeterminate growth. The northern form has a definite tendency to have a single cone (var. *canadense* Vict.). This species shows affinities with *L. alpinum*, from which it can be distinguished by its much greater size, loose elongated branches, and much-reduced ventral leaves with only their tips free.

Lycopodium inundatum L.

Sp. Pl. 1102, 1753; Macoun, Cat. Can. Pl. pt. 5, 288, 1890; Henry, Fl. S. British Columbia 12, 1915; Piper & Beattie, Fl. Northwest Coast 13, 1915; Maxon [in] Abrams, Ill. Fl. Pacific States 1:44, *Fig. 93*, 1923; Victorin, Contr. Lab. Bot. Univ. Montreal No. 3:98, 1925; Frye, Ferns Northwest 18, *Fig. III*, 1934; Peck, Man. Higher Pl. Ore. 56, 1941; Morton [in] Gleason, Ill. Fl. 1:2, *Fig. 1952*; Anderson, Fl. Alaska 21, 1959; Hultén, Fl. Alaska 26, 1968; Calder & Taylor, Fl. Queen Charlotte Islands pt. 1, 133, 1968.
Lepidotis inundata (L.) C. Borner, Volksflora. Eine Flora für das deutsche Volk. Leipzig. 1912.

Horizontal stem above ground, prostrate or arching, rooting at frequent intervals, particularly at the tip, flattened, up to 15 cm long (but usually much less), somewhat branched. Leaves spirally arranged in 8 or 10 ranks, those of the lower side twisted to an ascending position, soft, linear- or lance-attenuate, gradually long-acuminate, entire. Fertile stems few,

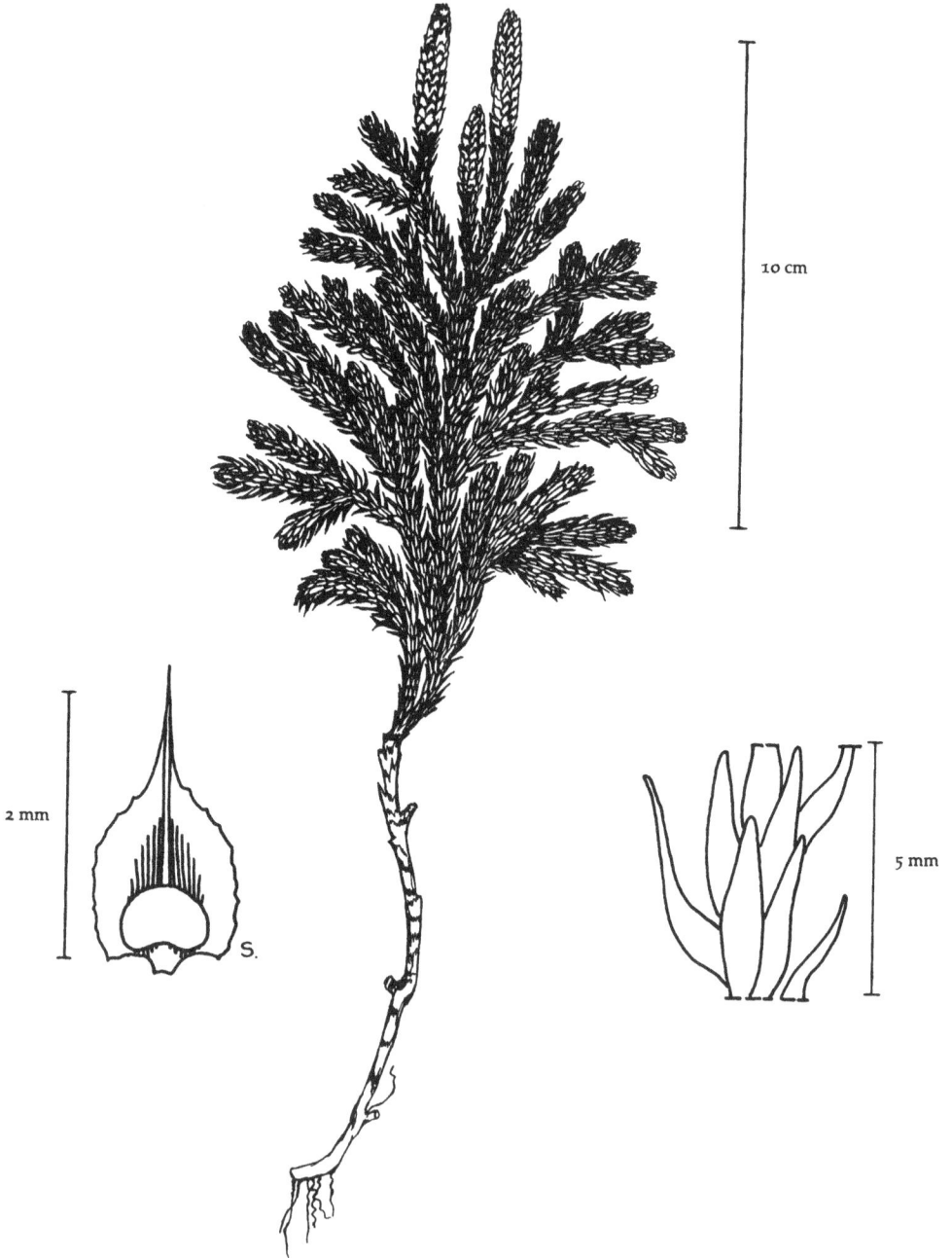

2 mm

10 cm

5 mm

S.

Lycopodium obscurum

usually only one, erect; leaves similar to those of sterile stems, spiralled, ascending or slightly incurved. Cones single, sessile. Sporophylls green, similar to the foliage leaves, much more spreading, narrowly ovate or lanceolate at base. Sporangia subglobose. ($n = 78$)

HABITAT Cool, acid bogs and damp sandy slopes or shores.

RANGE Alaska to New-foundland, south to Oregon, Idaho, Minnesota, northern Illinois, northern Ohio, Pennsylvania, in mountains to western Virginia and West Virginia; Eurasia.

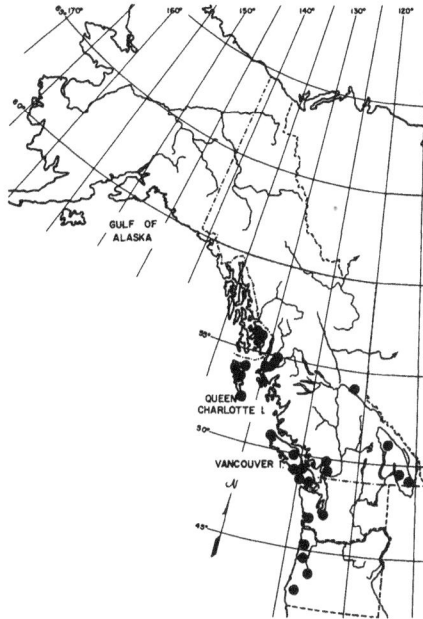

COMMENTS In eastern North America three intergrading varieties have been named; in the West only var. *inundatum* is present. The small size of this species makes it apt to be overlooked by collectors. It is to be expected in suitable habitats throughout although it seems to prefer coastal conditions in our area; it is probably much more common than the rather limited number of records would suggest.

Lycopodium obscurum L.

Sp. Pl. 1102, 1753; Macoun, Cat. Can. Pl. pt. 5, 288, 1890; Henry, Fl. S. British Columbia 12, 1915; Piper & Beattie, Fl. Northwest Coast 13, 1915; Maxon [in] Abrams, Ill. Fl. Pac. States 1:44, Fig. 95, 1923; Victorin, Contr. Lab. Bot. Univ. Montreal No. 3:99, 1925; Frye, Ferns Northwest 23, *Fig.* IV, 1934; Morton [in] Gleason, Ill. Fl. 1: 4, *Fig.*, 1952; Anderson, Fl. Alaska 21, 1959; Hultén, Fl. Alaska 28, 1968; Calder & Taylor, Fl. Queen Charlotte Islands pt. 1, 134, 1968. *L. dendroideum* Michx., Fl. Bor. Am. 2:282, 1803.

Rhizome deeply subterranean, wide-creeping, forking at intervals to give off aerial stems up to 30 cm tall, these mostly unbranched at the base but quite densely bushy-forked upwards, becoming treelike. Sterile branchlets ascending, more or less erect or spreading or recurving at the tips. Leaves very numerous, shining, green, 6-ranked, linear-subulate, tips free, entire, acuminate, sharp-pointed, the lower series usually appressed, the lateral spreading or incurved-ascending. Cones cylindric, sessile, terminating erect, leafy branches. Sporophylls yellowish-brown, broadly cordate, abruptly short-acuminate, margins hyaline. Sporangia reniform. ($2n = 68$)

HABITAT Acid soil of clearings, damp woods, or bog margins.

RANGE Alaska to New-foundland, south to northern Oregon, Montana, Minnesota, Michigan, West Virginia, and southern North Carolina; Asia.

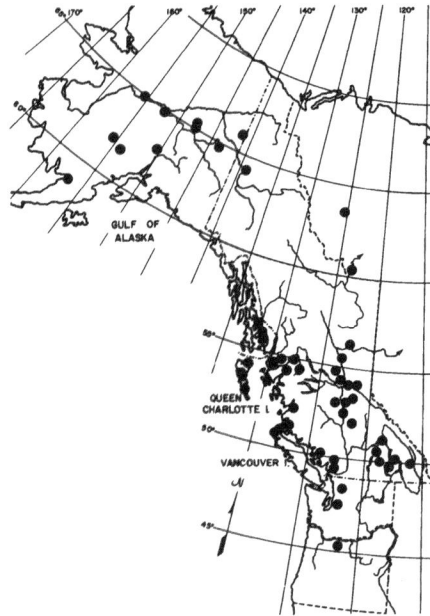

COMMENTS Common in British Columbia but apparently rare both north and south of this province. This species shows variation in the shape of the branchlets and the arrangement of the leaves. A form with rounded branchlets and leaves spreading uniformly in all planes has been distinguished as var. *dendroideum* (Michx.) D. C. Eaton. The typical form, which is commonest with us, has flattened branchlets with the leaves of the upper and lower planes smaller and subappressed. Intergradations between the two extremes are complete, however, so they are not considered to be taxonomically significant. Distinguishing features are the deeply subterranean rhizomes (usually lacking in herbarium specimens), the treelike growth of the erect portion, and the conspicuous sessile cones.

Lycopodium selago L.

Sp. Pl. 1102, 1753; Macoun, Cat. Can. Pl. pt. 5, 287, 1890; Henry, Fl. S. British Columbia 12, 1915; Piper & Beattie, Fl. Northwest Coast 13, 1915; Maxon [in] Abrams, Ill. Fl. Pac. States 1: 43, *Fig. 92*, 1923; Victorin, Contr. Lab. Bot. Univ.

Montreal No. 3:94, 1925; Frye, Ferns Northwest 15, *Fig.* 11, 1934; Peck, Man.
Higher Pl. Ore. 56, 1941; Morton [in] Gleason, Ill. Fl. 1:2, *Fig.* 1952; Anderson,
Fl. Alaska 20, 1959; Wiggins & Thomas, Fl. Alaskan Arctic Slope 34, 1962;
Hultén, Fl. Alaska 25, 1968; Calder & Taylor, Fl. Queen Charlotte Islands pt. 1,
134, 1968.
Huperzia selago (L.) Bernh., J. f.d. Bot. 2:121–35, 1801.

Horizontal rooting stems short, marcescent. Erect stems usually tufted,
several times dichotomously branched. Branches densely or loosely ascending, often forming tight, more or less flat-topped tufts. Leaves in 8
ranks, free, lanceolate, gradually acuminate, usually hollow at the base,
nearly entire, lustrous, broadest at the base, often bearing dilated gemmae
in their axils. Cones not obvious, sporophylls like the vegetative leaves,
borne in regularly alternating zones, each season's growth having a basal
sterile and an apical fertile zone. (Many and varied reports, n = ca 45;
$2n$ = ca 68, ca 88, ca 90, ca 264)

HABITAT Sphagnum bogs, shaded acid rocks, alpine and arctic regions.

RANGE Alaska to New-
foundland, south to Washington, Montana, Michigan,
northern New York, and
northern New England, south
in the mountains to North
Carolina; Greenland; Eurasia.

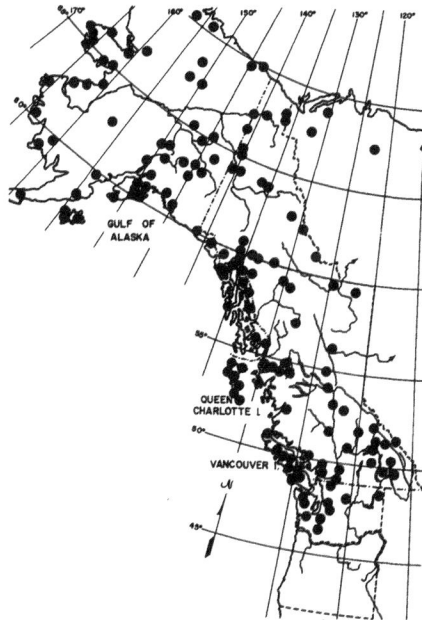

COMMENTS Another variable species, found in some form or another in
suitable habitats from Washington northward. Numerous varieties have
been described; the extremes appear quite distinct but are connected by a
complete series of intergrading forms. In var. *selago* the leaves are lance-
attenuate, ascending; in var. *appressum* Desv. they are shorter, firmer,
ovate-lanceolate, and appressed or incurved; var. *patens* (Beauv.) Desv. is
characterized by strongly divergent or reflexed lance-attenuate leaves and
loosely ascending stems; in var. *myoshianum* Makino the leaves are very

5 mm

5 mm

10 cm

g

1 mm

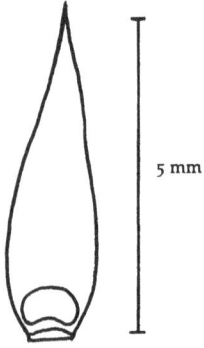

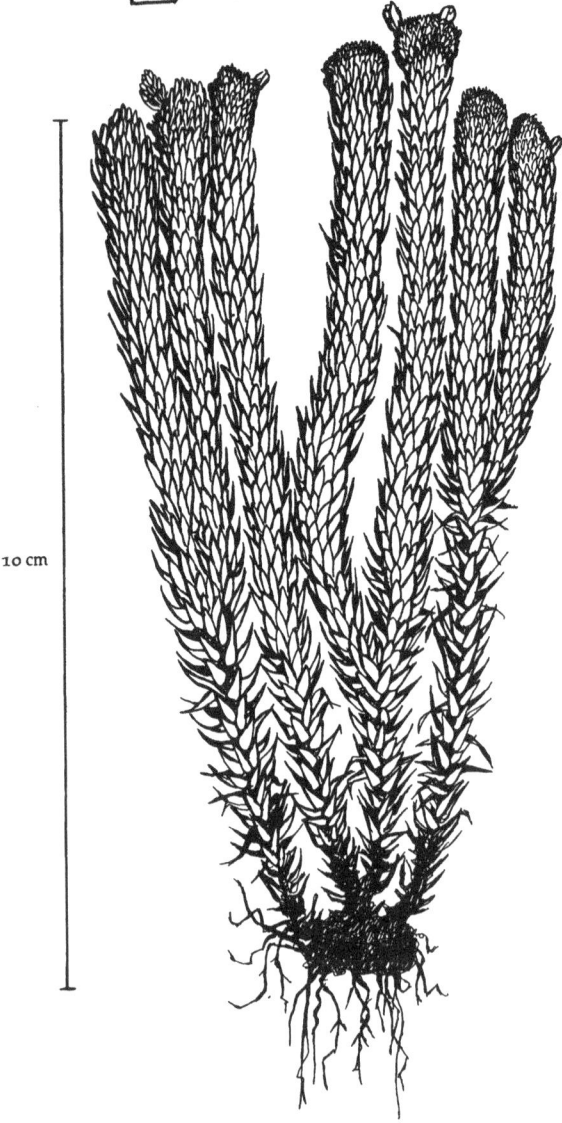

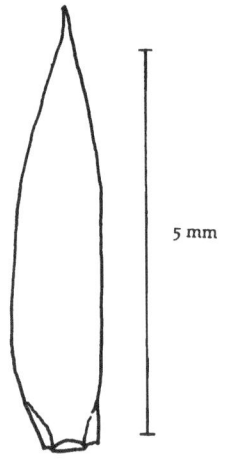

Lycopodium selago

fine, narrowly lanceolate, and somewhat divergent. It is not known if these varieties have a genetic basis or if they are merely environmental responses. Field observations tend to support the latter view. This is the only species in our area in which the cones are not clearly evident, the sporophylls being like the vegetative leaves. Another unique character is development of deciduous vegetative buds (gemmae) in the axils of the upper leaves.

Lycopodium sitchense Rupr.

Beitr. Pflanzenk. Russ. Reich. **3**:30, 1845; Henry, Fl. S. British Columbia 12, 1915; Piper & Beattie, Fl. Northwest Coast 14, 1915; Maxon [in] Abrams, Ill. Fl. Pac. States **1**: 45, *Fig.* 96, 1923; Victorin, Contr. Lab. Bot. Univ. Montreal No. **3**:105, 1925; Frye, Ferns Northwest 23, *Fig.* 11, 1934; Peck, Man. Higher Pl. Ore. 56, 1941; Morton [in] Gleason, Ill. Fl. **1**: 4, *Fig.*, 1952; Anderson, Fl. Alaska 22, 1959; Hultén, Fl. Alaska 30, 1968; Calder & Taylor, Fl. Queen Charlotte Islands pt. 1, 134, 1968.
L. sabinaefolium Willd. var. *sitchense* (Rupr.) Fern., Rhodora **25**: 166, 1923.

Horizontal stems nearly superficial, elongate, rooting at intervals. Erect stems numerous, several times dichotomously branched forming compact tufts; sterile branches not at all flattened. Leaves uniform, subulate, free tips usually longer than the decurrent base, sharply attenuate, entire, generally in 5 ranks, spreading or mostly incurved-ascending. Cones solitary at the ends of short pedunclelike stems with small appressed, scale-like leaves. Sporophylls yellow-green, ovate-deltoid, acuminate or subulate; the margins scarious and erose.

H A B I T A T Subalpine woods, alpine meadows, and barren slopes.

R A N G E Alaska to Newfoundland, south to northern Oregon, Idaho, Lake Superior region, and New Hampshire; east Asia.

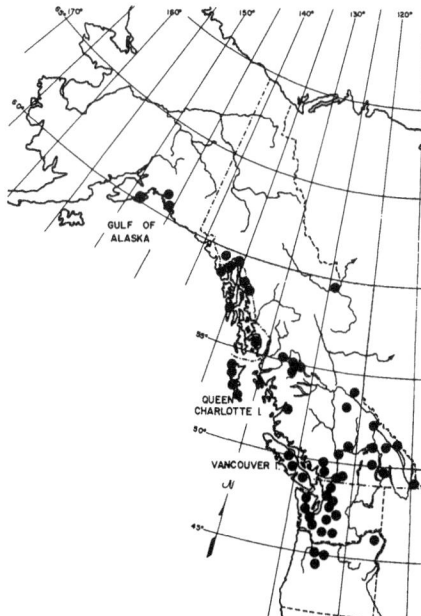

COMMENTS Fairly common at high altitudes. May be found growing with *L. alpinum* and is apt to be confused with it. It can be readily identified by its generally terete, rather than flattened, branches without trowel-shaped ventral leaves; by its leaves in 5–6 rather than 4 ranks; and by the free tips of the leaves being longer than the adnate portion. Some authors regard this as being var. *sitchense* (Rupr.) Fern. of the eastern American *L. sabinaefolium* Willd. The differences, however, appear to be consistent and sufficient to justify the recognition of two species.

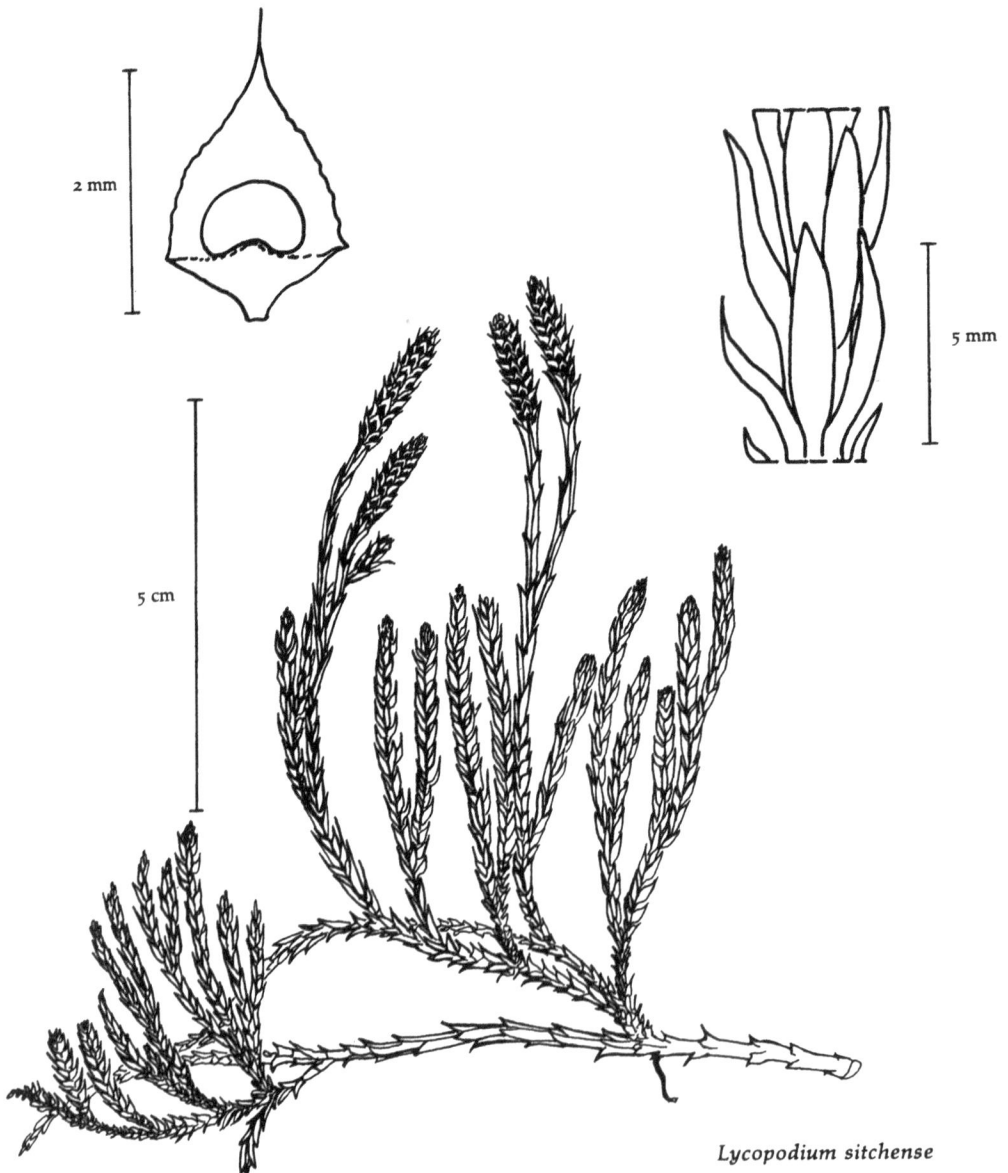

Lycopodium sitchense

Marsileaceae

Perennial herbaceous plants rooting in mud. Rhizome spreading, on or just beneath the surface; roots developing at the nodes. Leaves alternate in two rows, long-petioled, blade divided into four separate leaflets (in ours) that are folded together at night, or filiform and lacking a blade. Sori developed within hard, ovoid, or subhemispheric, pedunculate sporocarps, arising from the rhizome near the base of the petioles, or more or less consolidated with them. Spores tetrahedral in shape; microspores numerous but megaspores only one to a sporangium.

A family of three genera containing about 75 species found chiefly in the Old World. Only two genera occur in our area.

Leaves clearly differentiated into a 4-foliate blade and an elongated petiole.
Marsilea
Leaves simple, filiform, not differentiated into blade and petiole. *Piluaria*

MARSILEA L., Sp. Pl. 1099, 1753

Blades 4-foliate, cloverlike. Sporocarps ovoid, thick-walled with 2 teeth near the base, 2 loculate vertically, dehiscing into 2 valves and emitting a mucilaginous band of tissue bearing megasporangia on the top and microsporangia on the sides. (Named for Marsigli, an Italian botanist, 1658–1730.)

A genus of about 70 species, widely distributed but mainly in the Old World. Only one is recognized in our area.

Marsilea vestita Hook. & Grev.

Icon. Fil. **2**: *plate 159, 1831;* Macoun, Cat. Can. Pl. pt. 5:293, 1890 (as to b.c. record); Piper & Beattie, Fl. Northwest Coast 10, 1915; Maxon [in] Abrams, Ill. Fl. Pac. States **1**: 34, *Fig. 69, 1923;* Frye, Ferns Northwest 157, *Figs.* LIX, LX, 1934; Peck, Man. Higher Pl. Ore. 52, 1941; Weatherby, J. Arn. Arb. 24:325, 1943; Mason, Fl. Marshes Calif. 29, *Fig. 4, 1957;* St. John, Fl. Southeast. Wash. 10, 1963.
M. *mucronata* of Western Am. auth., not R. Br.
M. *oligospora* Gooding, Bot. Gaz. 33: 66, 1902.

Rhizomes wide-creeping, clothed at the nodes with reddish-brown or tawny, silky hairs. Leaves up to 20 cm tall; petioles slender, hairy at first, later glabrous; blades up to 3 cm broad, leaflets broadly cuneate, margins entire or nearly so, hairy at first. Peduncles short, mostly quite free from the petiole. Sporocarps solitary, at first very densely hairy, later more or less glabrous; upper tooth conspicuous, acute, often curved; lower tooth short, less conspicuous.

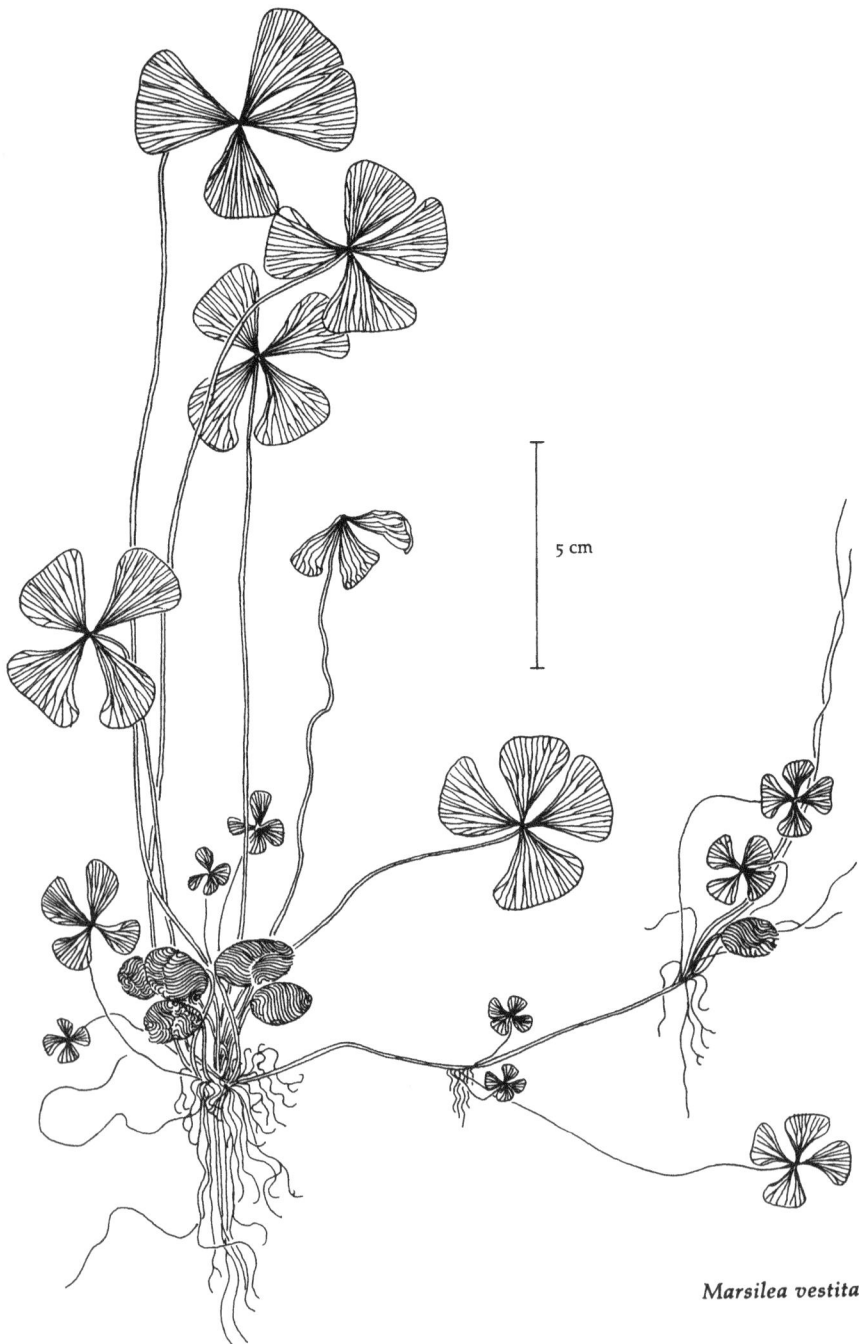

Marsilea vestita

HABITAT Edges of ponds and rivers, marshy places.

RANGE Southern British
Columbia, south to southern
California, east to southern
Alberta, South Dakota, Kansas, Oklahoma, and Texas.

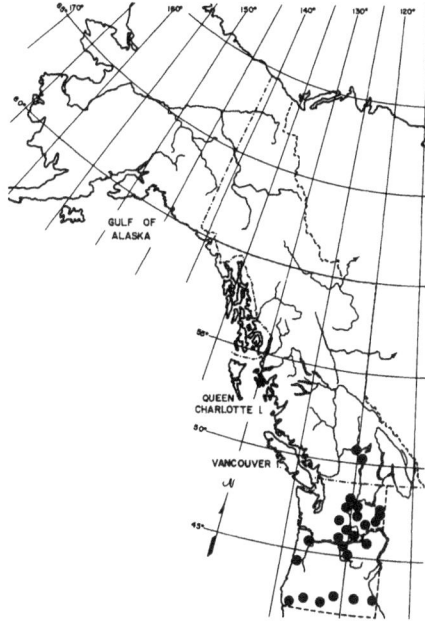

COMMENTS This species apparently reaches its northern limit at Kamloops in southern British Columbia. Gupta (1957) considers that *M. oligospora* should be recognized as a distinct species; this writer, however, is not convinced of this and prefers to include it with the present species, for the time being at least.

PILULARIA L., Sp. Pl. 1100, 1753

Small plants of wet, muddy situations. Rhizome delicate, extensively creeping. Leaves subulate, entire, borne one or several at a node. Sporocarps solitary, divided into 2–4 locules, each containing a single sorus, at maturity splitting longitudinally releasing the sporangia in a mass of mucilage; microsporangia above with numerous microspores, megasporangia below with a solitary megaspore. (Latin *pilula*, a little ball, referring to the sporocarps.)

A genus of 6 species in Europe and Mediterranean region, the Americas (temperate and tropical mountains), New Zealand, and Australia.

Pilularia americana A. Br.

Monatsb. Kon. Akad. Wiss. Berlin **1863**:435, 1864; Maxon [in] Abrams, Ill. Fl. Pac. States **1**:34, *Fig.* 71, 1923; Frye, Ferns Northwest 159, *Fig.* LXI, 1934; Peck, Man. Higher Pl. Ore. 54, 1961.

3 cm

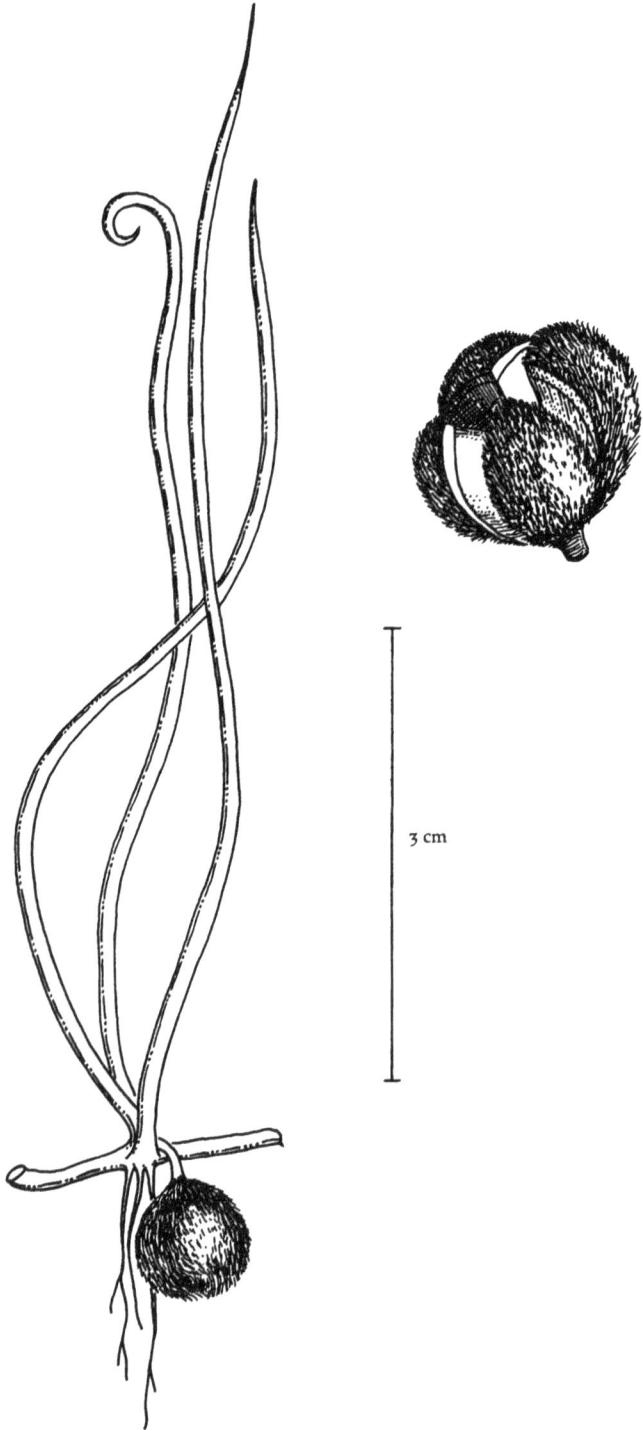

Pilularia americana

Rhizome threadlike, widely creeping and much branched, forming dense mats. Leaves very slender, almost bristlelike, occurring singly or severally at a node, glabrous, up to 5 cm tall, frequently much shorter. Sporocarps usually developed just below the surface of the ground, solitary at nodes, attached laterally to a short descending peduncle, orbicular, 2–4 mm in diameter, short, appressed, hairy, 2–4-celled, commonly 3-celled.

HABITAT Shallow, temporary pools in mud; springy hillsides.

RANGE Oregon to southern California and Texas; isolated in western Arkansas and central Georgia.

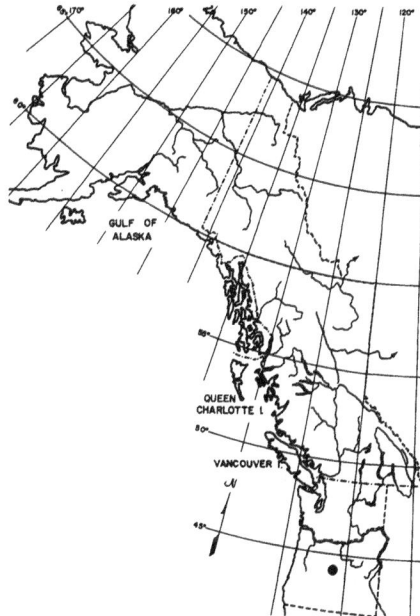

COMMENTS This species is included on the basis of a collection made in Crook County, Oregon, in 1894. As no later collections are known to the writer it may possibly be extinct.

Ophioglossaceae

Herbaceous, leafy, often fleshy plants with a short vertical rhizome bearing many fibrous, usually fleshy, roots, and a single frond. Frond erect consisting of a simple, or variously compounded, sterile blade and a simple, or paniclelike, compound fertile portion. Sporangia large, superficial or sunken in leaf tissue, opening by two valves, annulus lacking. Spores thick-walled, very numerous, all of one type.

A small family of 3 genera and about 90 species in tropical and temperate regions, some of the tropical species are epiphytes.

Sterile blade 1 to 4 times pinnately divided; veins free.	*Botrychium*
Sterile blade simple and unlobed; veins reticulate.	*Ophioglossum*

BOTRYCHIUM Sw. [in] Schrad., J. f.d. Bot. **2**: 110, 1801

Fleshy terrestrial plants. Rhizome short and thick, erect, bearing a few, unbranched, coarse, often corrugated roots. Leaves usually solitary, somewhat fleshy, glabrous or, rarely, with simple hairs; base of the leaf stalk containing the bud of next year's frond. Sterile blade sessile or long-stalked, 1–4 times pinnately or ternately compound. Fertile portion erect, a more or less long-stalked simple spike or variously compound panicle. Sporangia large, globose, free, mostly sessile, opening transversely by two valves. Spores very numerous, sulphur-coloured.

A cosmopolitan genus of about 25 species, usually preferring subacid soil in open or partially shaded situations. Some species are readily overlooked on account of their small size. (Named from the Greek *botryos*, a bunch of grapes.)

1	Sterile blades sessile, large, deltoid, much divided, often rather thin in texture; bud hairy, partially exposed by the sheathing base of the stalk which is open on one side.	*B. virginianum*
1	Sterile blades various, but not at one and the same time lax and membranous and broadly deltoid and sessile; bud hairy or glabrous, completely enclosed by the sheathing base of the stalk.	
2	Sterile blades rather large, ternately decompound, long-stalked to sessile, usually inserted towards the base of the plant; buds very hairy; leaves evergreen and somewhat fleshy.	*B. multifidum*
2	Sterile blades usually small, pinnately or ternately divided, sessile or short-stalked, inserted at various heights on the plant; buds glabrous.	

3   Sterile blade usually sessile, deltoid; both fertile and sterile segments completely reflexed in bud.   *B. lanceolatum*

3   Sterile blade sessile or stalked, oblong or ovate in outline, very rarely deltoid; the fertile and sterile segments in the bud either erect or with their tips variously inclined, but never both completely reflexed.

4   Fertile segments usually erect in bud or with the tip slightly inclined towards the sterile segments, the upper part of the sterile portion commonly bent down over and covering the fertile segment; blade almost sessile, ovate or ovate-oblong, inserted above the middle of the plant; divisions of the blade pinnately divided, oblong, obtuse at apex.   *B. boreale*

4   Both segments of the leaf erect in bud, or the fertile segment erect and the extreme tip of the sterile segment just slightly inclined over it; blade simple or once pinnate, sometimes with the basal division again divided, thus appearing ternate.

5   Sterile blade sessile or, at most, short-stalked, the divisions usually flabellate, generally similar except at the tip of the blade, basal lobes rarely again divided.

6   Pinnae and lobes approximate; breadth of pinnae 11 mm; penultimate divisions of the leaf coarse and angular; angle between the lower and upper margins of basal pinna about 150°.   *B. lunaria*

6   Pinnae and lobes distant; breadth of pinnae 4 mm; penultimate divisions of leaf rounded and small; angle between lower and upper margins of basal pinna about 90°.   *B. minganense*

5   Sterile blade usually stalked, rarely sessile; divisions usually obovate or oblong, seldom flabellate, generally dissimilar in shape; basal lobes often again divided giving the blade a ternate appearance.

7   Sterile blades simple or pinnate, sometimes subternately divided, stalked, inserted at various heights.   *B. simplex*

7   Sterile blade pinnately divided with the basal divisions again divided giving a ternate appearance; blade sessile, inserted above the middle of the plant.   *B. pumicola*

Botrychium boreale (Fr.)

Milde, Bot. Zeit. **15**:880, 1857; Frye, Ferns Northwest 68, *Fig.* XVII, 1934; Clausen, Mem. Torrey Bot. Club **19**(2):80, 1938; Peck, Man. Higher Pl. Ore. 44, 1941; Anderson, Fl. Alaska 7, 1959; Hultén, Fl. Alaska 40, 1968.
B. pinnatum St. John, Am. Fern J. **19**:11, 1929.

Plant relatively stout and fleshy, up to 26 cm tall. Bud glabrous with fertile segment erect and the sterile with its apex bent over the tip of the fertile segment. Blade inserted above the middle of the plant, usually sessile, oblong-triangular to broadly oblong-lanceolate, pinnate-pinnatifid to 2-pinnate. Lower pinnae more or less stalked and decurrent, segments oblong and obtuse at the apex. Fertile segment exceeding the sterile.

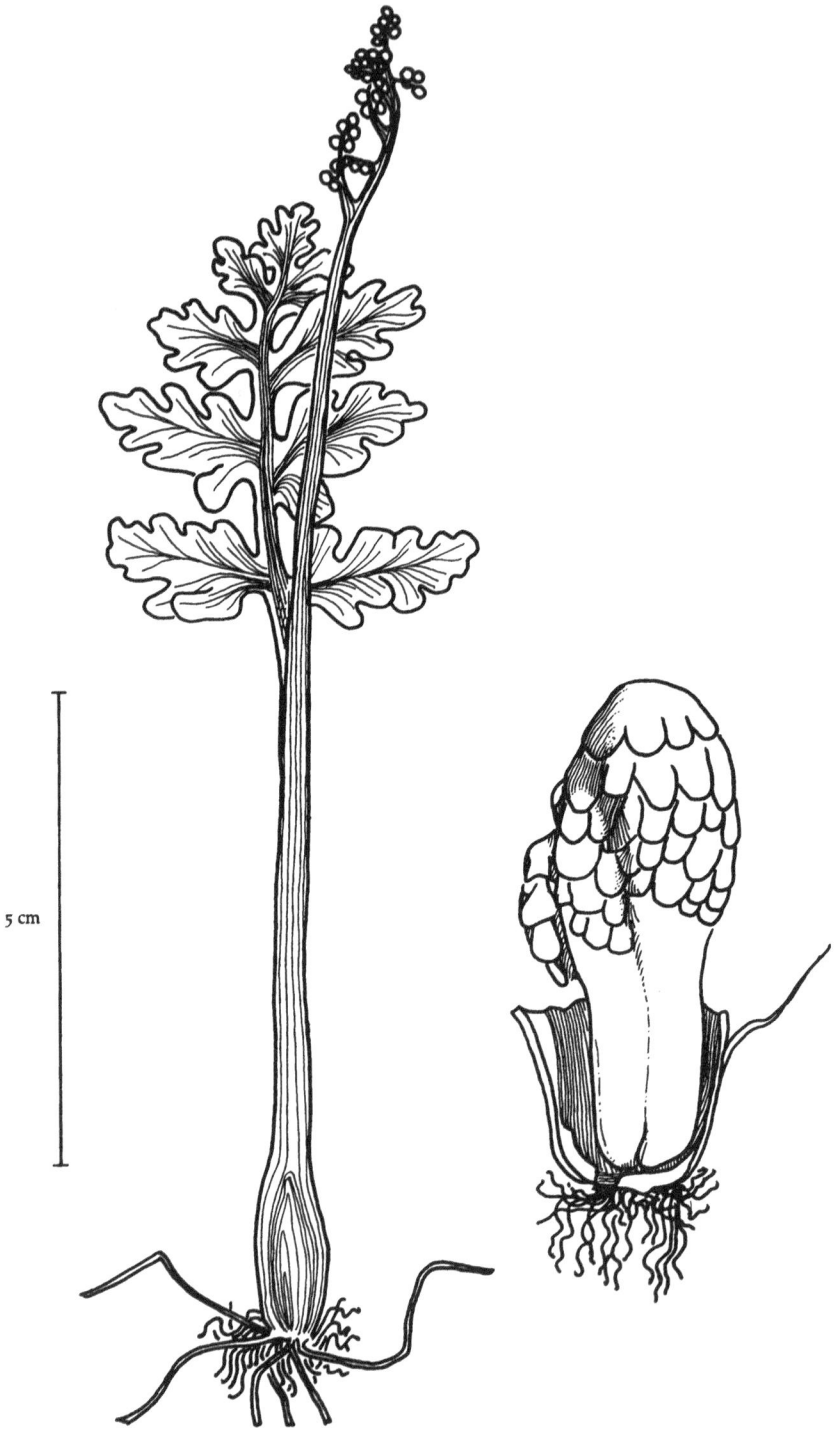

5 cm

Botrychium boreale

Stalk of the fertile segment mostly shorter than the blade, segment simple or paniculate. ($2n = 90$)

HABITAT Open grassy places, alpine meadows, peaty ground.

RANGE Alaska to western Alberta, south to Washington and Montana; Eurasia.

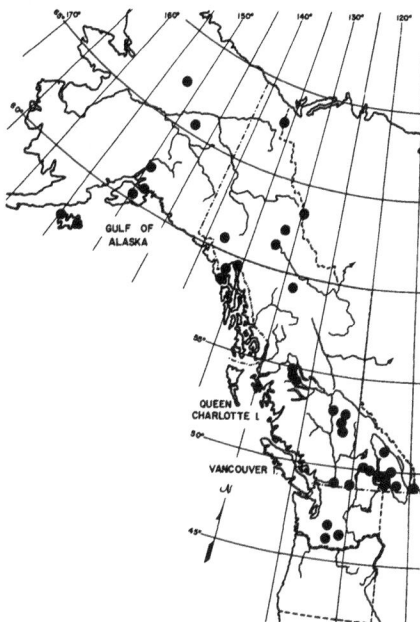

COMMENTS A boreal and alpine species met with infrequently, although of wide occurrence. Our form has been named ssp. *obtusilobum* (Rupr.) Clausen. Its ecological requirements are apparently similar to those of *B. lunaria* as the two species are often found growing together. The present species can be recognized by its almost sessile, ovate or oblong blade inserted above the middle of the plant, and by its pinnae which are lobed or divided into oblong and obtuse segments. In the south, at least, it is an inland species, apparently absent from the Queen Charlotte Islands.

Botrychium lanceolatum (Gmel.) Ångstr.

Bot. Not. **1854**:68, 1854; Macoun, Cat. Can. Pl. pt. 5, 254, 1890; Henry, Fl. S. British Columbia 2, 1915; Piper & Beattie, Fl. Northwest Coast 10, 1915; Maxon [in] Abrams, Ill. Fl. Pac. States 1:4, *Fig. 6*, 1923; Frye, Ferns Northwest 70, *Fig.* XIV, 1934; Clausen, Mem. Torrey Bot. Club 19(2):90, 1938; Peck, Man. Higher Pl. Ore. 44, 1941; Morton [in] Gleason, Ill. Fl. 1:18, *Fig.*, 1952; Anderson, Fl. Alaska 8, 1959; Hultén, Fl. Alaska 41, 1968.

Plant fleshy, up to 40 cm tall. Bud glabrous, both sterile and fertile segments completely reflexed. Common stalk up to 15 cm long. Sterile blade sessile or nearly so, inserted near the summit of the plant, glabrous, del-

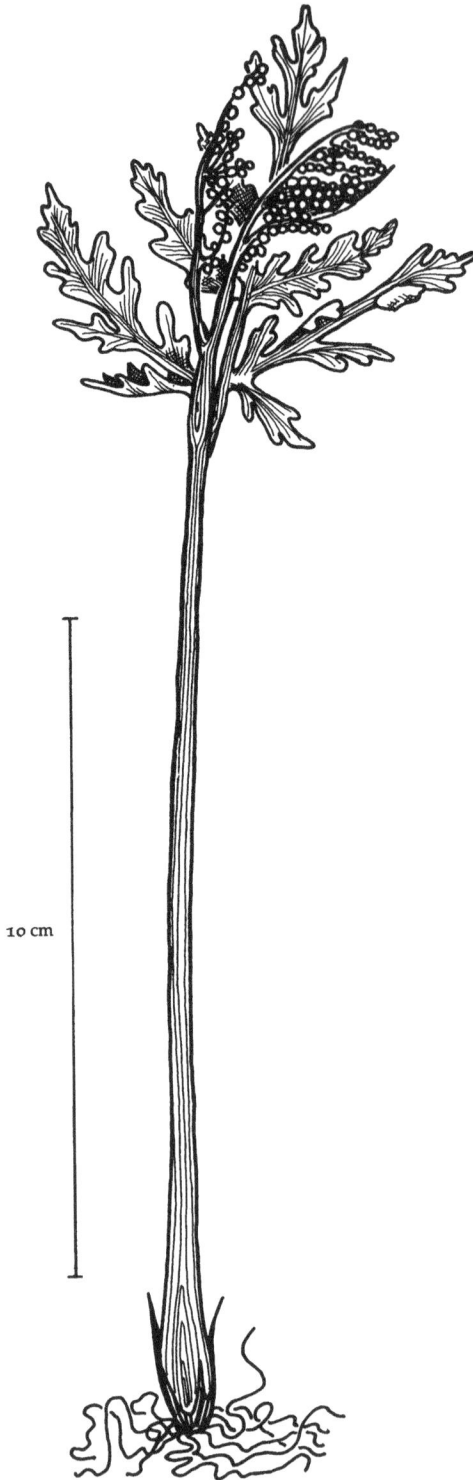

10 cm

Botrychium lanceolatum

toid with dark green, pinnate-pinnatifid, lanceolate segments of which the lowest pair is longest. Stalk of the fertile segment much shorter than the sterile blade. Fertile segment paniculate, rather compact with ascending branches, longer than the sterile blade. ($n = 45$)

HABITAT Peaty slopes, open sandy places, alpine meadows, and open woods.

RANGE Alaska to Newfoundland, south to northern Oregon, Colorado, Wyoming, Wisconsin, Ontario, Pennsylvania, and West ╱ Virginia; Greenland; Iceland; Eurasia.

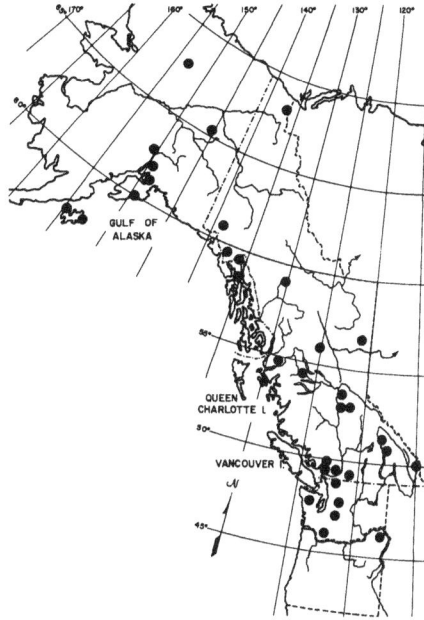

COMMENTS Rare but apparently quite widely spread, the species is one that can be expected anywhere. The stalk of the fertile segment, being very much shorter than the deltoid sterile blade, is a distinguishing feature. All of our specimens are referable to var. *lanceolatum* with thick, fleshy blades, the lobes of the pinnae rounded at the apex and broad at the base.

Botrychium lunaria (L.) Sw.

[In] Schrad., J. f.d. Bot. **2**:110, 1800; Macoun, Cat. Can. Pl. pt. 5, 253, 1890; Henry, Fl. S. British Columbia 1, 1915; Piper & Beattie, Fl. Northwest Coast 9, 1915; Maxon [in] Abrams, Ill. Fl. Pac. States **1**:3, *Fig.* 5, 1923; Frye, Ferns Northwest 67, *Fig.* XIV, 1934; Clausen, Mem. Torrey Bot. Club **19**(2):62, 1938; Peck, Man. Higher Pl. Ore. 44, 1941; Morton [in] Gleason, Ill. Fl. **1**:17, *Fig.*, 1952; Wagner & Lord, Bull. Torrey Bot. Club **83**:261–80, 1956; Anderson, Fl. Alaska 7, 1959; Wiggins & Thomas, Fl. Alaskan Arctic Slope 40, 1962; Hultén, Fl. Alaska 40, 1968; Calder & Taylor, Fl. Queen Charlotte Islands pt. 1, 140, 1968.

Plant up to 28 cm tall. Bud glabrous, both the sterile and fertile segments erect, apex of the sterile segment longer and arching over the tip of the

10 cm

Botrychium lunaria

fertile, pinnae large, nearly covering the fertile segment and overlapping the midrib of the sterile segment. Sterile blade more or less oblong, essentially sessile, somewhat leathery, glabrous, pinnately divided with more or less opposite lobes. Lobes broadly fanshaped to semicircular, approximate and often overlapping except in shade forms where they may be remote; angle formed by the upper and lower margins of the basal lobe is about 150°. Lobes rounded at apex, entire or somewhat incised, maximum breadth of median lobe about 11 mm. Penultimate divisions of the sterile blade generally larger than the ultimate, more or less sharply cuneate in outline, 3–6 mm broad. Fertile segment racemose or paniculate. ($n = 45$)

H A B I T A T Open places, alpine meadows, turfy slopes, and fields.

R A N G E Alaska to Newfoundland, south to California, Arizona, Colorado, Minnesota, Michigan, and Maine; Greenland; Iceland; Eurasia; Patagonia; possibly New Zealand and Australia.

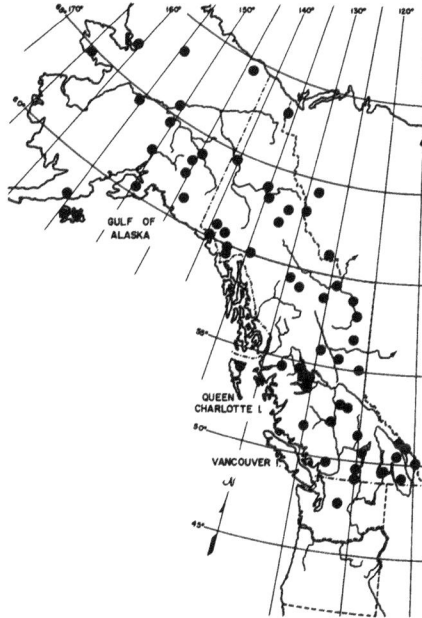

C O M M E N T S Generally distributed in suitable habitats, but on account of its small size it is likely to be overlooked. Wagner and Lord (1956) have drawn attention to numerous points of difference between the present species and *B. minganense* Victorin with which it is frequently confused. Certain of these characteristics are emphasized in the key to the species of *Botrychium* given above.

Botrychium minganense Victorin

Trans. Roy. Soc. Can. ser. III, **21**:331, 1927; Frye, Ferns Northwest 64, *Fig.* XVI, 1934; Clausen, Mem. Torrey Bot. Club **19**(2):67, 1938; Peck, Man. Higher Pl. Ore. 44, 1941; Wagner & Lord, Bull. Torrey Bot. Club **83**:261–80, 1956; Hultén,

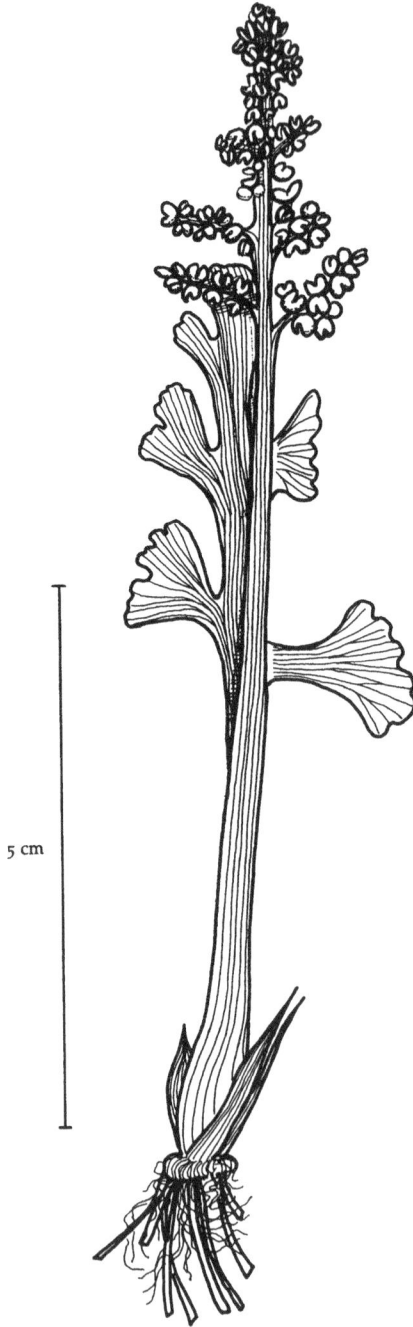

5 cm

Botrychium minganense

Fl. Alaska 40, 1968; Calder & Taylor, Fl. Queen Charlotte Islands pt. 1, 140, 1968.
B. lunaria var. *minganense* (Vict.) Dole [in] Fl. Vermont, 3rd ed., 1, 1937.

Plant up to 30 cm tall. Bud glabrous, both sterile and fertile segments erect, apex of the sterile segment equalling the tip of the fertile, pinnae small not covering the fertile segment nor overlapping the midrib of the sterile segment. Sterile blade inserted at about 1/3 total height of the plant or less, narrowly oblong, sessile to short-stalked, blade somewhat membranous, glabrous, usually pinnate, occasionally pinnate-pinnatifid below. Lobes opposite, oblique to the rachis, short-stalked, obovate, rhomboidal or oblong, frequently incised, cuneate, always remote, maximum breadth of median lobe about 4 mm, angle formed by upper and lower margins of basal lobe about 90°. Penultimate divisions of the sterile blade usually of the same size or smaller than the ultimate segment, rounded, obliquely-rhomboidal in outline, 1–3 mm broad. Fertile segment 1- to 2-pinnate. ($n = 90$)

H A B I T A T Open alpine areas, or in partial shade.

C O M M E N T S In the past *B. minganense* has been confused with both *B. lunaria* and *B. simplex*. Wagner and Lord (1956) have emphasized the differences between this species and *B. lunaria* and have also drawn attention to the difference in chromosome number. *B. minganense* may usually be distinguished from *B. simplex* by the sterile blade being inserted closer to the middle of the plant rather than near the base, by the blade being pinnate instead of tripartite, and by the usually greater height of *B. minganense*.

R A N G E Alaska to Labrador, south to California, Nevada, Idaho, Montana, Wisconsin, Michigan, New York, and Vermont.

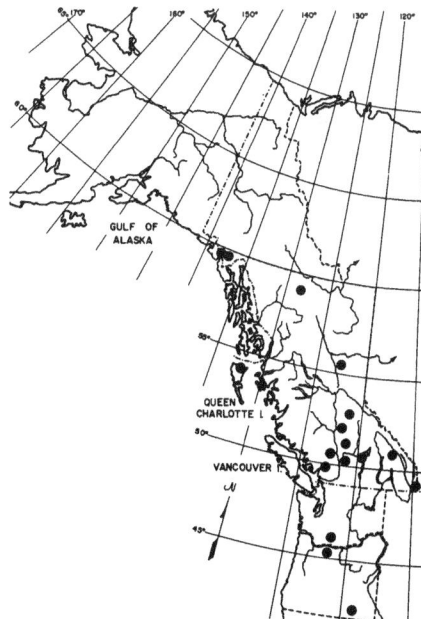

10 cm

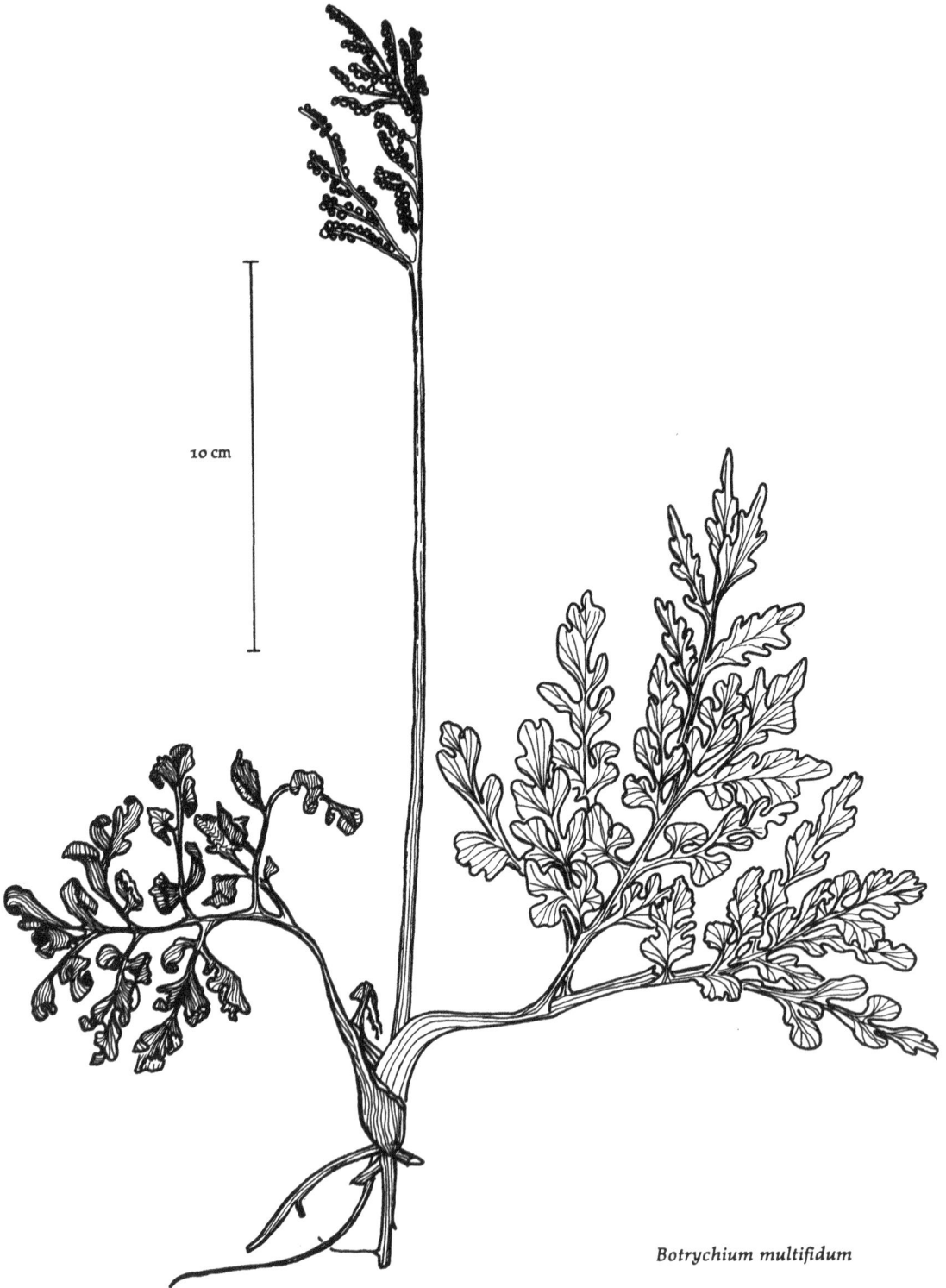

Botrychium multifidum

Botrychium multifidum (Gmel.) Rupr.

Beitr. Pflanzenk. Russ. Reich. **11**:40, 1859; Macoun, Cat. Can. Pl. pt. 5, 225, 1890; Henry, Fl. S. British Columbia 1, 1915; Piper & Beattie, Fl. Northwest Coast 9, 1915; Maxon [in] Abrams, Ill. Fl. Pac. States **1**:4, *Fig.* 8, 1923; Frye, Ferns Northwest 70, *Fig.* xiv, 1934; Clausen, Mem. Torrey Bot. Club 19(2):26, 1938; Peck, Man. Higher Pl. Ore. 5, 1941; Morton [in] Gleason, Ill. Fl. **1**:18, *Fig.*, 1952; Anderson, Fl. Alaska 8, 1959; St. John, Fl. Southeast. Wash. 3, 1963; Hultén, Fl. Alaska 41, 1968; Calder & Taylor, Fl. Queen Charlotte Islands pt. 1, 140, 1968.

B. ternatum var. *intermedium* Eaton [in] Gray's Man. 7th ed., 49, 1908.
B. silaifolium Presl, Rel. Haenk. 1:76, 1825.
B. occidentale Underwood, Bull. Torrey Bot. Club, **25**:538, 1898.

Rhizome short with numerous, long, wrinkled roots. Bud silky pilose, included within the closed stipe base, the sterile and fertile segments completely reflexed in bud. Stalk of sterile blade up to 13 cm long, stout, arising from near the base of the plant and usually underground. Blades evergreen, very fleshy and leathery, sparingly pilose, broad triangular, ternate, 2- to 4-pinnate, about as broad as long. Basal pinnae largest, obtuse, subentire to dentate, cuneate, decurrent. Fertile segment 2- to 4-pinnate, loose spreading. ($n = 45$)

HABITAT Fields and open woods in acid soil.

RANGE Alaska to Labrador, south of California, Montana, Iowa, Ohio, Pennsylvania, and New Jersey; Eurasia.

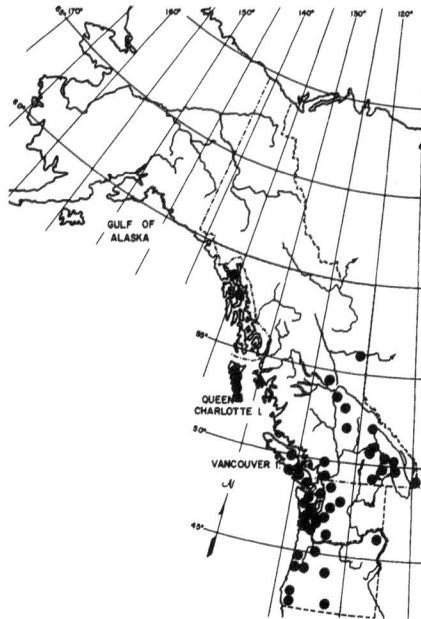

COMMENTS Common everywhere in suitable sites. Our plants nearly all belong to var. *intermedium* (D. C. Eaton) Farwell, characterized by its larger size, 15–40 cm tall, with blades 7–15 cm long and 3–21 cm wide, the

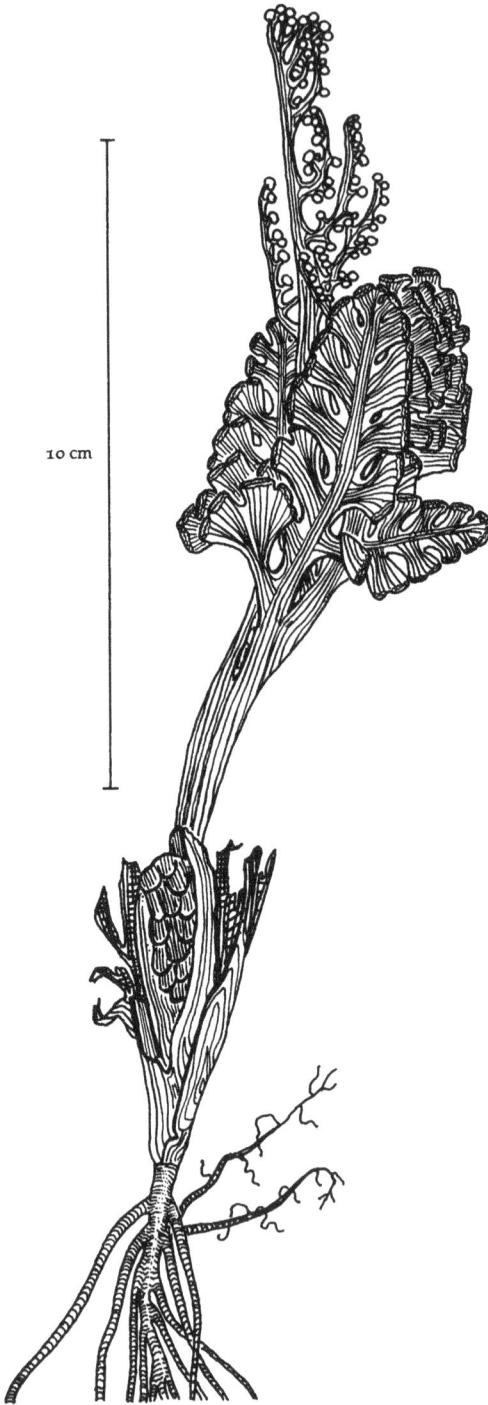

10 cm

Botrychium pumicola

ultimate divisions being rather remote and not imbricate. This species is late in starting into growth in the spring; the sterile frond of the previous year may be found persisting even until the September following. This feature, together with the fact that the stalk of the sterile blade arises from below ground, will distinguish it from *B. virginianum*.

Botrychium pumicola Colville

[In] Underw., Nat. Ferns, 6th ed., 69, 1900; Maxon [in] Abrams, Ill. Fl. Pac. States **1**:3, *Fig. 4*, 1923; Frye, Ferns Northwest 67, *Fig. xv*, 1934; Clausen, Mem. Torrey Bot. Club **19**(2):79, 1938; Peck, Man. Higher Pl. Ore. 4, 1941.

Plant stout and fleshy, about 10 cm tall. Bud glabrous; both sterile and fertile segments erect in bud or the apex of the sterile segment slightly over-topping the fertile. Leaf bases from previous years form a conspicuous brown sheath around the lower half of the plant. Sterile blade sessile or nearly so, leathery, glaucous, up to 3 cm long and nearly as wide, usually ternately divided with each of the major divisions ovate-deltoid; ultimate segments fan-shaped or obovate, markedly imbricate. Stalk of fertile segment very short; spike compact, paniculate. ($n = 45$)

H A B I T A T Fine, loose pumice gravel.

R A N G E Oregon.

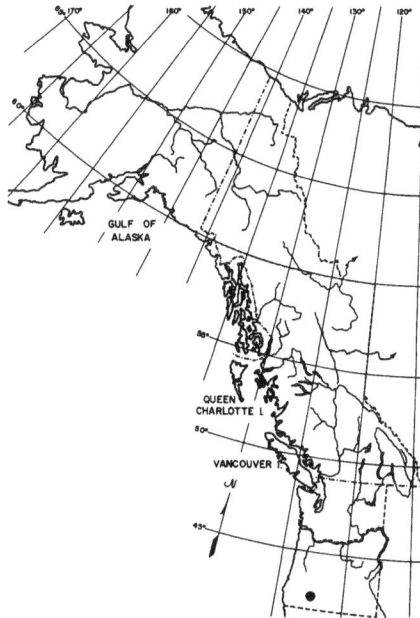

C O M M E N T S A very distinctive species, known only from Crater Lake, Klamath County, Oregon, where it has been collected many times.

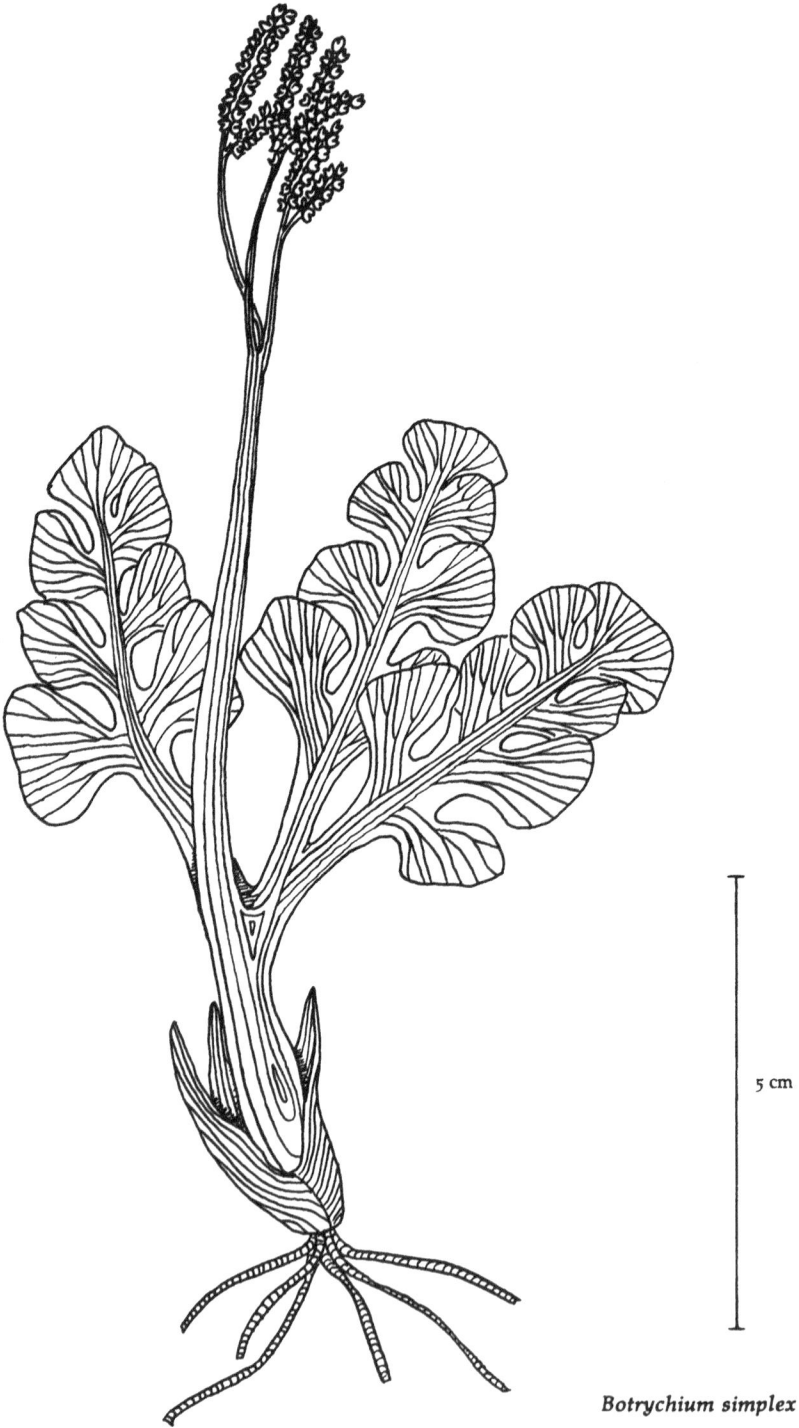

5 cm

Botrychium simplex

Botrychium simplex Hitchc.

Am. J. Sci. 6:103, *Plate 8*, 1823; Maxon [in] Abrams, Ill. Fl. Pac. States **1**:3, *Fig. 3*, 1923; Frye, Ferns Northwest 64, *Fig.* xv, 1934; Clausen, Mem. Torrey Bot. Club **19**(2):70, 1938; Peck, Man. Higher Pl. Ore. 44, 1941; Morton [in] Gleason, Ill. Fl. **1**:18, *Fig.*, 1952.

Plants slender, up to 18 cm tall. Bud glabrous with both the sterile and fertile segments erect or with the tip of the sterile segment slightly inclined over the top of the fertile. Blade short-stalked, inserted almost at the base of the common stalk or towards the middle of the stalk, simple, lobed or pinnately divided, lobes oblong, rhomboid, or reniform, sometimes overlapping; basal lobes may be again divided. Fertile segment long-stalked, simple or compound. ($n = 45$)

HABITAT Open grassy slopes and meadows, mossy woods.

COMMENTS Apparently limited to Oregon and Washington in our area. No specimens have been seen to support its reputed occurrence in British Columbia. Western material for the most part is separable as var. *compositum* (Lasch) Milde. This has the blade ternately divided and inserted basally on the common stalk; both the fertile and sterile segments are entirely erect in the bud.

RANGE Oregon to Newfoundland, south to California, New Mexico, Minnesota, Wisconsin, Michigan, New York, Pennsylvania, and Massachusetts; Europe; Japan.

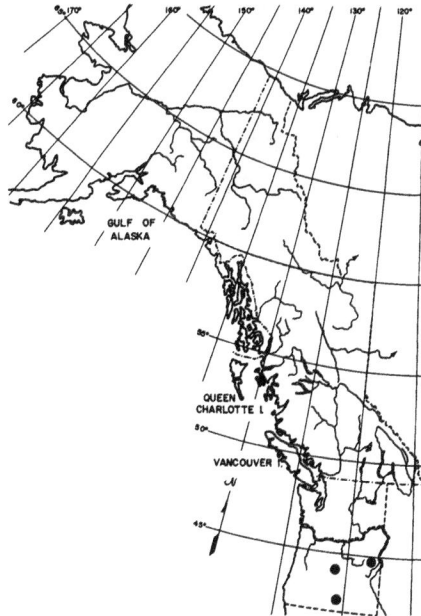

Botrychium virginianum (L.) Sw.

[In] Schrad., J. f.d. Bot. **2**:111, 1800; Macoun, Cat. Can. Pl. pt. 5, 256, 1890; Henry, Fl. S. British Columbia 1, 1915; Piper & Beattie, Fl. Northwest Coast 9,

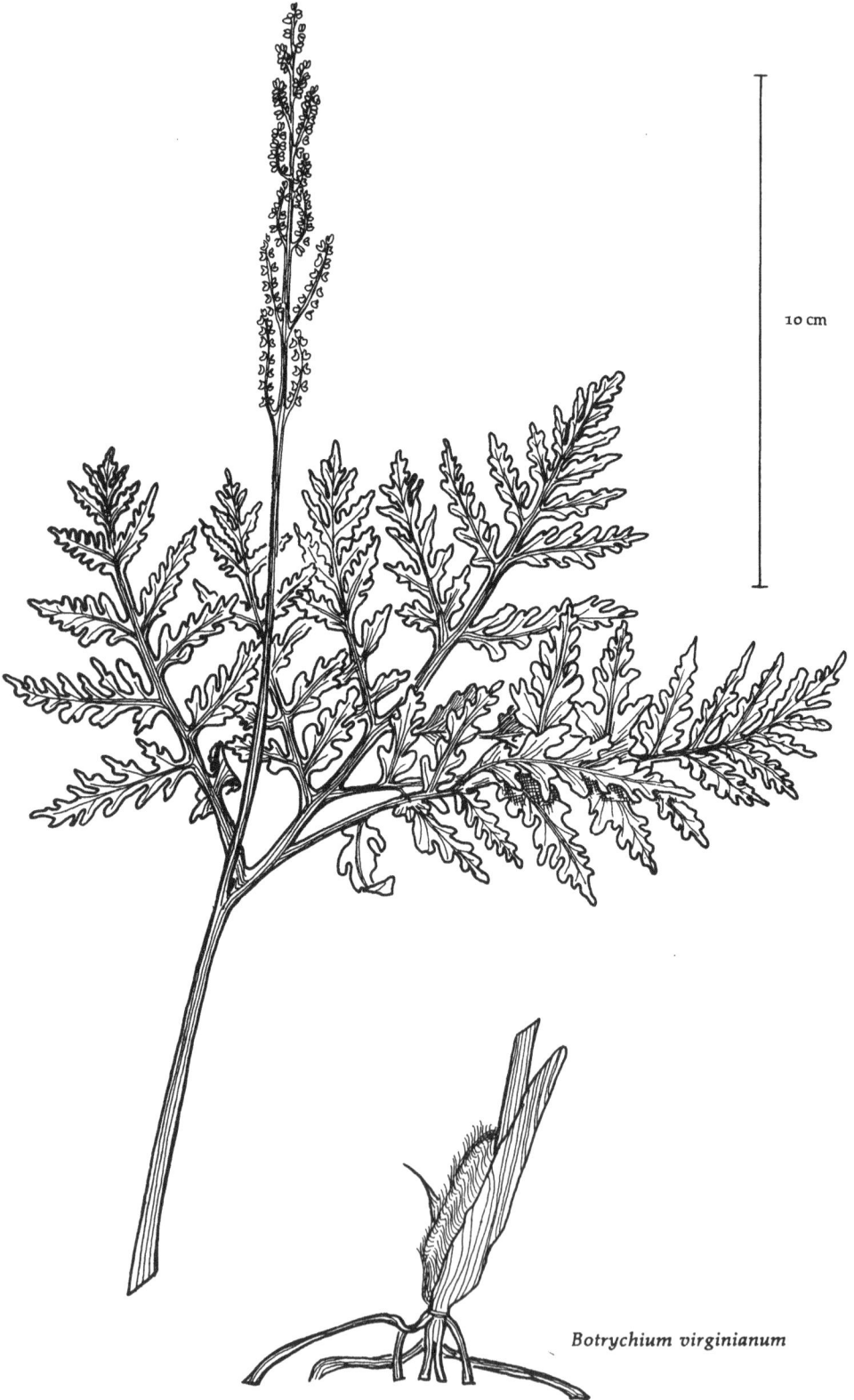

10 cm

Botrychium virginianum

1915; Maxon [in] Abrams, Ill. Fl. Pac. States **1**:4, *Fig. 7*, 1923; Frye, Ferns Northwest 71, *Fig.* XIV, 1934; Clausen, Mem. Torrey Bot. Club **19**(2):97, 1938; Peck, Man. Higher Pl. Ore. 5, 1941; Morton [in] Gleason, Ill. Fl. **1**:18, *Fig.*, 1952; Anderson, Fl. Alaska 8, 1959; Hultén, Fl. Alaska 42, 1968.

Plant erect, up to 76 cm tall, glabrous or sparsely pubescent. Bud pilose and, at least late in the season, partially exposed by the sheathing base of the leaf stalk which opens on one side. Fertile and sterile segments both completely reflexed in vernation. Blade deciduous, thin and membranous, sometimes leathery, deltoid, large, nearly sessile, ternately decompound, much divided; basal pinnae largest, ultimate divisions oblong-lanceolate, variously toothed or lobed, blunt or acutish. Spike 2- or 3-pinnately compound. ($n = 92$)

HABITAT Woods, meadows, and damp places.

COMMENTS Widely spread and easily recognized by its thin, triangular blade inserted well above the ground. Two subspecies occur in our region: ssp. *virginianum* and ssp. *europaeum* (Ångstr.) Clausen. The latter has a more compact, leathery blade, with the pinnae not so deeply or finely divided, the ultimate divisions often being crowded or imbricate; the valves of the sporangia are usually not wide-spreading or recurved in dehiscence. It is possible that broad ecological preferences determine the distribution of these two subspecies. It appears that ssp. *europaeum* is more northern and is a plant of coniferous forests and damp open spaces. The typical subspecies, on the other hand, is found in dry or somewhat moist, deciduous woods.

RANGE Alaska to Newfoundland, south of California, Arizona, and Florida; Eurasia.

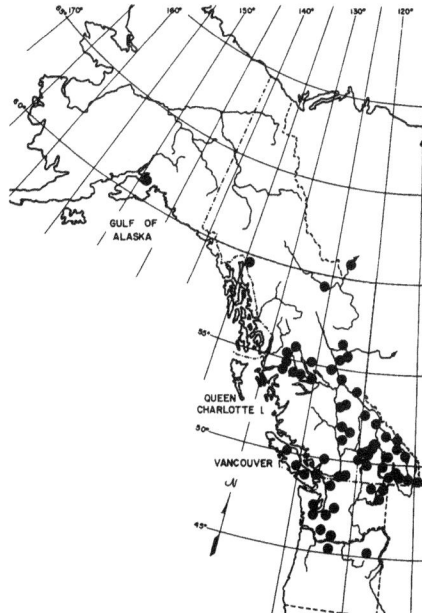

OPHIOGLOSSUM L., Sp. Pl. 1063, 1753

Small, perennial herbs, ours terrestrial. Rhizome short, erect, with an exposed apical bud. Roots rather thick and fleshy. Leaves one to several arising at the side of the apical bud. Sterile segment sessile or short-

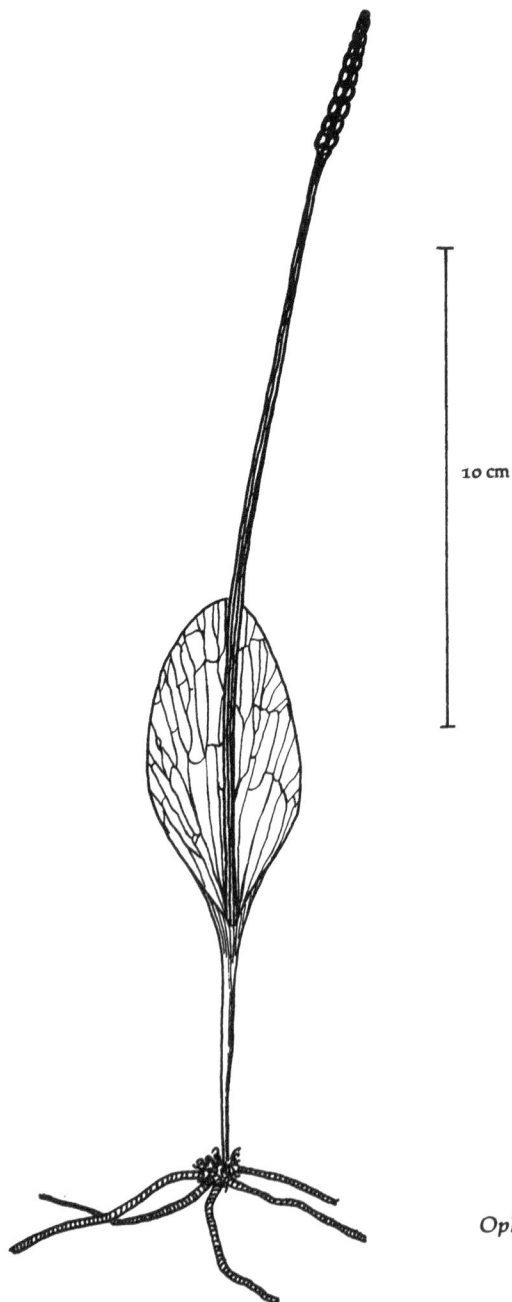

10 cm

Ophioglossum vulgatum

stalked, simple, linear-lanceolate to ovate, venation reticulate. Fertile segment consisting of a stalk and simple spike. (From *ophis*, serpent, and *glossa*, a tongue, perhaps in allusion to the narrow leaves of some species.)

A genus of about 30 species widely distributed in a variety of habitats in both hemispheres.

Ophioglossum vulgatum L.

Sp. Pl. 1062, 1753; Piper & Beattie, Fl. Northwest Coast 9, 1915; Maxon [in] Abrams, Ill. Fl. Pac. States **1**:2, *Fig.* 1, 1923; Frye, Ferns Northwest 72, *Fig.* xiv, 1934; Clausen, Mem. Torrey Bot. Club **19**(2):123, 1938; Peck, Man. Higher Pl. Ore. 43, 1941; Morton [in] Gleason, Ill. Fl. **1**:20, *Fig.*, 1952; Anderson, Fl. Alaska 7, 1959; Hultén, Fl. Alaska 39, 1968.
O. alaskanum Britton, Bull. Torrey Bot. Club **24**:556, 1897.

Plant up to 35 cm tall from a short rootstock with numerous rather fleshy roots and usually only a single leaf. Bud apical, not enclosed by the base of the leaf of the season. Both the length of the common stalk and that of the blade extremely variable. Blade sessile, variable in shape from lanceolate to ovate, blunt at apex, reticulate venation quite apparent. Fertile segment with stalk longer, often considerably so, than the spike. (Several chromosome numbers reported, $n =$ ca 150–160, ca 250; $2n =$ ca 300, 480)

H A B I T A T Variety of open places or in semi-shade, fields, thickets, woods, and swamps.

C O M M E N T S Very rare in our area, but collections have been seen from Mason and Kittitas Counties, Washington, and another reported from Skamania County, Washington, by Clausen (1938). The only station known in Alaska is from the island of Unalaska at the extreme west end of the Aleutian chain. The affiliation of the form there is apparently westward with that of northern Japan.

R A N G E Prince Edward Island to Ontario, south to northern Florida, Gulf states; Washington; Alaska; Iceland; Eurasia; Africa.

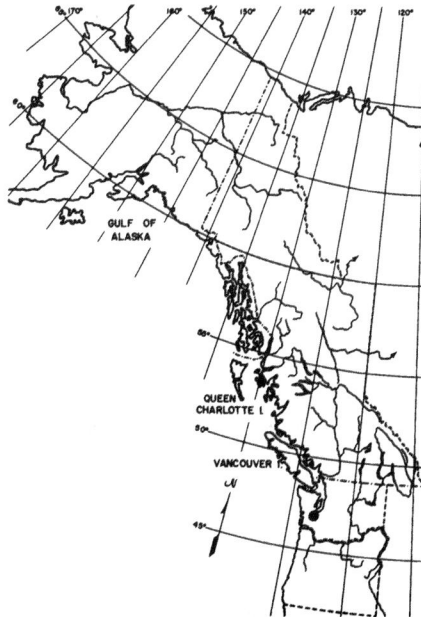

Polypodiaceae

Leafy plants of various habits. Rhizomes creeping or erect, often clothed with scales. Fronds usually rather large, stalked, erect, spreading or drooping. Blades circinate in bud, simple or variously compound, glabrous or pubescent, sometimes glandular; fertile and sterile fronds alike, or in some cases dissimilar. Sporangia grouped together in sori; these are dorsal and various in shape and size, sometimes enclosed or covered by a membranous indusium. Sporangia usually long-stalked, somewhat helmet-shaped (the annulus being the crest), elastically dehiscent through the action of the hygroscopic cells of the annulus.

A very large family of over 150 genera and probably more than 7,000 species. World wide in distribution but most abundant in damp tropical regions; in the sense taken in this book the family includes most of the plants familiarly known as ferns.

1		Sori on or near the margin of the lower surface of the frond, covered by the inrolled indusiumlike margin, or apparent margin.
	2	Sori clearly separate, borne on the underside of reflexed margins of pinnae; segments of frond fan-shaped. *Adiantum*
	2	Sori continuous along the margin of the reflexed frond segment; frond segments more or less oblong or linear-lanceolate.
	3	Rhizome coarse, much elongated, with hairs only, lacking scales; fronds scattered, very large and coarse. *Pteridium*
	3	Rhizome very short, scaly but often with hairs as well; fronds mostly clustered, much smaller, sometimes quite delicate.
	4	Stipes herbaceous, green except at base; fertile and sterile fronds very dissimilar, glabrous. *Cryptogramma*
	4	Stipes wiry, dark and shining.
	5	Blades glabrous (except *Pellaea atropurpurea* and then the fronds coriaceous and dimorphic, the segments 0.5–1.0 cm wide); indusia linear, more or less marginal; margins of segments revolute.
	6	Fronds deltoid not much longer than wide; segments linear, narrowly elongate, upper surface striate, shining; indusium erose-denticulate. *Aspidotis*
	6	Fronds considerably longer than broad; upper surface of segments neither striate nor shining, linear or often broader; indusium at most crenulate. *Pellaea*

5   Blades variously hairy and scaly; segments small, oblong to oval or round, reflexed margins scarcely modified; sori interrupted.   *Cheilanthes*

1   Sori on the lower surface of the frond, not covered by the margin (in *Blechnum* the indusium may appear marginal).

7   Sori continuous along both sides of the midrib of the linear pinnae; fertile fronds with pinnae much narrower than those of the sterile fronds.   *Blechnum*

7   Sori not as above; sterile and fertile fronds much alike, or if markedly dissimilar then the fertile fronds linear, stiffly erect.

8   Indusium present, at least on young fertile fronds.

9   Indusium inferior, borne in part at least beneath the sporangia, cup-shaped, hoodlike or deeply lacerate into filamentous segments.

10   Fertile and sterile fronds very dissimilar, coarse and stiff; fertile segments podlike.   *Matteuccia*

10   Fertile and sterile fronds alike, delicate, not stiff; sori flat or convex.

11   Indusium borne symmetrically under the sorus, splitting into filamentous segments; veins not reaching the margin.   *Woodsia*

11   Indusium hoodlike, attached at one side, partly under the sorus; veins reaching the margin.   *Cystopteris*

9   Indusium superior, either centrally attached at a point or with a lateral linear attachment.

12   Indusium peltate or attached at its centre, more or less circular or kidney-shaped.

13   Indusium centrally attached, shield-shaped, without a deep sinus.   *Polystichum*

13   Indusium laterally attached, kidney-shaped or with a deep sinus.

14   Rhizome thick, short-creeping to erect; veins not reaching the margin.   *Dryopteris*

14   Rhizome slender, creeping; veins reaching the margin.   *Thelypteris*

12   Sori oblong, linear, hooked, or horseshoe-shaped.

15   Sori very large in chainlike rows parallel to midrib of the frond.   *Woodwardia*

15   Sori small, oblique, not in chainlike rows.

16   Blades small, 1-pinnate, evergreen; sori linear to oblong, straight or nearly so.   *Asplenium*

16   Blades large, at least 2-pinnate; sori mostly hooked or horseshoe-shaped, often crossing veins.   *Athyrium*

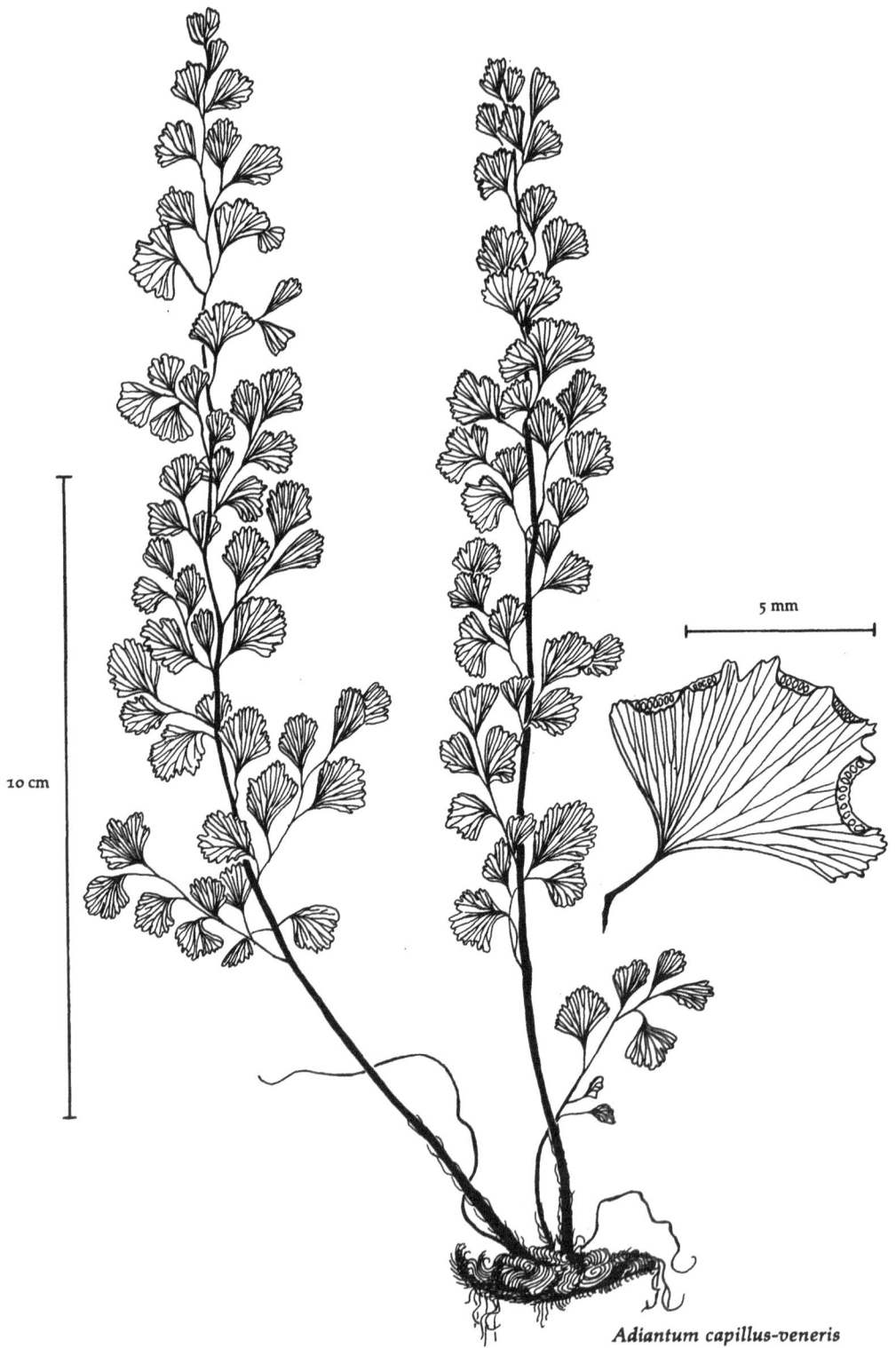

10 cm

5 mm

Adiantum capillus-veneris

8	Indusium lacking; sori roundish or elongated.
17	Sori elongated following the veins, usually confluent; backs of fronds covered with white or yellowish powder. *Pityrogramma*
17	Sori roundish, as a rule not confluent.
18	Fronds coriaceous, evergreen, simply pinnatifid; stipes jointed to the superficially, creeping rhizome. *Polypodium*
18	Fronds herbaceous, not evergreen, at least 2-pinnate; stipes not jointed to the rhizome.
19	Rhizome very stout, short, crowned by fronds and the remains of stipe bases; fronds lanceolate, at least 3-pinnate. *Athyrium*
19	Rhizome very slender, elongate, and branched, fronds scattered or solitary, deltoid or broader.
20	Fronds more or less ternate, 2-pinnate-pinnatifid; lower pinnae articulate at the base; glabrous and scaleless, sometimes glandular. *Gymnocarpium*
20	Fronds not ternate, 2-pinnatifid; lower pinnae not articulated at the base, hairy and scaly throughout. *Thelypteris*

ADIANTUM L., Sp. Pl. 1094, 1753

Sori marginal, not continuous, the edge of the frond being reflexed to form the indusium. Midrib of the pinnules wanting (in ours) or close to the lower edge. Stipe black and shining. Pinnules distinct and short-stalked. (The classical name, from the Greek *a,* not, and *diaine,* to wet, hence unwetted—the foliage sheds rain.)

A large genus of ferns of rich, damp soil in tropical and temperate regions. Many species are cultivated in greenhouses or out of doors in damp mild climates.

1	Rachis branched into two, more or less equal, parts; pinnae only on one side of the rachis except towards the end. *A. pedatum*
1	Rachis essentially unbranched.
2	Pinnules roundish or semicircular, only slightly lobed; sori mostly linear or somewhat arcuate. *A. jordanii*
2	Pinnules cuneate-obovate to rhombic, deeply cleft; sori mostly oblong-lunate. *A. capillus-veneris*

Adiantum capillus-veneris L.

Sp. Pl. 1096, 1753; Maxon [in] Abrams, Ill. Fl. Pac. States 1:24, *Fig. 46, 1923*;
Morton [in] Gleason, Ill. Fl. 1:30, *Fig.,* 1952.
A. modestum Underw., Bull. Torrey Bot. Club 28:46, 1901.
A. rimicola Slosson, Bull. Torrey Bot. Club 41:308, Plate 7, 1914.

Rhizome slender, densely covered with brown scales, horizontal, producing a row of fronds. Fronds drooping, up to 50 cm long, more or less ever-

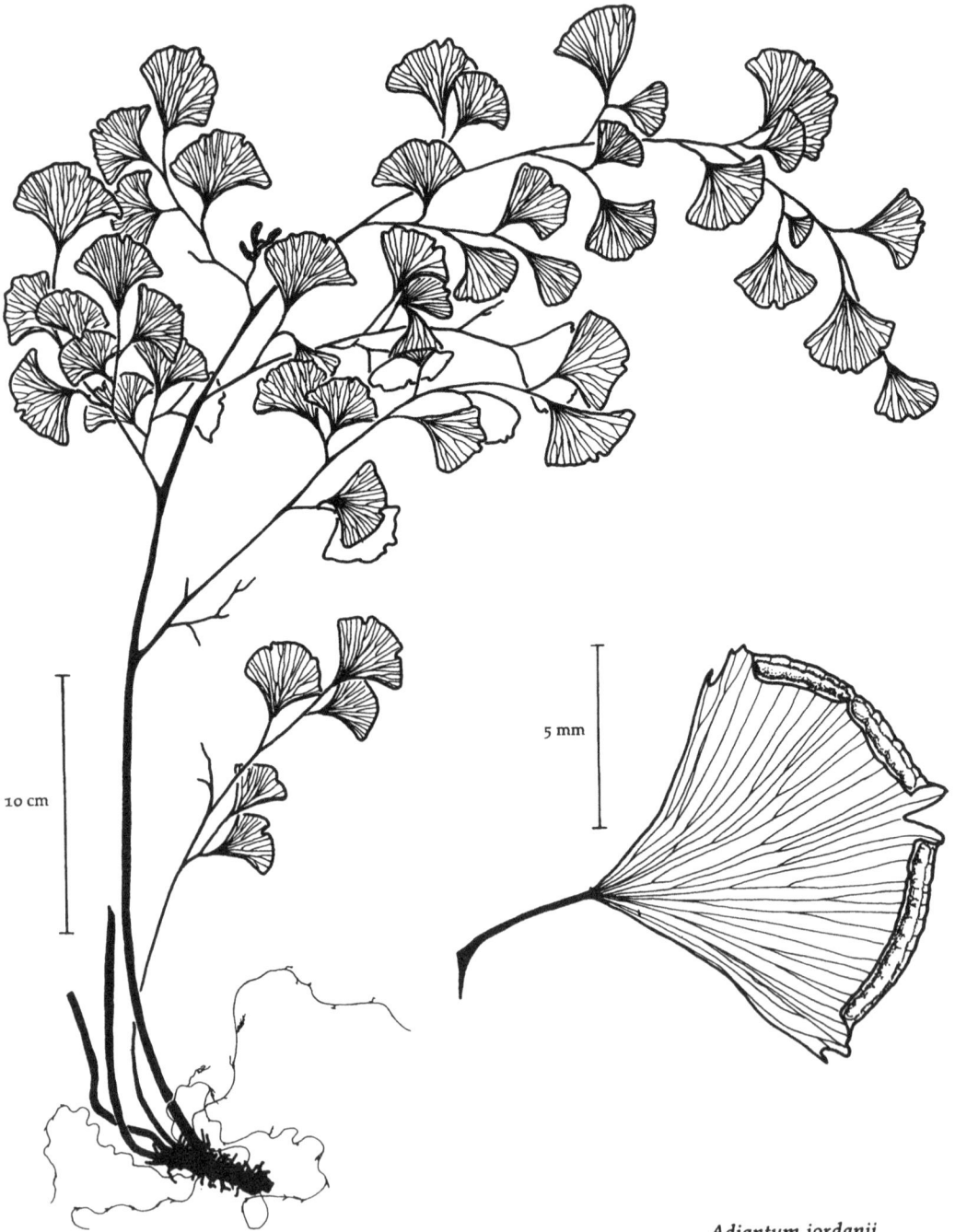

10 cm

5 mm

Adiantum jordanii

green, glabrous. Stipe lustrous, reddish-brown to black, continuing without branching as the main rachis of the frond, as long as or shorter than the blade. Blade ovate-lanceolate, 2- to 3-pinnate at the base, becoming simply pinnate in the upper portion. Pinnules very thin, fan-shaped, rather irregularly jagged-lobed, some of the lobes bearing the transversely oblong, or lunate, indusia. Veins dichotomously branched, free, in sterile pinnules ending in the apices of sharp teeth; midrib lacking. Sori on the margins of recurved lobules. ($n = 30, 60$)

H A B I T A T Damp, usually shaded, cliffs of alkaline rocks or limey soil; in the northern part of its range sometimes associated with hot springs.

R A N G E Southeastern British Columbia, South Dakota, Virginia, south to Florida, west through Kentucky, to Utah, New Mexico, Arizona, and California; subtropical regions of both hemispheres.

C O M M E N T S Widely distributed in tropical and warm-temperate regions of both hemispheres. With us it is known only from the runnels of hot springs at Fairmont, British Columbia, north of Columbia Lake. Despite the fact that the nearest known colony is some hundreds of miles to the south, there is very strong circumstantial evidence to support the view that it is indigenous here and not an escape from cultivation. Reports of the tetraploid form from Tropical West Africa as well as from Florida suggest that it is quite widespread throughout the range of the species.

Adiantum jordanii C. Müll.

Bot. Zeit. **1864**:26, Plate I, *Fig.* 1, 1864; Maxon [in] Abrams, Ill. Fl. Pac. States **1**:24, *Fig.* 45, 1923; Frye, Ferns Northwest 86, 1934; Peck, Man. Higher Pl. Ore. 51, 1941.

A. emarginatum D. C. Eaton, Ferns N. Am. **1**:285, Plate 38; *Figs*. 1–3, 1879, not Bory, 1810.

Rhizome slender, creeping, densely scaly; scales dark, large, rigid, lance-attenuate, entire. Fronds several, often close and clustered, erect, up to 50 cm long; stipe about as long as the blade, wiry, dark brownish-black, shiny. Blade broadly ovate to deltoid, 2- to 3-pinnate below, 1-pinnate at apex; pinnae obliquely spreading, stalked, lower ones about half as long as the blade. Pinnules long-stalked, roundish, semicircular or reniform, entire below, upper margin shallowly 2–5-lobed; if sterile, sharply denticulate with a veinlet ending in each tooth. Sori linear, straight or somewhat arcuate, almost continuous.

H A B I T A T Shaded rocks and rocky soil.

C O M M E N T S Very local in our area being known only from Curry and Josephine Counties in Oregon. For some account of a natural hybrid with *A. pedatum* see under the latter species.

R A N G E Southern Oregon to northern Mexico.

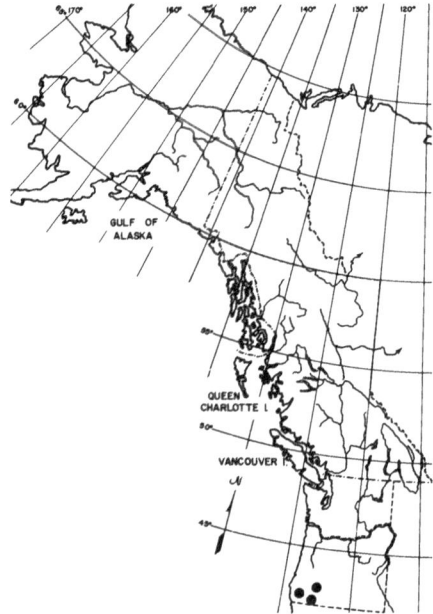

Adiantum pedatum L.

Sp. Pl. 1095, 1753; Macoun, Cat. Can. Pl. pt. 5, 263, 1890; Henry, Fl. S. British Columbia 4, 1915; Piper & Beattie, Fl. Northwest Coast 4, 1915; Maxon [in] Abrams, Ill. Fl. Pac. States **1**:24, *Fig.* 44, 1923; Frye, Ferns Northwest 84, *Fig.* xx, 1934; Peck, Man. Higher Pl. Ore. 51, 1941; Morton [in] Gleason, Ill. Fl. **1**:30, *Fig.*, 1952; Anderson, Fl. Alaska 11, 1959; St. John, Fl. Southeast. Wash. 3, 1963; Hultén, Fl. Alaska 42, 1968; Calder & Taylor, Fl. Queen Charlotte Islands pt. 1, 145, 1968.

Rhizome slender, short-creeping, dark-coloured, producing a row of several fronds in one season. Fronds erect, up to 10 dm tall, glabrous, deciduous. Stipes blackish to deep reddish-brown, lustrous, forking at the top into two, more or less equal, rachises that are widely divergent and arched-recurved. Frond fan-shaped, 2-pinnate; pinnae borne on the upper side of the rachises. Pinnules numerous, petioled, obliquely triangular-oblong, lobed on the upper margin. Sori marginal, covered by indusia formed by the reflexed tips of the lobes; indusium transversely linear, linear-oblong, or lunate. ($n = 29$)

HABITAT Shady situations in damp soil rich in humus.

COMMENTS Found in suitable habitats throughout. Our western form has been described as var. *aleuticum* Rupr. It differs by being more glaucous, with merely spreading to strongly ascending, not recurved branches, and pinnae with the upper margins often deeply and unequally cleft. Intergrading individuals are common, however.

Wagner (1956) has named as *A.* × *tracyi* C. C. Hall a natural hybrid between *A. pedatum* and *A. jordanii*. This is apparently of very local occurrence but should be searched for wherever the two parents grow together. It is intermediate in its morphology between the parental species, has abortive spores, and Wagner (1962) reports $2n = 59$. In the genus *Adiantum* both $n = 29$ and $n = 30$ occur and although there is no report of the chromosome situation in *A. jordanii* it is very likely to be $n = 30$.

RANGE Alaska to Newfoundland, south to California, Utah, Oklahoma, and Georgia; northeast Asia.

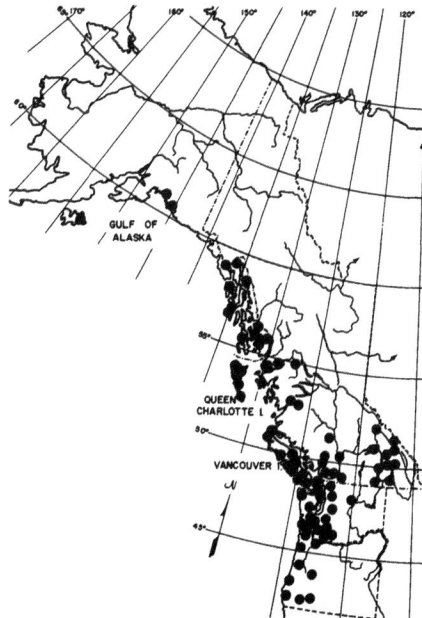

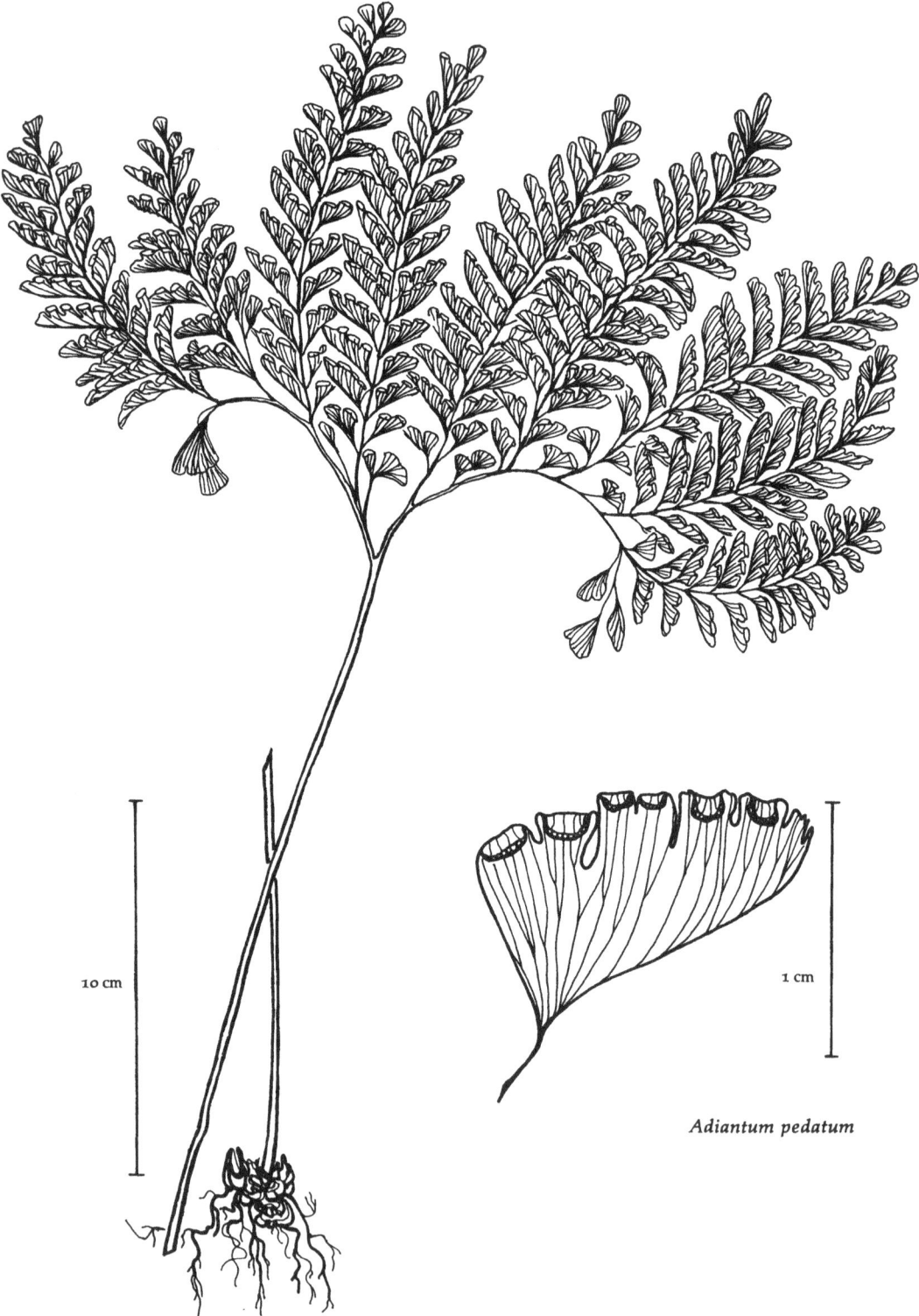

Adiantum pedatum

ASPIDOTIS (Nutt. ex Hook. & Baker) Copel

Small terrestrial ferns with short, creeping rhizome clothed with black setaceous scales. Fronds tufted, glabrous, subcoriaceous; stipe shiny, brown. Blade deltoid-ovate, pinnately divided with narrow, linear, mucronate-rostrate, distantly toothed segments, upper surface striate and shiny. Sori small in the sinuses of the teeth; indusia broad, scarious.

A small genus of three North American species related to, and often included in, *Cheilanthes*. (Name from Greek *aspis*, shield.)

Aspidotis densa (Brack. [in] Wilkes) Lellinger

Am. Fern J. **58**:141, 1968; Macoun, Cat. Can. Pl. pt. 5, 261, 1890; Henry, Fl. S. British Columbia 5, 1915; Piper & Beattie, Fl. Northwest Coast 4, 1915; Maxon [in] Abrams, Ill. Fl. Pac. States **1**:26, *Fig. 48*, 1923; Frye, Ferns Northwest 95, *Fig.* xxxii, 1934; Peck, Man. Higher pl. Ore. 50, 1941; Morton [in] Gleason, Ill. Fl. **1**:33, *Fig.*, 1952; St. John, Fl. Southeast. Wash. 9, 1963.
Pellaea densa (Brack.) Hook., Sp. Fil. **2**:150, 1858.
Cryptogramma densa (Brack.) Diels [in] Engl. & Prantl., Pflanzenfam. **1**:280, 1899.
Cheilanthes siliquosa Maxon, Am. Fern J. **8**:116, 1918.
Cheilanthes densa (Brack.) St. John, Am. Fern J. **19**:14, 1924.

Rhizome with several crowns, cespitose; scales abundant, very narrow, dark chestnut-brown. Fronds erect, densely tufted, up to 30 cm tall; stipes dark chestnut-brown, stout, glabrous, lustrous. Blades broadly ovate, deltoid-oblong, or somewhat pentagonal, 3-pinnate at least below, glabrous, mostly fertile. Pinnae few, close together, definitely oblique, the basal broadly triangular, inequilateral; the lower pinnae longest. Segments narrowly linear, mucronate; margins sharply revolute bearing a thin, continuous, erose-denticulate indusium. Sterile fronds with broader segments, margins callous, serrate. ($n = 30$)

HABITAT Exposed talus slopes and crevices of calcareous or serpentine rocks.

COMMENTS As can be judged by the number of generic synonyms, the relationship of this species has been a matter of considerable debate. Of recent years it has usually been included in *Cheilanthes* but, as Lellinger (1968) points out, it differs from species of this genus in a number of ways. The present writer agrees with his decision to place it in *Aspidotis* and it is treated accordingly. The species is easily recognized by its tufted habit with long stipes and short, ascending, somewhat feathery blades.

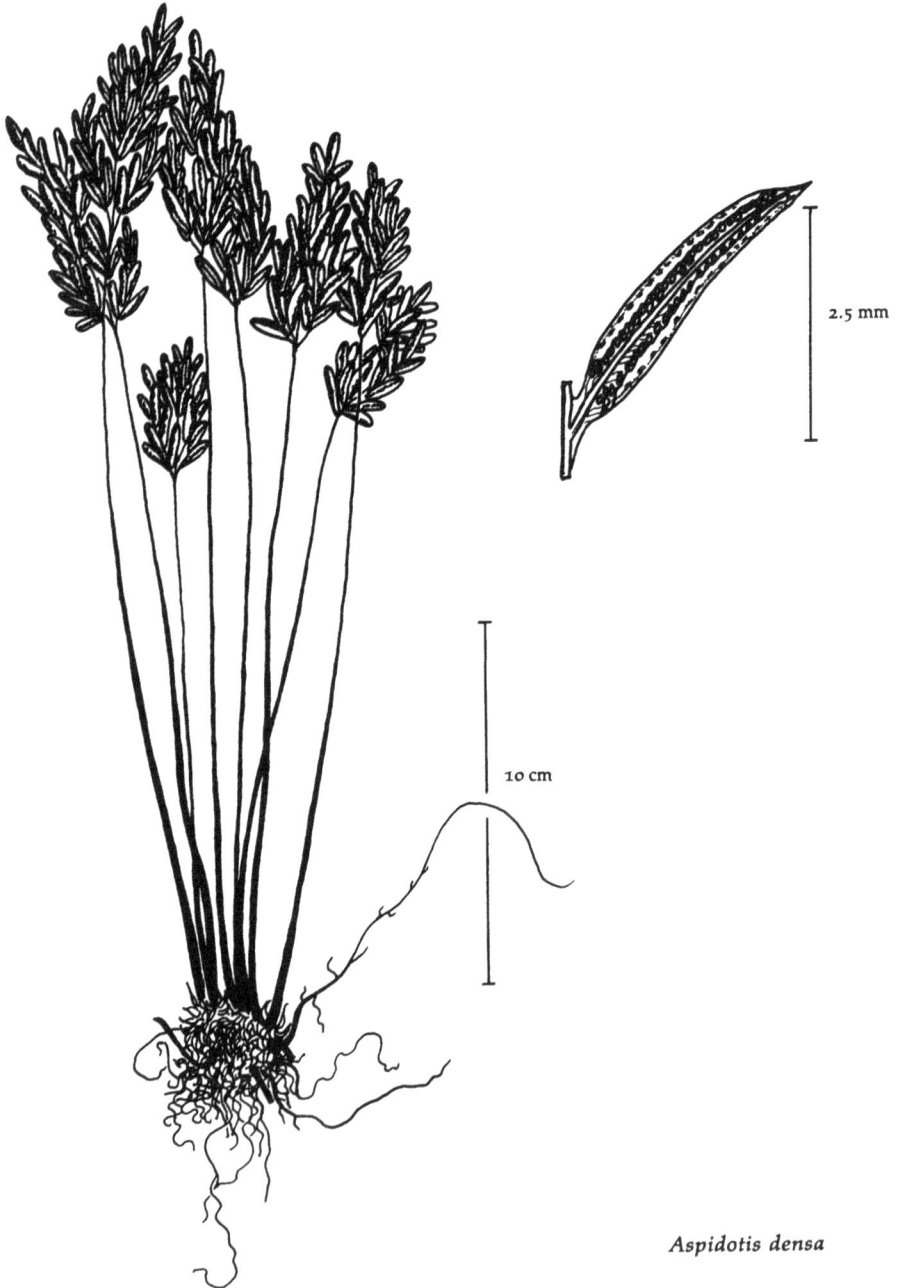

2.5 mm

10 cm

Aspidotis densa

RANGE Southern British Columbia to California, east to northern Montana, Wyoming, and Utah; Gaspé Peninsula, Quebec.

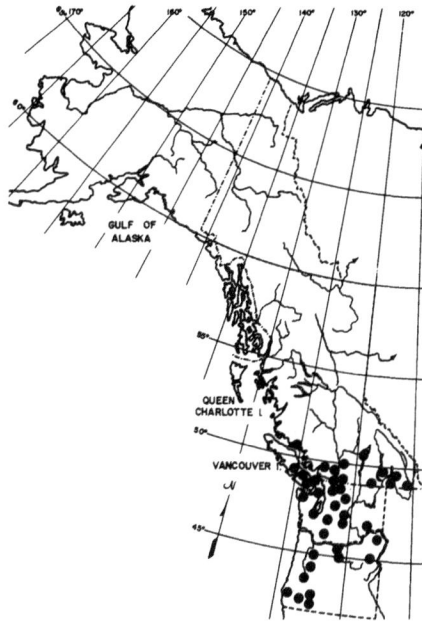

ASPLENIUM L., Sp. Pl. 1078, 1753

Small to large, often evergreen ferns of moist, particularly rocky places. Fronds somewhat crowded, varying from rosettes to erect-spreading, with creeping or erect rhizomes. Stipes slender to filiform. Blades variously compound (simply pinnate in ours), glabrous, variously pubescent or slightly scaly; rachises green and soft to hard and lustrous. Sori oblong to linear, oblique; indusia always present, usually membranous, straight or slightly curved, attached lengthwise to the upper side of an ultimate veinlet.

A very large genus mainly of tropical and subtropical regions in both hemispheres. (Classical Greek name *Asplenon* given to some fern supposed to cure diseases of the spleen.)

1   Plants grasslike, densely tufted; stipe longer than the blade; pinnae few, narrowly linear to narrowly cuneate.   *A. septentrionale*

1   Plants not grasslike; stipe much shorter than the blade; pinnae numerous, oval or oblong.

2   Stipe and rachis dark brown or purplish-brown throughout; glossy, somewhat stiff.   *A. trichomanes*

2   Stipe sometimes reddish-brown below, the upper part and rachis green, lax, herbaceous.   *A. viride*

Asplenium septentrionale

Asplenium septentrionale (L.) Hoffm.

Deutsch. Fl. **2**:12, 1795.

Rhizome erect, densely tufted, short, creeping. Fronds up to 14 cm long; stipe slender, considerably longer than the blade, green but becoming abruptly dark brown at the base. Blade much reduced, more or less dichotomously forked 1–3 times, dark green, glabrous. Segments linear to narrowly triangular-cuneate, about 1 mm wide, often somewhat curved, minutely forked at the tip, decurrent down the rachis. Sori 2–3 to a segment, elongate running the length of the segment. ($n = 72$)

H A B I T A T   Siliceous rock crevices and ledges, often at high altitudes.

R A N G E   South Dakota, south to Colorado, western Oklahoma, and New Mexico; Oregon; Eurasia.

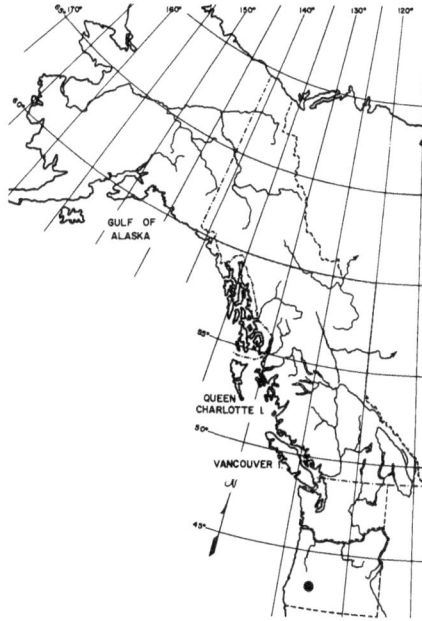

C O M M E N T S   This little fern which when dry looks like a small tuft of grass or a sedge is known in our area only from Douglas County, Oregon, where it has recently been collected by Lang.

Asplenium trichomanes L.

Sp. Pl. 1080, 1753; Macoun, Cat. Can. Pl. pt. 5, 265, 1890; Henry, Fl. S. British Columbia 5, 1915; Piper & Beattie, Fl. Northwest Coast 6, 1915; Maxon [in] Abrams, Ill. Fl. Pac. States **1**:18, *Fig.* 34, 1923; Frye, Ferns Northwest 143, *Fig.* LI, 1934; Peck, Man. Higher Pl. Ore. 49, 1941; Morton [in] Gleason, Ill. Fl. **1**:39, *Fig.*, 1952; Anderson, Fl. Alaska 16, 1959; Hultén, Fl. Alaska 46, 1968; Calder & Taylor, Fl. Queen Charlotte Islands pt. 1, 146, 1968.

Rhizomes erect or decumbent, very short, forming dense tufts with many

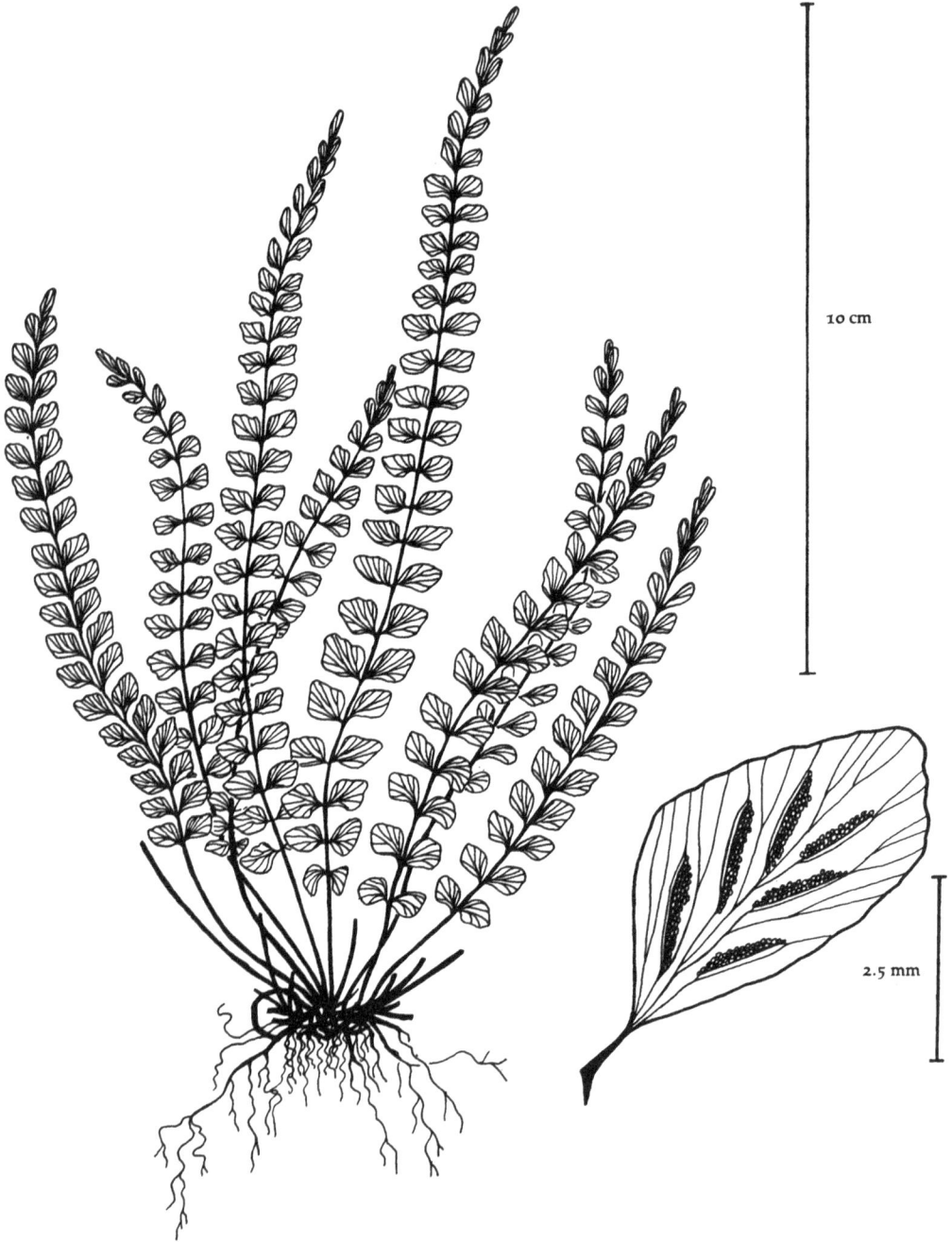

10 cm

2.5 mm

Asplenium trichomanes

denuded old rachises. Fronds up to 22 cm tall, wide-spreading; stipe slender, curved, like the rachis dark chestnut to purplish-brown, lustrous, glabrous. Blades linear, once pinnate, somewhat narrower below. Pinnae evergreen, roundish-oblong or oval, inequilateral at the cuneate to rounded base, lower ones progressively more distant, crenulate or shallowly crenate; veins mostly once forked. Sori 3–5 pairs, linear-oblong, medial; indusia ample, more or less entire. ($n = 36, 72, 108$)

H A B I T A T Moist, often calcareous, crevices and rock ledges.

C O M M E N T S General across southern British Columbia and up the coast to about 58° north latitude. In Washington and Oregon it is found mostly in the Cascades. The bare stipes and rachises persist for several years after the pinnae have been shed and may be even more numerous than the living fronds. Manton (1950) reports two races of this species in western Europe, one has $n = 36$ and the other $n = 72$. Britton (1953) also reports diploid and tetraploid races in North America; Meyer (1952) reports a triploid with $2n = 108$ from Europe.

R A N G E Southern Alaska to Nova Scotia, south to Oregon, Arizona, North Dakota, Wisconsin, Alabama, and Georgia; Eurasia; Hawaii; Africa.

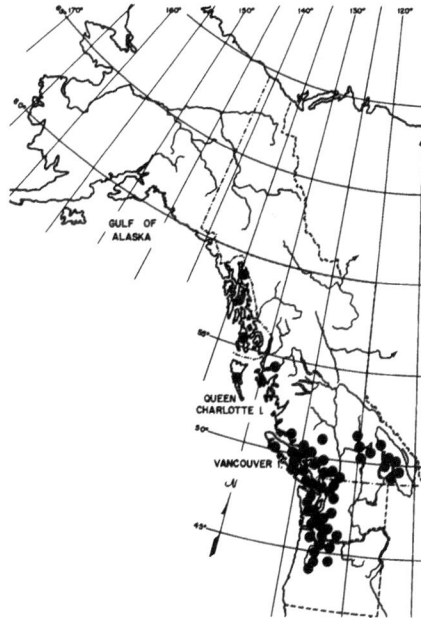

Asplenium viride Huds.

Fl. Angl. 385, 1762; Macoun, Cat. Can. Pl. pt. 5, 265, 1890; Henry, Fl. S. British Columbia 5, 1915; Piper & Beattie, Fl. Northwest Coast 6, 1915; Maxon [in] Abrams, Ill. Fl. Pac. States 1:19, *Fig.* 36, 1923; Frye, Ferns Northwest 144, *Fig.* LI, 1934; Morton [in]Gleason, Ill. Fl. 1:38, *Fig.*, 1952; Anderson, Fl. Alaska 17, 1959; Hultén, Fl. Alaska 47, 1968; Calder & Taylor, Fl. Queen Charlotte Islands pt. 1, 147, 1968.

Rhizome with blackish scales, short but creeping. Fronds numerous, tufted, up to 13 cm tall, laxly ascending. Stipes pale straw-coloured to green from a reddish-brown base. Blades narrow, linear to linear-lanceolate, somewhat attenuate, once pinnate; rachis delicate green. Pinnae soft, herbaceous; the lower pinnae opposite and somewhat remote, becoming progressively closer together and more alternate higher up; round or rhombic-ovate, stalked, broadly cuneate at the inequilateral base, crenate or crenately lobed. Sori in 2–4 pairs, close to the rather indistinct midvein, soon confluent and concealing the delicate, subentire indusium. ($n = 36$)

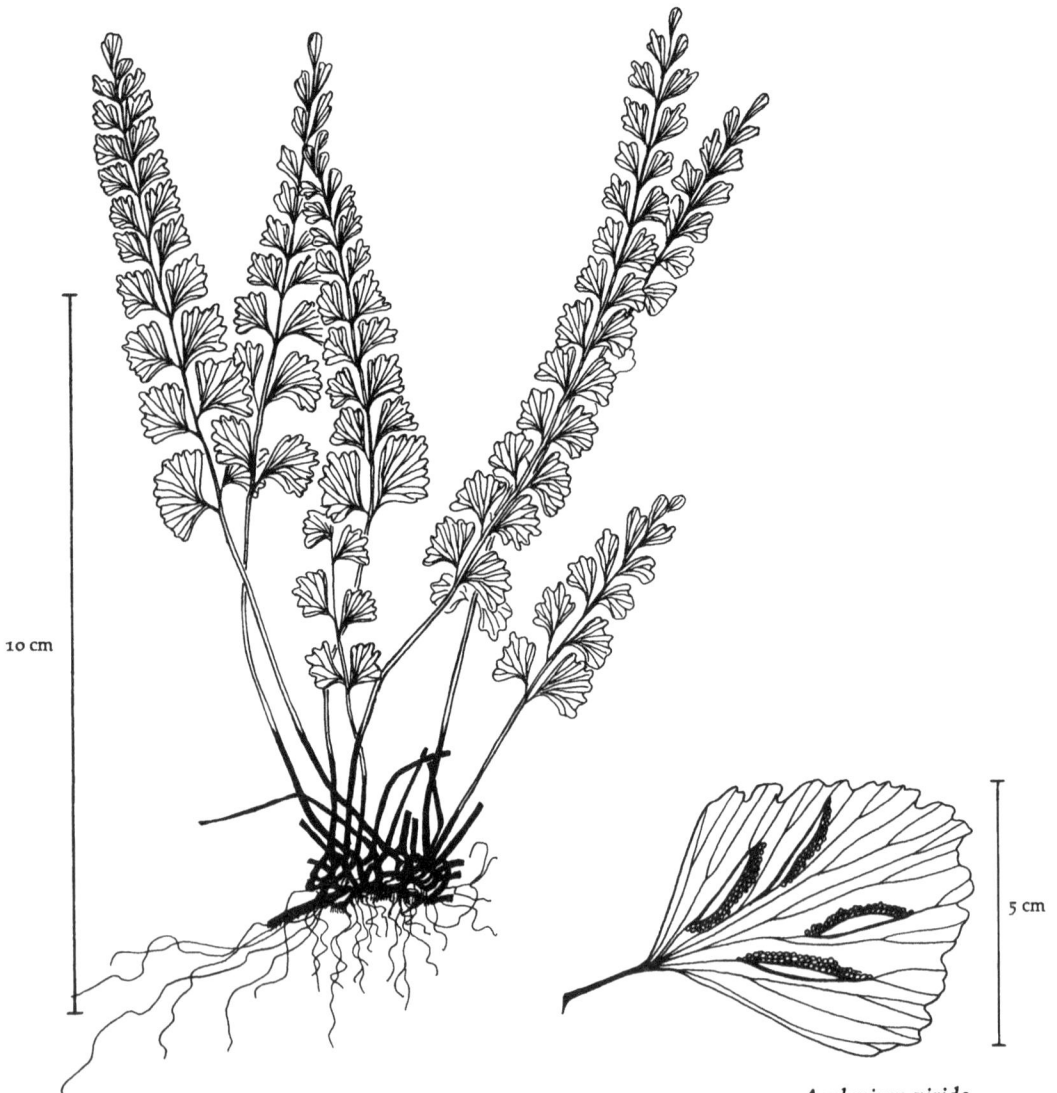

10 cm

5 cm

Asplenium viride

HABITAT Cool, shaded crevices in limestone or other basic rocks.

RANGE Alaska to New-
foundland, south in moun-
tains to Washington, Colorado, and Vermont; Green-
land; Eurasia.

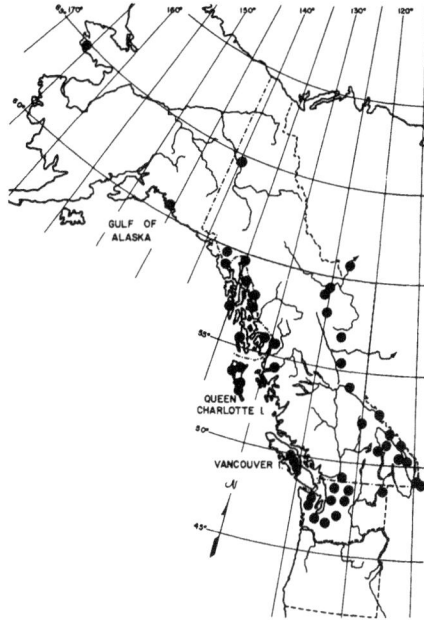

COMMENTS This delicate little fern can always be recognized by its
tufted habit, narrow linear fronds, and green to straw-coloured rachis. It
extends from the Cascades of central Washington north to about 50°
north latitude.

ATHYRIUM Roth, Fl. Germ. 3:58, 1799

Medium to large ferns of erect habit. Rhizome stout and short, ending in
a stipe-covered crown. Fronds herbaceous and soft, usually large, erect-
spreading; stipe long to almost wanting. Blades ample, 1- to 3-pinnate,
elongate; pinnae variously pinnately incised, membranous. Sori oblique
to the midvein, narrowly oblong, either along one side of a veinlet or curving and crossing it to become somewhat horseshoe-shaped. Indusia thin,
straight or curved, laterally attached, the free edge usually toothed, sometimes minute or obsolete.

  A large genus, mainly of tropical regions. (Name from Greek *athyros*,
literally shieldless or without a door, perhaps with reference to the growth
of the sporangia forcing back the indusium.)

Blades much dissected, almost skeletonlike, segments very narrow and distant;
sori roundish, indusia obsolete or wanting.   *A. distentifolium*

Blades ample, leaflike, segments broad and close; indusia and sori mostly cres-
cent-shaped or broadly hooked.   *A. filix-femina*

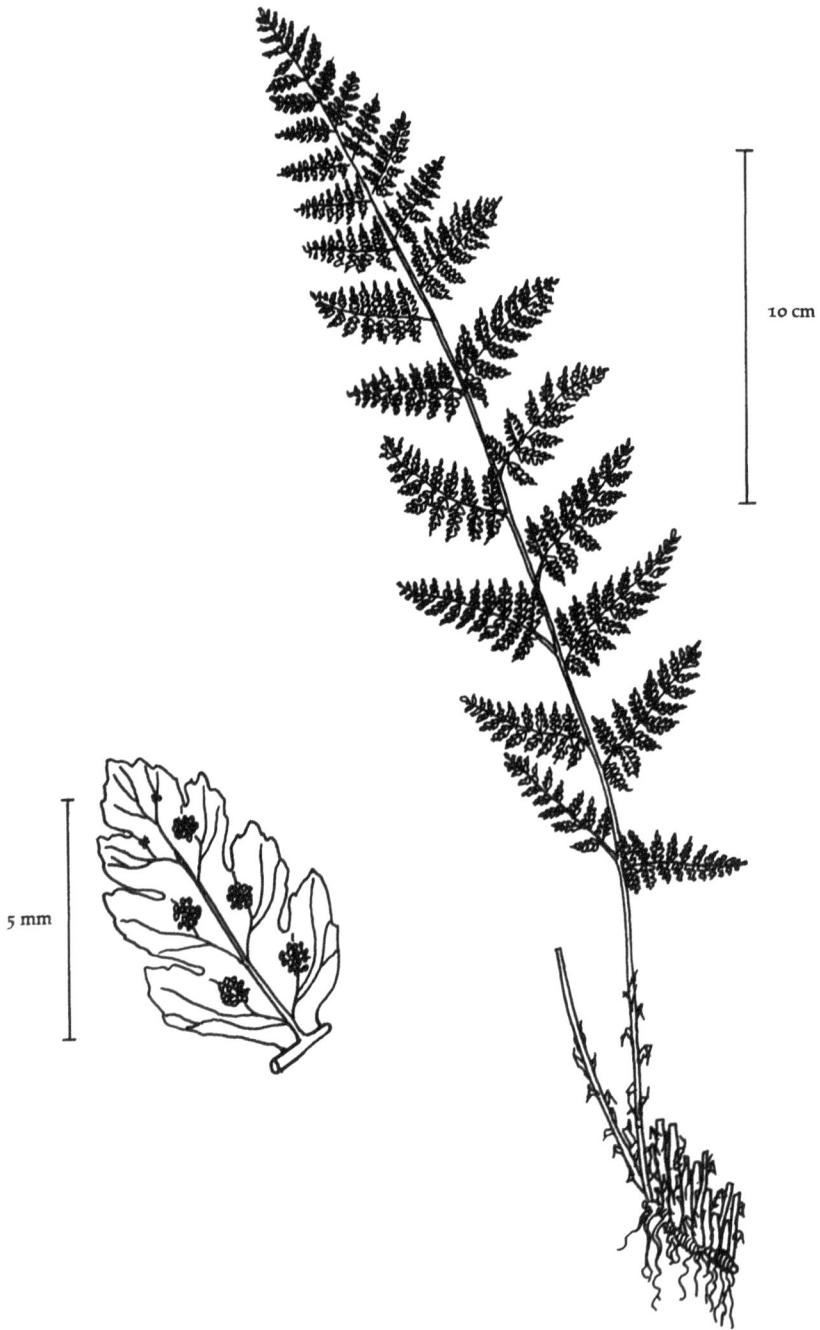

10 cm

5 mm

Athyrium distentifolium

Athyrium distentifolium Tausch

Opiz, Tent. Fl., Crypt Boem. 1:14, 1820; Macoun, Cat. Can. Pl. pt. 5, 271, 1890; Henry, Fl. S. British Columbia 4, 1915; Piper & Beattie, Fl. Northwest Coast 3, 1915; Maxon [in] Abrams, Ill. Fl. Pac. States 1:20, *Fig.* 38, 1923; Frye, Ferns Northwest 140, *Fig.* xlviii, 1934; Peck, Man. Higher Pl. Ore. 48, 1941; Morton [in] Gleason, Ill. Fl. 1:43, *Fig.*, 1952; Anderson, Fl. Alaska 17, 1959; Hultén, Fl. Alaska 47, 1968.
Phegopteris alpestris (Hoppe) Mett., Fil. Hort. Bot. Lips. 83, 1856.
A. alpestre (Hoppe) Rylands [in] Moore, Ferns of Gt. Brit. & Ire. Nat. Printed *Plate* 7, 1857, *non* Clairv.
A. americanum (Butters) Maxon, Am. Fern J. 8:120, 1918.

Rhizome very stout, erect or somewhat decumbent, branching, forming large round clumps, copiously covered with brown scales and coarse stipe bases. Fronds numerous in vaselike arrangement at the end of the rhizome, up to 80 cm tall; stipe short, sparsely scaly, dark straw-coloured from a darker base. Blades glabrous, subcoriaceous, linear to oblong-lanceolate, acuminate, usually 2-pinnate-pinnatifid, mostly 1/4 to 1/3 as broad as long. Pinnae oblique, rather narrowly deltoid, gradually acuminate, the lower ones distant, sometimes shorter than the median. Pinnules thin, stalked, somewhat oblique, broadest at the base, oblong-lanceolate to narrowly triangular, incised; the larger ones deeply pinnatifid. Sori numerous, very small, round; indusia not evident, rudimentary or lacking. ($n = 40$)

HABITAT Wet talus slopes, rocky hillsides, and alpine meadows.

RANGE Southeastern Alaska, south to California, Nevada, Colorado; Gaspé Peninsula, Quebec; New-foundland; Iceland; Eurasia.

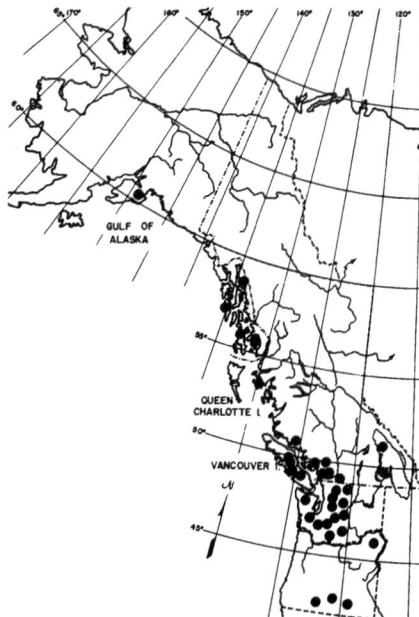

1 cm

15 cm

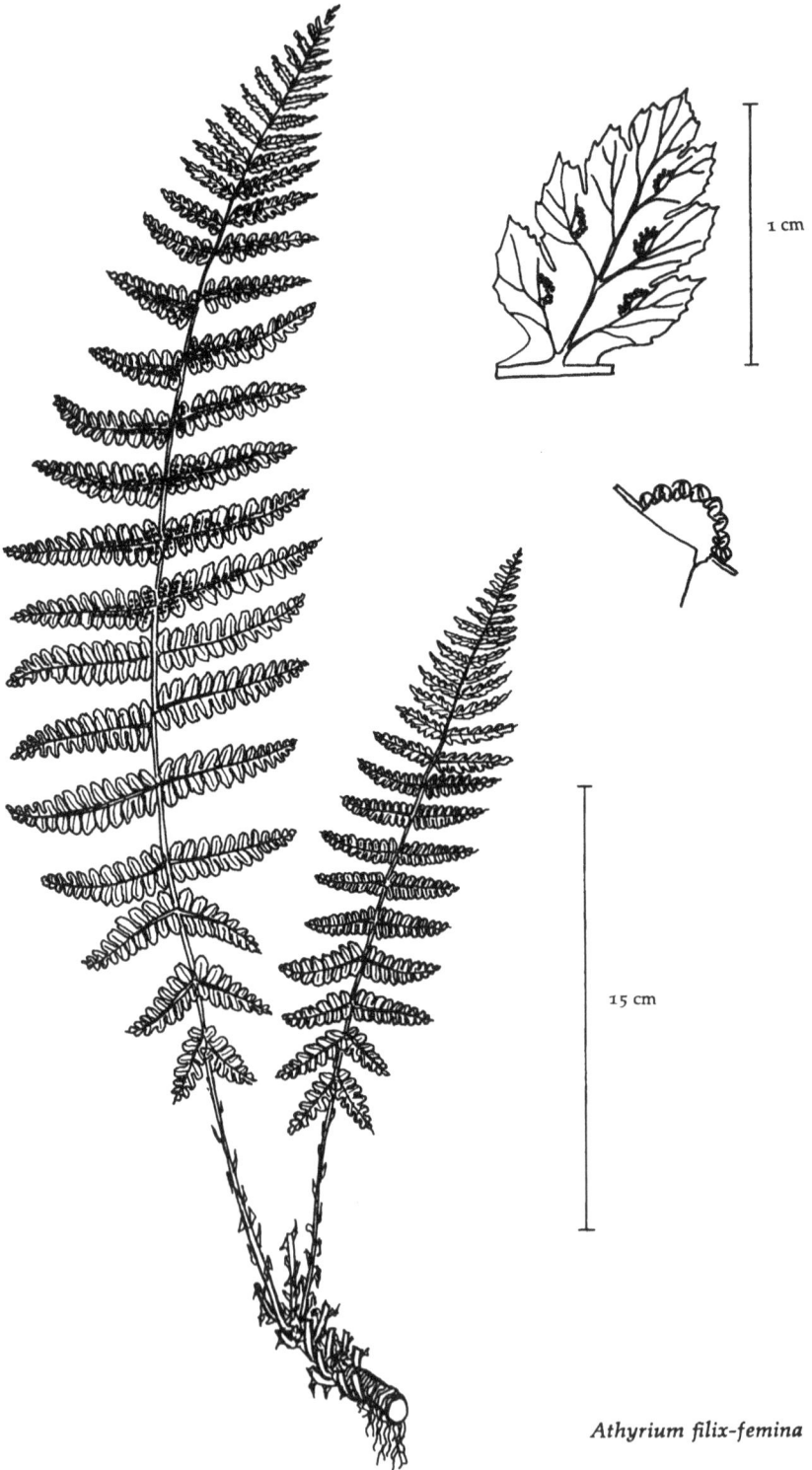

Athyrium filix-femina

COMMENTS This widely ranging alpine species occurs in several geographic varieties. The North American form is regarded by some as a distinct species *A. americanum* (Butters) Maxon. It is, however, so close to the Eurasian form that it seems best to treat the two as conspecific.

Athyrium filix-femina (L.) Roth

[ex] Mertens, Arch. Bot. 2:106, 1799; Macoun, Cat. Can. Pl. pt. 5, 267, 1890; Henry, Fl. S. British Columbia 5, 1915; Piper & Beattie, Fl. Northwest Coast, 6, 1915; Maxon [in] Abrams, Ill. Fl. Pac. States 1:19, *Fig.* 37, 1923; Frye, Ferns Northwest 137, *Figs.* XLIX, LI, 1934; Peck, Man. Higher Pl. Ore. 48, 1941; Morton [in] Gleason, Ill. Fl. 1:43, *Fig.,* 1952; Anderson, Fl. Alaska 17, 1959; St. John, Fl. Southeast. Wash. 5, 1963; Hultén, Fl. Alaska 48, 1968; Calder & Taylor, Fl. Queen Charlotte Islands pt. 1, 147, 1968.
Asplenium filix-femina (L.) Bernh., Neu. J. Bot. Schrad. 1:26, 1806.
Athyrium cyclosorum Rupr., Beitr. Pflanzenk. Russ. Reich. 3:41, 1845.

Rhizome stout, erect or ascending, terminating in a tuft of fronds, chaffy, with numerous, thin, lanceolate, light to dark brown scales. Fronds up to 2 m tall, erect-spreading; stipes short, somewhat fragile, furrowed, straw-coloured, scaly at the dark base. Blades narrowly to broadly lanceolate, tapering both ways from about the middle, 2- to 3-pinnate. Pinnae linear to lanceolate-oblong, symmetrical, acuminate to attenuate, practically sessile, often decurrent on the narrowly winged rachis, glabrous. Sori oblong to lunate or horseshoe-shaped; indusia similar, laterally attached by their inner side to a veinlet, often toothed, septate-ciliate. ($n = 40$)

HABITAT Forests and damp shady places generally.

RANGE Alaska to Newfoundland, south to California, Texas, and Florida; Eurasia.

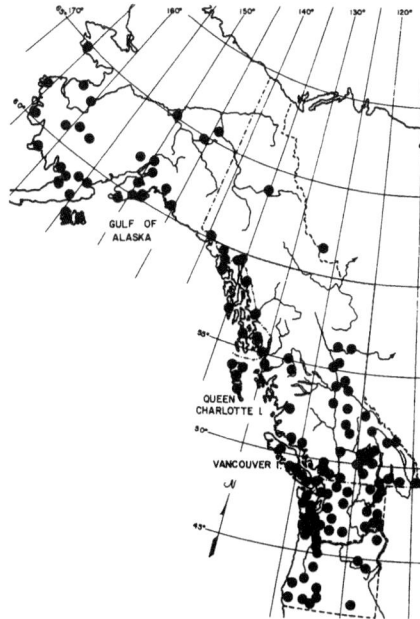

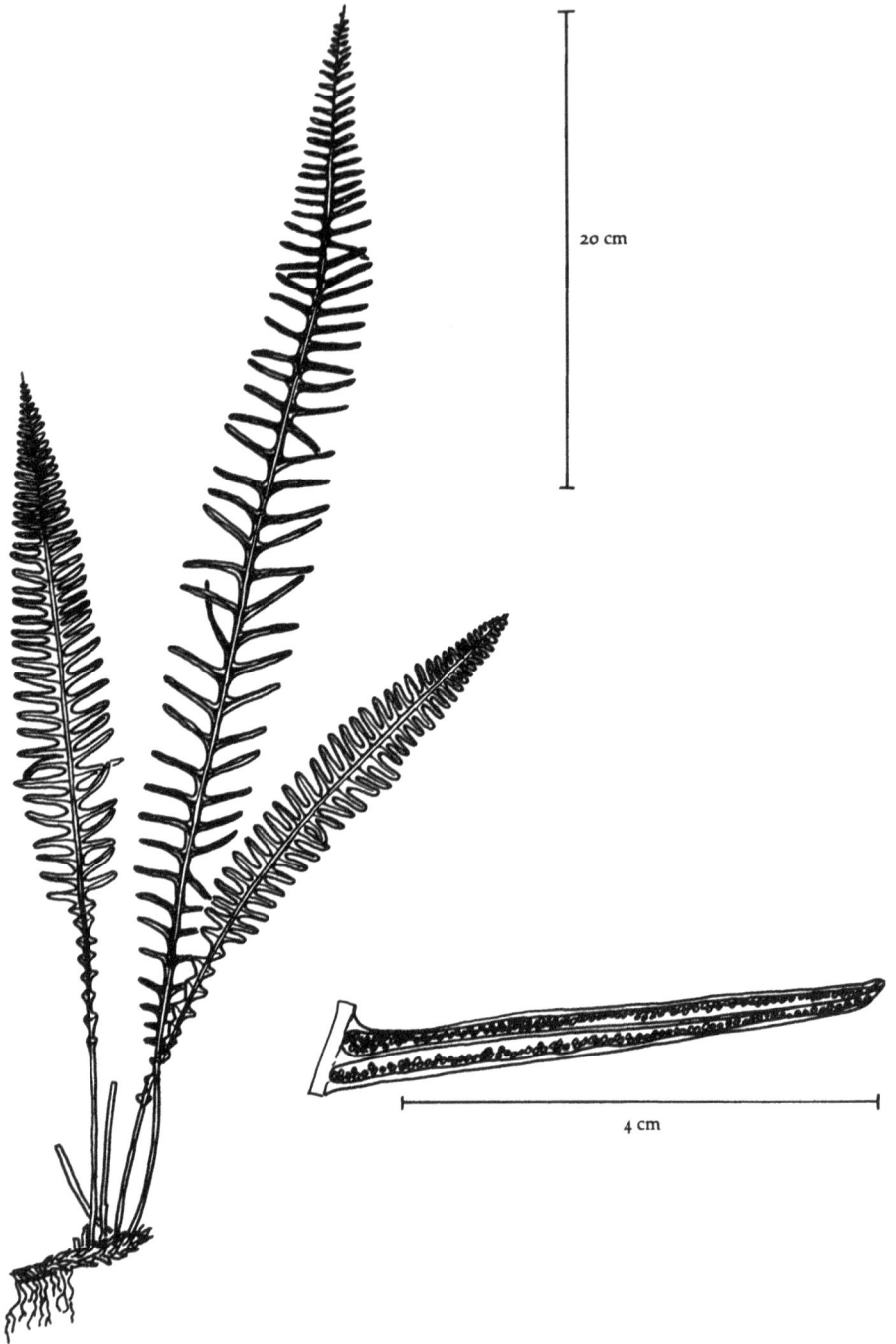

20 cm

4 cm

Blechnum spicant

c o m m e n t s To be expected throughout our area wherever conditions are suitable for its growth. A variable and widely distributed species that is found in one form or another throughout the northern hemisphere and on tropical mountains. The western North American form is often referred to var. *cyclosorum* (Ledeb.) Moore. It is a large, rather coarse variety with subcoriaceous fronds, the ultimate segments having broad, blunt teeth.

B L E C H N U M L. Sp. Pl. 1077, 1753

Chiefly forest ferns of various habitats, terrestrial or epiphytic. Rhizomes woody, climbing, short-creeping or erect and trunklike in some tropical species. Fronds spreading or ascending, sometimes of two kinds; blades pinnatifid or pinnate. Sori continuous, parallel to the midrib and near the margin of the segment; indusia membranous, attached to the margin and opening towards the midrib, entire or lacerate, often reflexed at maturity. (Greek name for a kind of fern.)

A large genus of tropical and mainly south temperate distribution. Only one species grows in our area.

Blechnum spicant (L.) J. Sm.

Mem. Acad. Roy. Sci. Turin 5:411, 1793; Macoun, Cat. Can. Pl. pt. 5, 263, 1890; Henry, Fl. S. British Columbia 3, 1915; Piper & Beattie, Fl. Northwest Coast 5, 1915; Maxon [in] Abrams, Ill. Fl. Pac. States 1:21, *Fig.* 40, 1923; Frye, Ferns Northwest 145, *Fig.* LI, 1934; Peck, Man. Higher Pl. Ore. 49, 1941; Anderson, Fl. Alaska 9, 1959; Hultén, Fl. Alaska 58, 1968; Calder & Taylor, Fl. Queen Charlotte Islands pt. 1, 148, 1968.
Struthiopteris spicant (L.) Weiss, Pl. Crypt. Gott. 287, 1770.
Lomaria spicant (L.) Desv., Gesell. Nat. Fr. Berlin 5:325, 1811.

Rhizome short-creeping, woody, with narrow, chestnut-coloured scales. Fronds of two kinds; the sterile evergreen, coriaceous, numerous in a rosette, spreading and appressed to the ground; stipes very short, yellowish to chestnut-brown. Blades linear-oblanceolate, pinnatifid but becoming pinnate at the long attenuate base, basal segments broader, often distant. Segments very numerous, mostly linear to linear-oblong, somewhat roundish at the apex, entire or crenulate, glabrous, somewhat paler beneath. Fertile fronds deciduous, less numerous, arising from the centre of the crown, erect, longer than the sterile fronds; stipes dark chestnut-coloured. Blades similar to the sterile ones in general outline, pinnate, with very much reduced basal segments; pinnae distant, mostly narrowly linear, somewhat dilated at the base; sori continuous, parallel to the midvein; indusia membranous. ($n = 34$)

h a b i t a t Damp, mostly coniferous forests, swampy places, and wet banks.

2 mm

3 cm

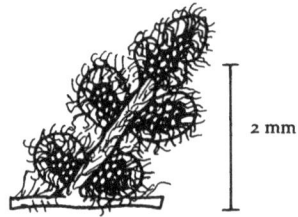

Cheilanthes feei

RANGE Coastal Alaska to
California; Iceland; Eurasia.

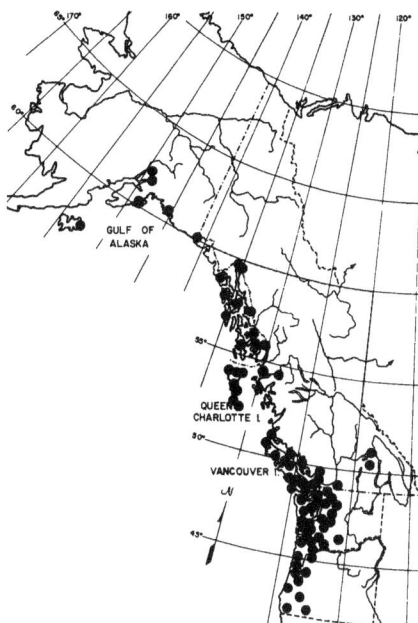

COMMENTS Very common, mainly growing not far from the coast,
except for two stations in the interior wet belt of southern British Columbia.

CHEILANTHES Sw., Syn. Fil. 126, 1806

Mostly small rock-loving ferns of dry regions. Fronds 1- to 4-pinnate,
variously pubescent or glabrous; segments usually very small and beadlike. Sori borne at the enlarged tips of veins, solitary on minute lobes, or
numerous and confluent. Indusia formed of the revolute, more or less
modified margin of the lobes or segments, or in some species the margin
closely reflexed and pouchlike, giving rise to a proper inward-opening
indusium.

About 125 species of dry temperate and tropical regions. (Name from
Greek *keilos*, lip or margin, and *anthos*, flower, referring to the marginal
sori.)

1   Blades wholly lacking scales but conspicuously long-pubescent;
     upper side of blade thinly villous.   *C. feei*

1   Blades scaly as well as villous or tomentose, devoid of hairs above.

  2   Blades mostly 2-pinnate; segments usually oblong, densely tomentose
       beneath, glabrous above.   *C. gracillima*

  2   Blades mostly 3-pinnate; segments rounded to oval, densely imbri-
       cate-scaly beneath, with a few pale, stellate scales above. *C. intertexta*

Cheilanthes feei Moore

Ind. Fil. xxxviii, 1857; Macoun, Cat. Can. Pl. pt. 5, 259, 1890; Henry, Fl. S. British Columbia 4, 1915; Maxon [in] Abrams, Ill. Fl. Pac. States 1:27, *Fig.* 51, 1923; Frye, Ferns Northwest 97, *Fig.* xxvii, 1934; Morton [in] Gleason, Ill. Fl. **1**:33, *Fig.*, 1952; St. John, Fl. Southeast. Wash. 9, 1963.
C. lanuginosa Nutt. [ex] Hook., Sp. Fil. **2**:99, 1858.

Rhizome with several crowns, short-creeping or ascending, cespitose; scales narrow, cinnamon- or darker brown with a distinct central blackish stripe. Fronds erect, somewhat flexuous, tufted, up to 25 cm tall; stipes slender, dark brown, deciduously soft hairy. Blades linear-oblong to ovate, acuminate, usually 3-pinnate. Pinnae oblique, deltoid to ovate-oblong, thinly villous above with soft whitish hairs, densely and coarsely tomentose beneath with pale brown hairs, which are also on the rachis; ultimate segments roundish, simple or crenately lobed or divided.

5 cm

5 mm

Cheilanthes gracillima

H A B I T A T Dry crevices of limestone cliffs.

C O M M E N T S Found in two limestone localities of southern British Columbia, where it is quite common, also in Whitman County, Washington. Recently a collection of a cheilanthoid fern has been made at the east end of the Olympic Peninsula. It may be the present species or it may prove to be closer to *C. lanosa* (Michx.) D. C. Eaton. At the present time insufficient material is available for adequate study and a proper decision as to its identity.

R A N G E Southern British Columbia to western Wisconsin, south to California, Texas, and Arkansas.

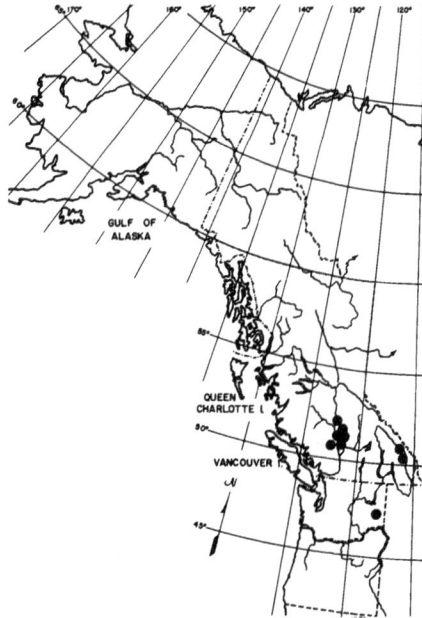

Cheilanthes gracillima D. C. Eaton

[In] Torrey, Bot. Mex. Bound. 234, 1859; Macoun, Cat. Can. Pl. pt. 5, 259, 1890; Henry, Fl. S. British Columbia 4, 1915; Piper & Beattie, Fl. Northwest Coast 5, 1915; Maxon [in] Abrams, Ill. Fl. Pac. States 1:28, *Fig.* 54, 1923; Frye, Ferns Northwest 96, *Fig.* xxvii, 1934; Peck, Man. Higher Pl. Ore. 50, 1941; St. John, Fl. Southeast. Wash. 9, 1963.

Rhizomes tufted with numerous, very scaly, short branches; scales rather light brown, needlelike or slightly broader. Fronds very numerous, up to 24 cm tall, more or less erect; stipes dark brown, soon naked or nearly so. Blades linear to oblong-lanceolate, 2-pinnate; lower pinnae lance-oblong, pinnately divided, progressively less divided upwards; segments oblong to oval, dull green above, densely cinnamon tomentose beneath, a few minute stellate hairs above; margins deeply recurved. Rachis appressed scaly, the scales linear-attenuate from a broader ciliate base.

10 cm

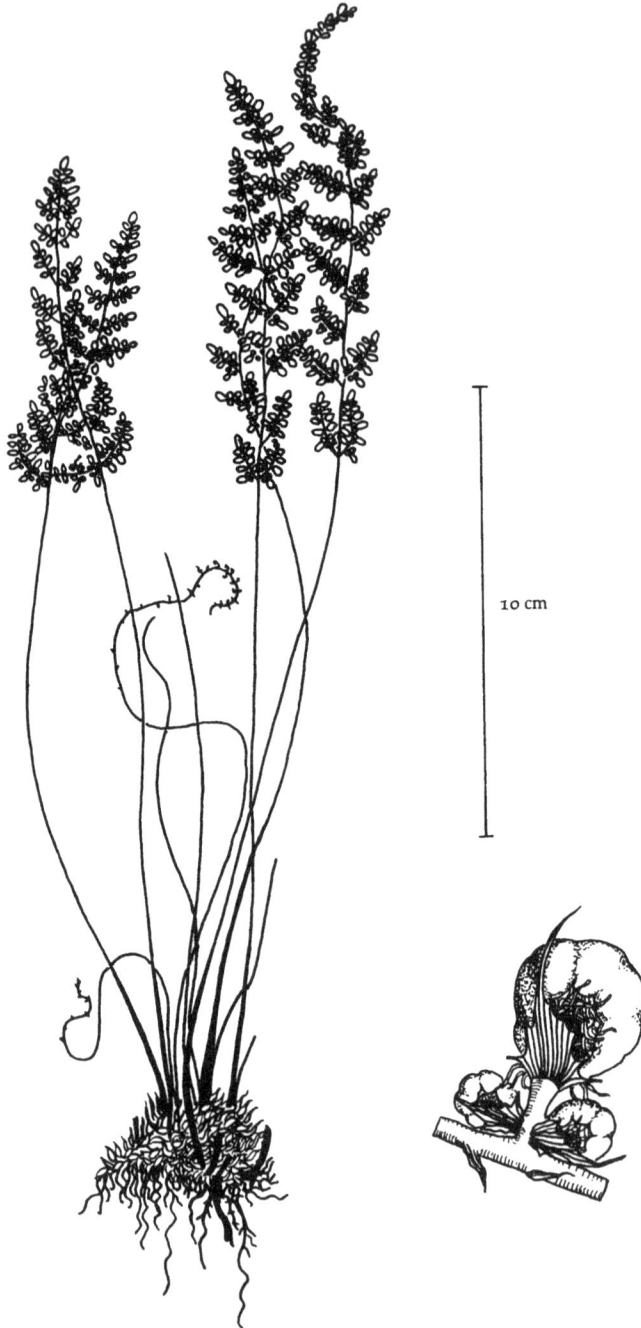

Cheilanthes intertexta

HABITAT Exposed crevices in cliffs and ledges of igneous rocks.

COMMENTS Generally distributed from the extreme southern part of British Columbia southward. A tufted little fern with delicate, wiry fronds and small segments that are glabrous above.

RANGE Southern British Columbia to Idaho and Montana, south to California and Nevada.

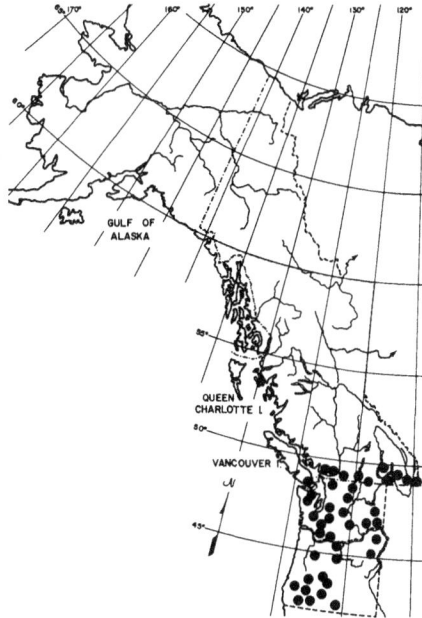

Cheilanthes intertexta Maxon

[In] Abrams, Ill. Fl. Pac. States 1:28, 1923.
C. covillei Maxon var. *intertexta* Maxon, Proc. Biol. Soc. Wash. 31:149, 1919.

Rhizome short, creeping, somewhat nodulose, appressed scaly; scales long, linear-lanceolate, brownish. Fronds numerous, up to 28 cm tall. Stipes brownish-purple, somewhat wiry, with scattered scales, at least when young. Blades ovate-deltoid to oblong, 3-pinnate, dark green above with a few small, whitish stellate scales above, densely clothed below with numerous, bright chestnut or cinnamon-coloured, imbricate scales; larger scales deltoid-lanceolate, long attenuate, margins sinuate denticulate, long ciliate at the cordate base, smaller scales progressively more ciliate; ultimate segments small and beadlike with strongly recurved margins. Sori more or less covered by the recurved segment border.

HABITAT Dry rock crevices.

COMMENTS Known in our area only from Jackson County, Oregon. By some authors this is regarded as a variety of the more widely ranging *C. covillei* Maxon. In general appearance it suggests *C. gracillima* but

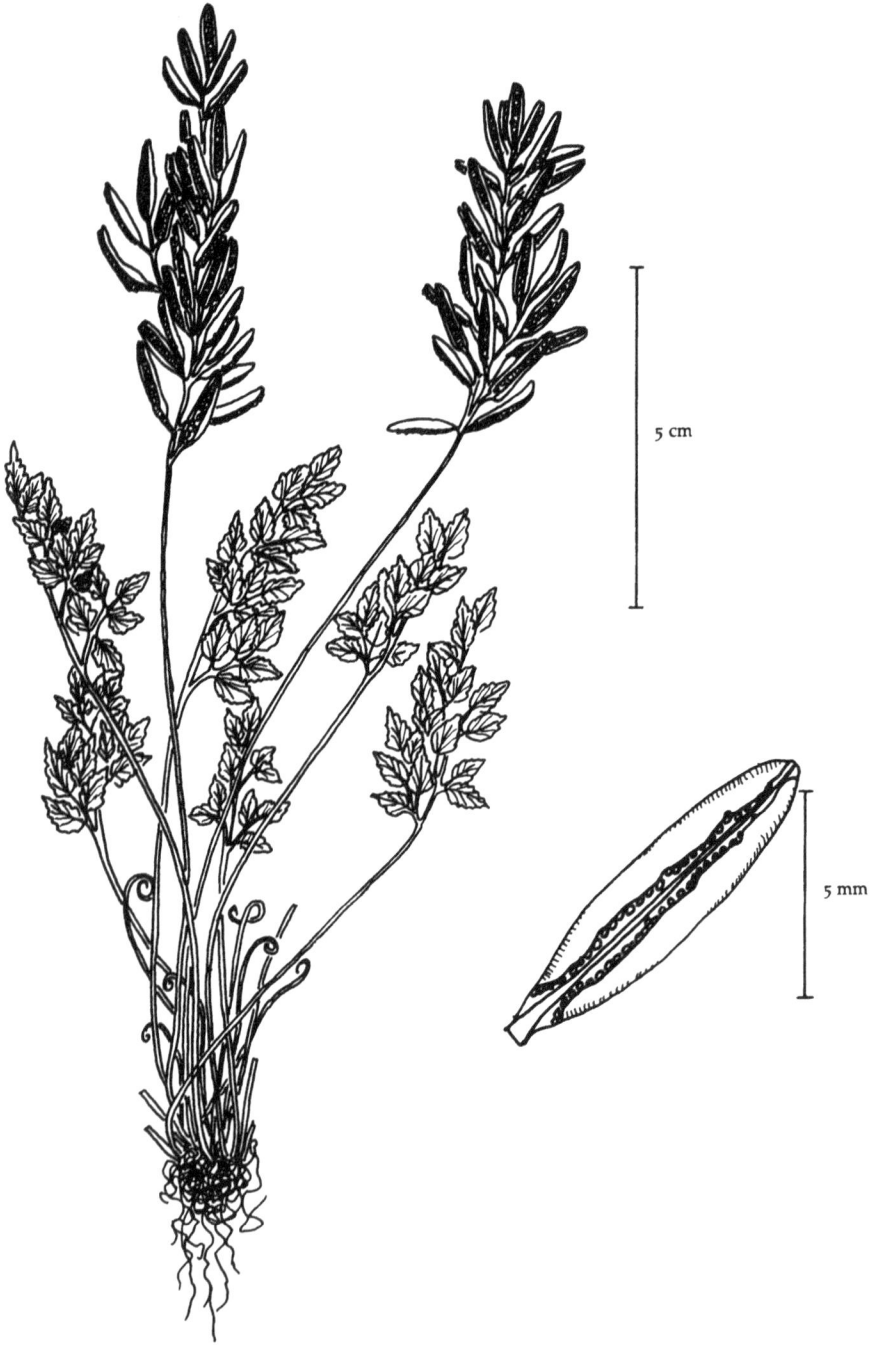

Cryptogramma crispa

may be distinguished by its 3-pinnate fronds that are densely imbricate scaly below.

R A N G E Southern Oregon, western Nevada, and northern California.

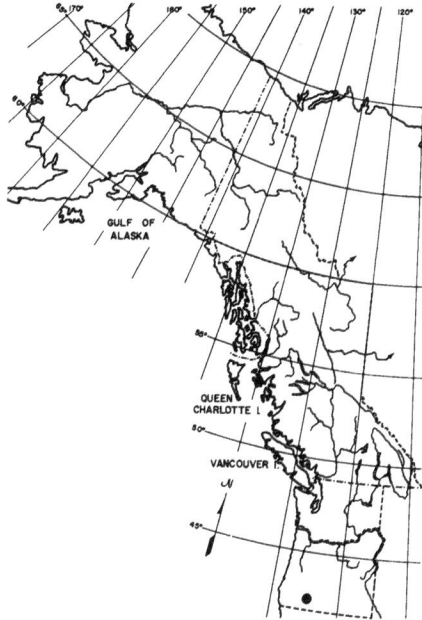

CRYPTOGRAMMA R. Br. [in] Franklin, Narr. Journey 767, 1823

Fronds evergreen, markedly dimorphic, numerous, densely clustered from the erect or somewhat creeping rhizome, not articulated at the base. Sterile fronds thin, membranous, spreading, glabrous. Fertile fronds longer than the sterile ones, segments narrow, entire, margins revolute forming a false indusium, a true indusium not developed.

A small genus of delicate, boreal or alpine ferns of rocky situations. (Name from *kryptos*, hidden, and *gramme*, line, referring to the line of sporangia hidden under the reflexed margin.)

Rhizome relatively stout, ascending; fronds numerous, clustered, blades thick and opaque. *C. crispa*

Rhizome quite slender, creeping; fronds fewer, scattered, blades thin and membranous. *C. stelleri*

Cryptogramma crispa (L.) R. Br.

[In] Richards., Bot. App. Franklin Narr. 767, 1823; Macoun, Cat. Can. Pl. pt. 5, 261, 1890; Henry, Fl. S. British Columbia 2, 1915; Piper & Beattie, Fl. Northwest Coast 4, 1915; Maxon [in] Abrams, Ill. Fl. Pac. States 1:22, *Fig.* 41, 1923; Frye, Ferns Northwest 87, *Fig.* xxvii, 1934; Peck, Man. Higher Pl. Ore. 51, 1941;

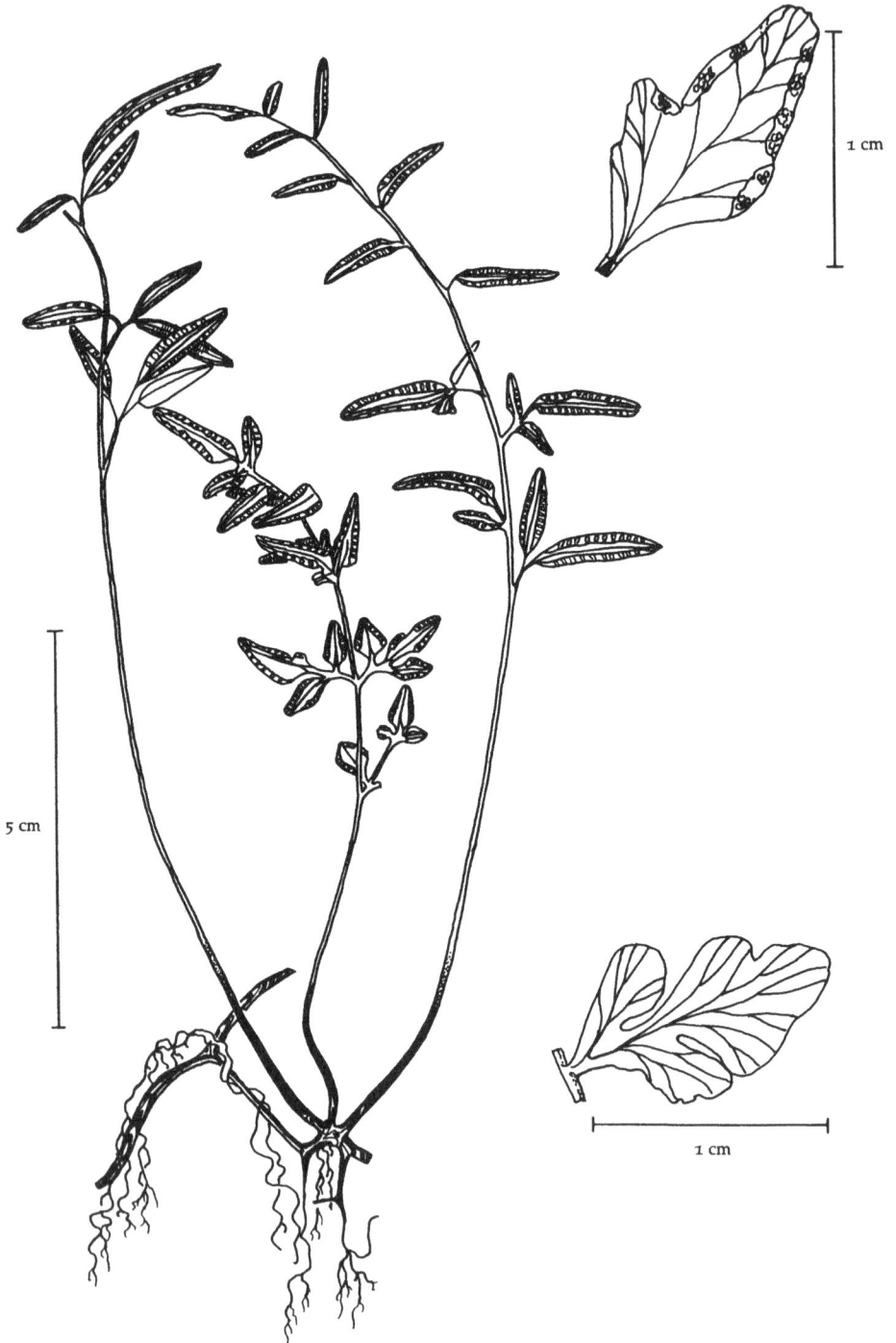

1 cm

5 cm

1 cm

Cryptogramma stelleri

Morton [in] Gleason, Ill. Fl. **1**:31, *Fig.*, 1952; Anderson, Fl. Alaska 10, 1959; Hultén, Fl. Alaska 44, 1968; Calder & Taylor, Fl. Queen Charlotte Islands pt. 1, 149, 1968.
C. acrostichoides R. Br. [in] Richards., Bot. App. Franklin Narr. 767, 1823.

Rhizomes short-creeping but mostly ascending, forming quite large tufts, chaffy; scales lance-ovate, attenuate, entire, brown with a darker central portion. Fronds very numerous, densely clustered, glabrous, markedly dimorphic. Sterile fronds spreading, up to 15 cm long; stipe straw-coloured, scaly at the base; blade ovate or ovate-lanceolate, 2-pinnate-pinnatifid. Rachis compressed, greenish; pinnae alternate or the lower subopposite, few, short-petioled; ultimate segments ovate, oblong or obovate, obtuse, crenate or incised, thick, opaque. Fertile fronds stiffly upright, simpler, segments fewer, linear-oblong, entire, the thin yellowish margins revolute. ($n = 30$)

H A B I T A T Rock crevices and ledges, talus slopes.

C O M M E N T S Common, particularly on talus slopes. The North American form (var. *acrostichoides* (R. Br.) Clarke) differs from the Eurasian in having sterile fronds of a thicker texture and scales with a dark centre. In the north, var. *sitchense* (Rupr.) C. Chr. is a marked variation with broadly deltoid sterile fronds which are finely dissected having obovate, toothed tertiary lobes.

R A N G E Alaska to Labrador, south in mountains to California, and New Mexico; Michigan, Ontario, and Quebec; Eurasia.

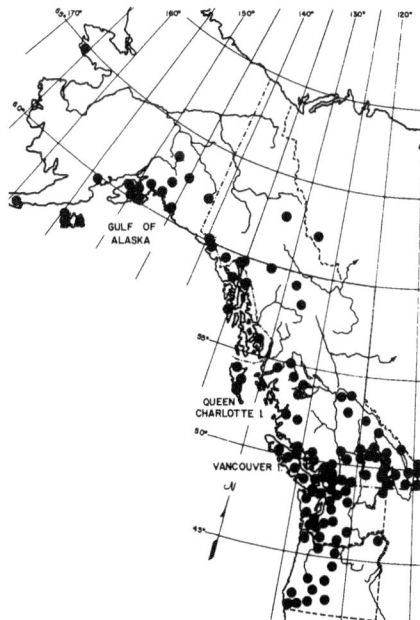

Cryptogramma stelleri (Gmel.) Prantl

[In] Engler, Bot. Jahrb. **3**:413, 1882; Macoun, Cat. Can. Pl. pt. 5, 259, 1890;
Henry, Fl. S. British Columbia 3, 1915; Maxon [in] Abrams, Ill. Fl. Pac. States
1:22, *Fig.* 42, 1923; Frye, Ferns Northwest 88, *Fig.* xxii, 1934; Morton [in]
Gleason, Ill. Fl. **1**:31, *Fig.*, 1952; Anderson, Fl. Alaska 10, 1959; Hultén, Fl.
Alaska 45, 1968.
Pellaea gracilis (Michx.) Hook., Sp. Fil. **2**:138, 1858.

Rhizome very slender, cordlike, creeping, scaly, pilose. Fronds few,
scattered, dimorphic, glabrous, lax. Sterile fronds up to 15 cm long, stipe
almost chaffless, slender, usually longer than the blade, yellowish-green
above, chestnut-brown below; blade ovate or oblong-ovate, 2-pinnate;
the rachis green. Pinnae few, membranous, the larger ones with pinnately
divided basal pinnules; ultimate segments ovate to obovate, cuneate,
crenate, petiolulate. Fertile fronds longer than the sterile, stipe longer than
the blade, the blades 2- to 3-pinnate, segments linear-oblong to lanceolate, mostly acute, rather widely separated, entire. Sori marginal and
continuous, false indusium formed by the delicate membranous, translucent, revolute margin. ($n = 30$)

HABITAT Moist, shaded, usually calcareous, rock crevices or cliffs.

COMMENTS A very attractive, delicate little fern that is easily overlooked. It occurs mostly in the northern Rockies but is known as far south
as northeastern Oregon.

RANGE Alaska to southeast
Labrador and Newfoundland,
south to Oregon, Utah,
Colorado, northeast Iowa,
northern Illinois, Michigan,
northern Pennsylvania, and
West Virginia; Asia.

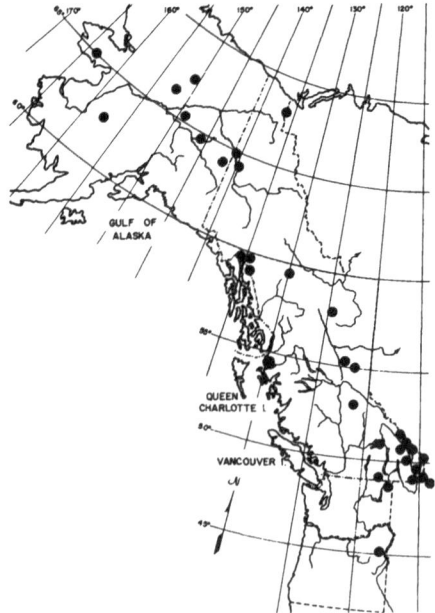

CYSTOPTERIS Bernh., Neu. J. Bot. Schrad. 1(2):26, 1806

Delicate ferns of woodlands or damp rocks with pinnately divided fronds. Rhizomes creeping, scaly. Fronds erect, remote or somewhat tufted, stipes straw-coloured, scaly only at the base, not jointed to the rhizome. Blades glabrous or with a few scattered, septate hairs on the rachis, thin and delicate, 1- to 4-pinnate. Sori dorsal on free, simple or forked veins, roundish, separate. Indusium very delicate, membranous, hoodlike or arched, attached by a broad base on the side towards the midrib and so partly under the sorus; soon thrown back and withering. Spores echinate or rugose. (From the Greek *kystis*, a bladder, an allusion to the inflated indusium.)

About a dozen species are recognized, mostly from temperate and boreal regions.

Fronds lanceolate or narrowly ovate lanceolate, tufted; rhizome thickish, unbranched. *C. fragilis*

Fronds deltoid-ovate, scattered, rhizome slender, branched. *C. montana*

Cystopteris fragilis (L.) Bernh.

Neu. J. Bot. Schrad. 1:27, 1806; Macoun, Cat. Can. Pl. pt. 5, 279, 1890; Henry, Fl. S. British Columbia 7, 1915; Piper & Beattie, Fl. Northwest Coast 8, 1915; Maxon [in] Abrams, Ill. Fl. Pac. States 1:7, *Fig.* 12, 1923; Frye, Ferns Northwest 106, *Fig.* XXXI, 1934; Peck, Man. Higher Pl. Ore. 46, 1941; Morton [in] Gleason, Ill. Fl. 1:47, *Fig.*, 1952; Anderson, Fl. Alaska 13, 1959; Wiggins & Thomas, Fl. Alaskan Arctic Slope 43, 1962; St. John, Fl. Southeast. Wash. 6, 1963; Calder & Taylor, Fl. Queen Charlotte Islands pt. 1, 150, 1968; Hultén, Fl. Alaska 49, 1968. *Filix fragilis* (L.) Gilib., Exerc. Phyt. 558, 1792.

Rhizome light brown, short-creeping, thickish, unbranched, scaly towards the apex; scales light brown, ovate, attenuate, entire. Fronds up to 30 cm tall, erect-spreading, usually clustered. Stipe shorter than the blade, slender glabrous, straw-coloured to light chestnut. Blades variable, bright green, lanceolate to narrowly lance-ovate, 2- to 3-pinnate; rachis green, glabrous except for a few hairs towards the base of the pinnae. Pinnae deltoid to deltoid-oblong, mostly acutish, petiolulate, the lowermost pair somewhat shorter than the second and rather distant from it. Pinnules spreading, ovate to oblong-lanceolate, incised, decurrent on the margined rachis, membranous. Sori rather small, often numerous and even coalescing; indusium thin and fragile, deeply convex, tapering towards the apex which is free and often somewhat lacerate-toothed or lobed. ($n = 42, 84, 126$)

HABITAT Damp shady woods, rocky slopes, and crevices in circumneutral soil.

COMMENTS Found very commonly throughout our area wherever suitable conditions are present. It is extremely variable in the size and shape

Cystopteris fragilis

7.5 cm

of its fronds, yet all plants are best placed in var. *fragilis*. In eastern North America some variants are sufficiently consistent to warrant special names. One form of particular interest in the West, where it is quite common, has rugose rather than finely echinate spores. In this respect it comes close to the European *C. dickieana* Sim but it does not have the foliage characteristic of that species. Further study is required to determine the proper status of this interesting form.

RANGE Alaska to Labrador and Newfoundland, south to southern California, Texas, Missouri, northern Illinois, northern Ohio, and Virginia; Eurasia; on mountains in the tropics and southern hemisphere.

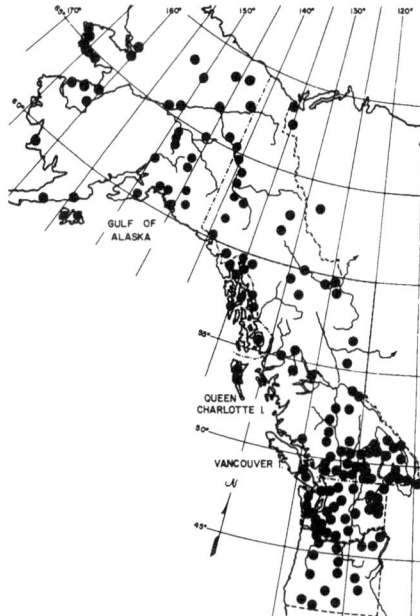

Cystopteris montana (Lam.) Bernh.

Neu. J. Bot. Schrad. 1:26, 1806; Macoun, Cat. Can. Pl. pt. 5, 280, 1890; Henry, Fl. S. British Columbia 7, 1915; Frye, Ferns Northwest 107, *Fig.* xxxiv, 1934; Morton [in] Gleason, Ill. Fl. 1:46, *Fig.*, 1952; Anderson, Fl. Alaska 13, 1959; Hultén, Fl. Alaska 50, 1968.

Rhizome black, slender, cordlike, long, widely creeping and branching; scales conspicuous, brown, ovate, acuminate, denticulate. Fronds few, scattered, arising singly from the spreading rhizome. Stipes dark brown at the base, pale above, considerably longer than the blades, sparingly glandular with a few septate hairs. Blade deltoid, somewhat ternate, up to 16 cm long and about as wide, sparsely glandular beneath, 2-pinnate below, progressively less compound upwards, the lowermost basal pinnule of each pinna the largest, ultimate segments ovate, rounded often somewhat cleft at the apex. Indusia whitish, roundish, irregularly toothed, not conspicuous. ($n = 84$)

HABITAT Damp woods and calcareous or basic rocky slopes.

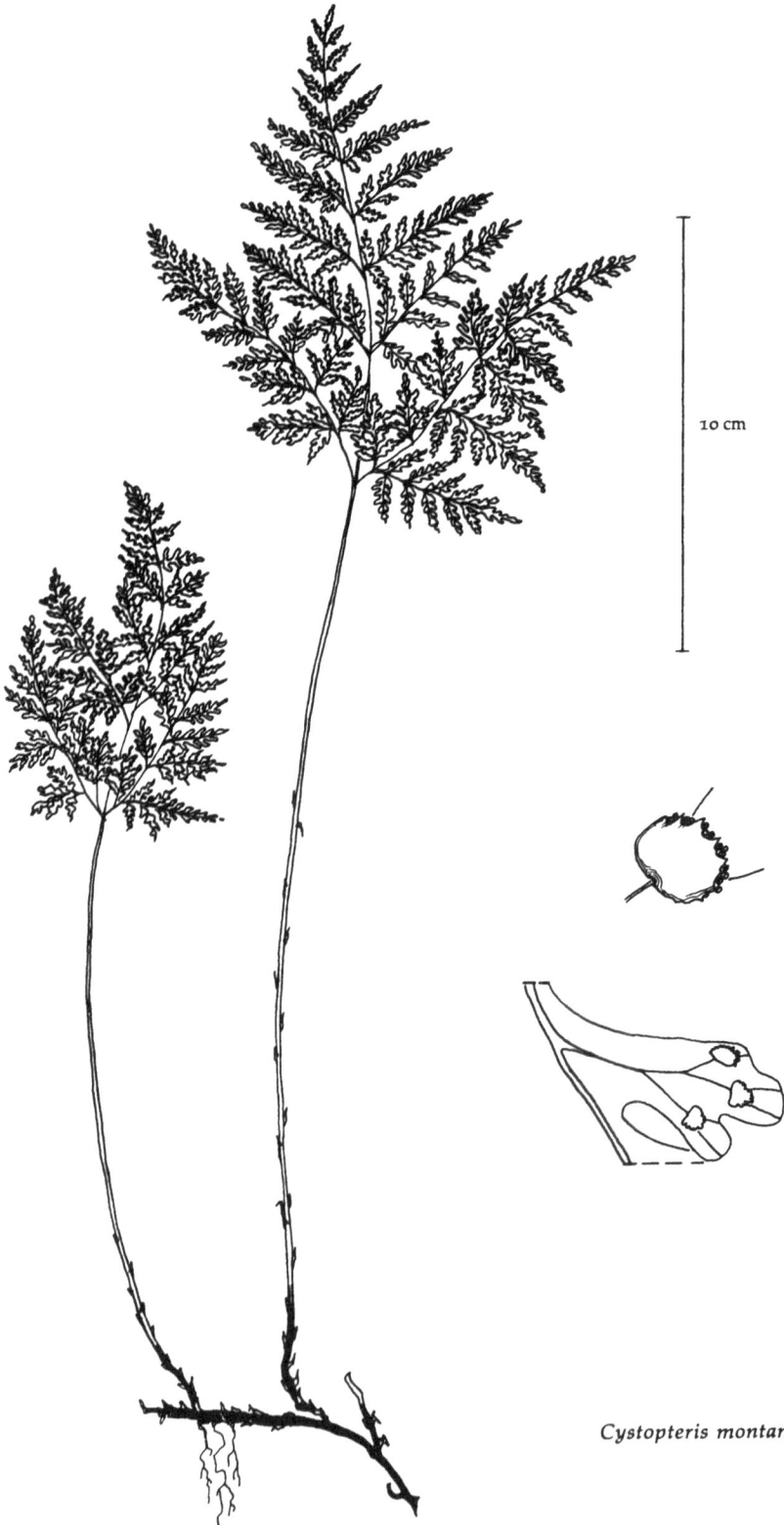

10 cm

Cystopteris montana

RANGE Alaska to Labrador, south to British Columbia, Colorado, northern shore of Lake Superior, and Gaspé, Quebec; Eurasia.

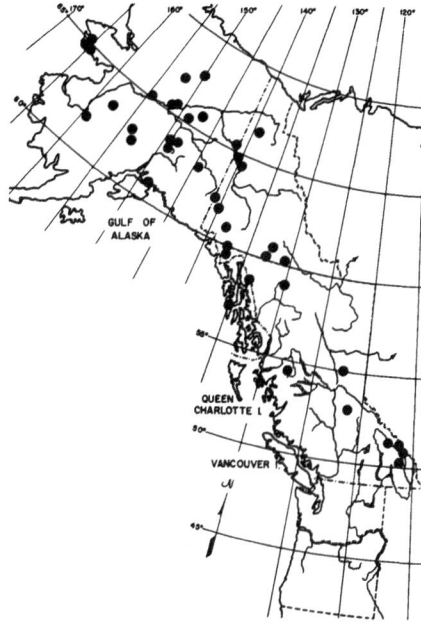

COMMENTS Found only in the more northern parts of our area and limited by its preference for alkaline soil conditions. This very attractive fern with its delicately compound frond and long spreading rhizome is very distinctive in appearance.

DRYOPTERIS Adans., Farn. Pl. 2:20, 550, 1763

Mainly woodland ferns of upright habit. Rhizomes stout, short-creeping, erect and copiously chaffy. Fronds borne in a crown; fertile and sterile blades alike (except in *D. cristata*), 1- to 3-pinnate, glabrous or glandular or variously pubescent, conspicuously chaffy only in a few species. Sori mostly roundish, dorsal, the indusium (in ours) roundish-reniform, largish, persistent, attached at its sinus, glabrous or sometimes glandular.

This very large, mainly tropical, woodland genus has been variously treated by different authorities. The modern tendency is to break it up into a number of smaller genera, e.g., *Gymnocarpium*, *Thelypteris*, and others. (Greek *drys*, oak, and *pteris*, the name given to a particular kind of fern.)

1   Blades subtripinnate, triangular, widest at or near the base; the lowest pinnae markedly unequal-sided.   *D. austriaca*

1   Blades pinnate-pinnatifid, not triangular, widest at or near the middle (except *D. arguta*); lowest pinnae almost equal-sided.

2   Blades copiously capitate-glandular; indusia glandular-margined; blades narrow, about 5 cm wide, aromatic.   *D. fragrans*

2   Blades and indusia glandless; blades wider, not aromatic.

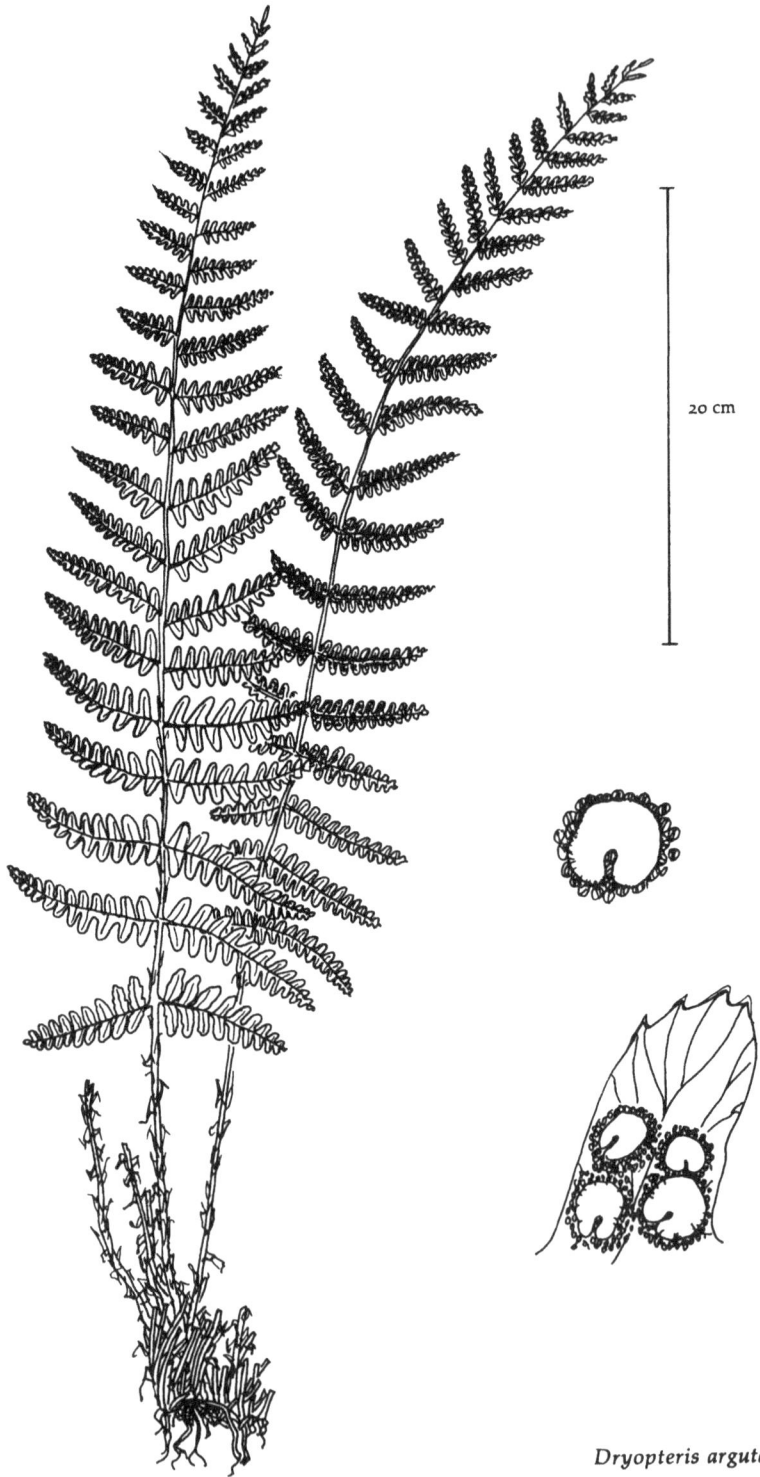

20 cm

Dryopteris arguta

3   Fronds strongly dimorphic, fertile much taller and narrower than the sterile; pinnae markedly reduced towards the base.   *D. cristata*

3   Fronds essentially alike; pinnae not reduced towards the base.

    4   Pinnae subsessile, oblong-lanceolate; the lower basal pinnule semi-cordate, usually overlying the main rachis; veinlets spreading, all ending in salient spinelike teeth.   *D. arguta*

    4   Pinnae clearly stalked, deltoid-lanceolate; basal pinnules symmetrical; veinlets oblique, ending in oblique, usually curved, acute teeth.
                                                    D. filix-mas

Dryopteris arguta (Kaulf.) Maxon

Am. Fern J. **11**:3, 1921; Macoun, Cat. Can. Pl. pt. 5, 274, 1890; Henry, Fl. S. British Columbia 7, 1915; Piper & Beattie, Fl. Northwest Coast 7, 1915; Maxon [in] Abrams, Ill. Fl. Pac. States **1**:16, *Fig.* 30, 1923; Frye, Ferns Northwest 129, *Fig.* XLIX, 1934; Peck, Man. Higher Pl. Ore. 47, 1941; Ewan, Am. Fern J. **34**:108, 1944.
Aspidium rigidum Am. auth.
D. rigida var. *arguta* (Kaulf.) Underw., Nat. Ferns, 4th ed., 116, 1893.

Rhizome stout, short-creeping, covered by abundant bright chestnut-coloured scales. Fronds evergreen, tufted, 2-pinnate, ovate to deltoid-lanceolate, acuminate, subcoriaceous, up to 70 cm tall. Stipe stout, up to 1/2 the length of the blade, scaly. Blade widest towards the base. Pinnae spreading, oblong-lanceolate, long-acuminate, the lower ones broadest. Pinnules somewhat distant, oblong, mostly rounded-obtuse, 2-serrate to pinnately incised; veinlets spreading, all ending in salient, often cartilaginous, spinelike teeth. Sori large; indusia pale greenish-yellow, orbicular, firm, persistent, flat to somewhat concave with a deep narrow sinus, glabrous but with somewhat glandular margins.

H A B I T A T Rocky ledges and woods along the coast.

R A N G E Southwestern British Columbia to California, locally inland to Arizona.

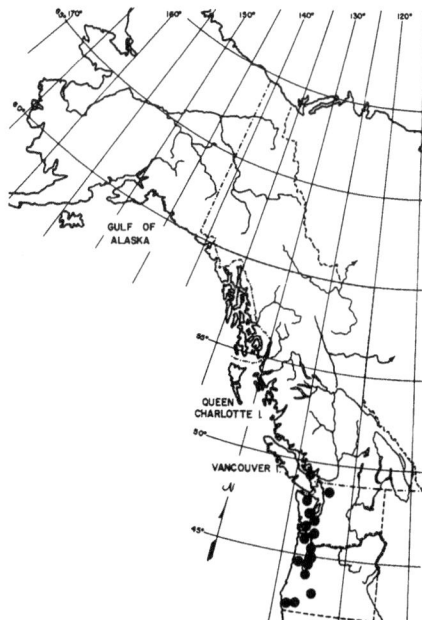

20 cm

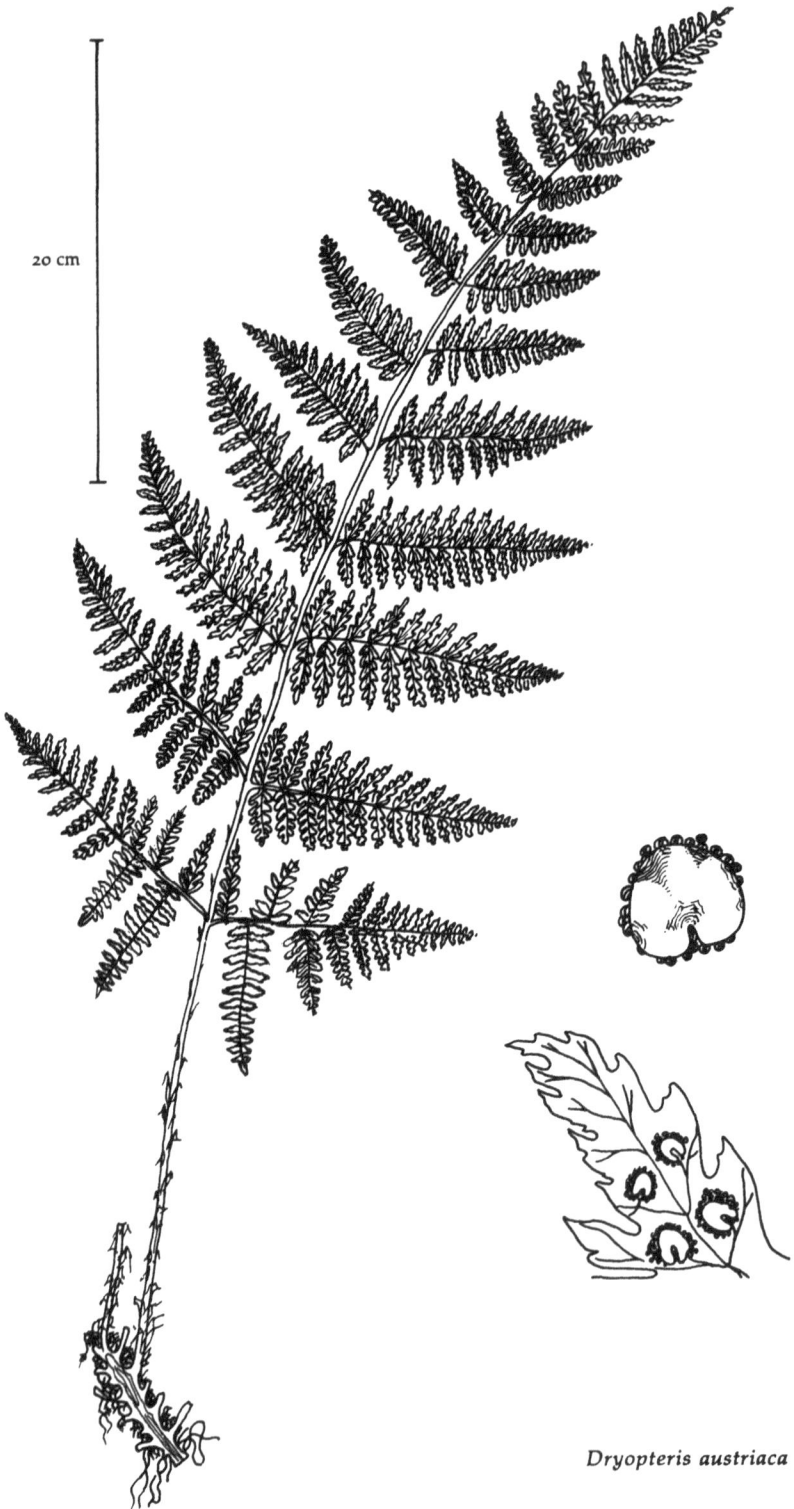

Dryopteris austriaca

COMMENTS Limited with us to a comparatively narrow coastal belt from Denman Island, British Columbia, southward. Very similar to the European *D. villarii* (Bell.) Woynar (*D. rigida* (Swartz) A. Gray) and, by some, considered conspecific with it.

Dryopteris austriaca (Jacq.) Woynar

[Ex] Schinz & Thell., Vierteljarssch. Naturf. Ges. Zurich **60**:339, 1915; Macoun, Cat. Can. Pl. pt. 5, 275, 1890; Henry, Fl. S. British Columbia 7, 1915; Piper & Beattie, Fl. Northwest Coast 7, 1915; Maxon [in] Abrams, Ill. Fl. Pac. States **1**:17, *Fig.* 32, 1923; Frye, Ferns Northwest 132, 1934; Peck, Man. Higher Pl. Ore. 47, 1941; Anderson, Fl. Alaska 16, 1959; St. John, Fl. Southeast. Wash. 4, 1963; Hultén, Fl. Alaska 55, 1968; Calder & Taylor, Fl. Queen Charlotte Islands pt. 1, 150, 1968.
D. dilatata Am. auth.
Aspidium spinulosum Sw. var. *dilatatum* Am. auth.

Rhizome stout, erect or ascending, chaffy. Fronds up to 10 dm tall, spreading, borne in a crown, deciduous or subpersistent. Stipes usually shorter than the blade, clothed with brownish, often darker-centred, entire, ovate-lanceolate scales. Blades broadly triangular to ovate or broadly oblong, abruptly acuminate, nearly or quite 3-pinnate, herbaceous. Pinnae short-stalked, acuminate; basal pinnae broadly ovate or triangular, markedly inequilateral, the upper ones lanceolate; margins of pinnules and lobes serrate, teeth mucronate, straight or falcate, somewhat spreading. Indusium glabrous or very sparsely glandular. ($n = 41, 82$)

HABITAT Cool, moist woods.

RANGE Southern Alaska possibly to Newfoundland, south to California, Montana, and New England; the mountains of North Carolina and Tennessee; Greenland; Eurasia.

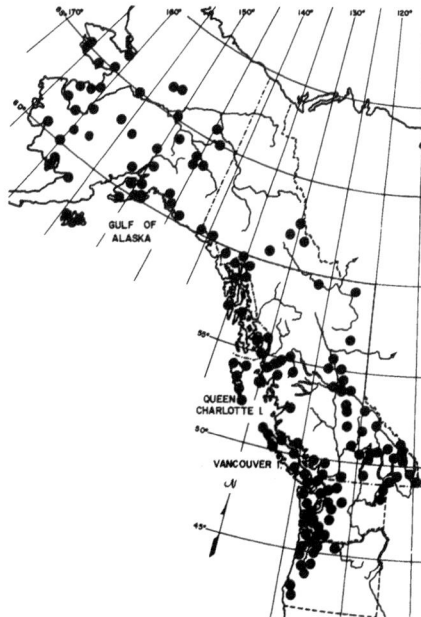

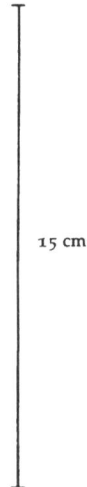

Dryopteris cristata

15 cm

COMMENTS One of our commonest and most widespread species complexes. In one form or another it is to be found over the entire forested area of the northern hemisphere. In our region it is represented largely by var. *austriaca*, characterized by the basal pinnae, the basal pinnules of which are 5–15 mm apart and the lowermost pinnule usually considerably more than twice as long as the one above it on the opposite side.

In eastern North America the investigations of Walker, Britton, and Wagner have shed much light on our understanding of the species and hybrids in this complex. Until similar studies are made in western America it seems best to retain the name *D. austriaca* for our representatives of the complex, i.e., for those plants that are morphologically very similar to *D. dilatata* (Hoffm.) Gray of Europe. This is being done despite the fact that Walker (1959, 1961) has reported both diploid and tetraploid forms in "*dilatata*" and that he considers the diploid to be a separate species *D. assimilis* Walker. He further (1961) reports that a British Columbian plant that he examined was diploid ($2n = 82$) and so may prove to be conspecific with *D. assimilis*.

Dryopteris cristata (L.) A. Gray

Man., 1st ed., 631, 1848; Henry, Fl. S. British Columbia 6, 1915; Frye, Ferns Northwest 126, 1934; Morton [in] Gleason, Ill. Fl. 1:54, *Fig.*, 1952.
Aspidium cristatum (L.) Sw., Schrad. Bot. J. **1800**:37, 1801.
Thelypteris cristata (L.) Nieuwl., Am. Midl. Nat. **1**:226, 1910.

Rhizome stout, decumbent or somewhat creeping, densely covered with old stipe bases. Fronds of two kinds; the sterile evergreen, shorter and laxer than the erect, deciduous fertile frond. Stipes with thin, ovate, uniformly coloured, cinnamon or pale brown scales. Blades glabrous, linear to lance-oblong, somewhat narrowed at the base, pinnate-pinnatifid. Pinnae on fertile fronds often twisted on the rachis to lie more or less horizontally, lower pinnae deltoid, the upper deltoid-oblong, mostly obtusish, deeply cut into deltoid-oblong, obtuse, appressed-serrate lobes, the inner basal segment of the lowest pinnae obviously longer than the second. Sori more or less medial; indusia round-reniform, glabrous, persistent. ($n = 82$)

HABITAT Marshes, bogs, and swampy thickets.

COMMENTS This distinctive fern is very rare and local in its distribution with us, being limited to the Columbia and Kootenay Valleys of British Columbia where it has been collected in the vicinity of Revelstoke, Nelson, and Kitchener.

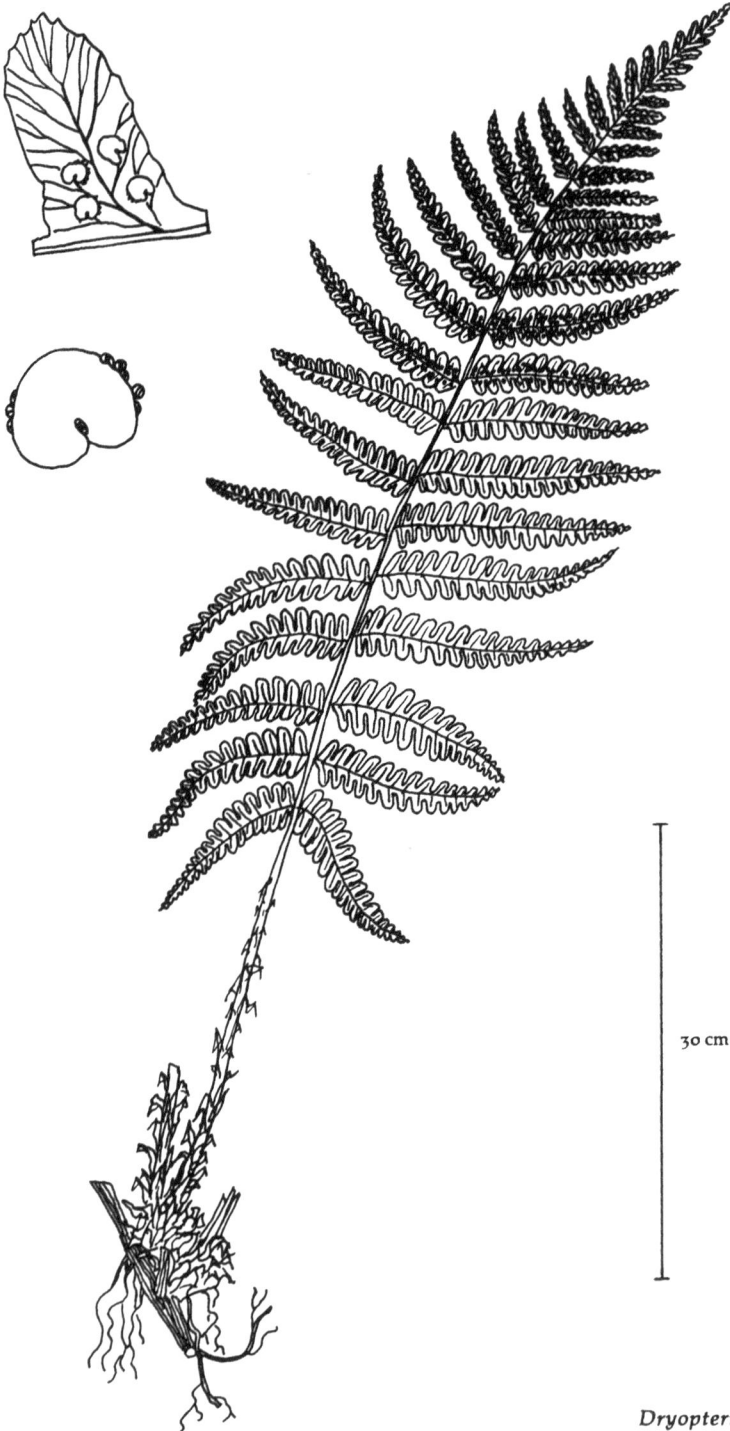

Dryopteris filix-mas

30 cm

RANGE Southeastern British
Columbia to Newfoundland,
south to Idaho, Montana,
Nebraska, Arkansas, and
Virginia; Europe.

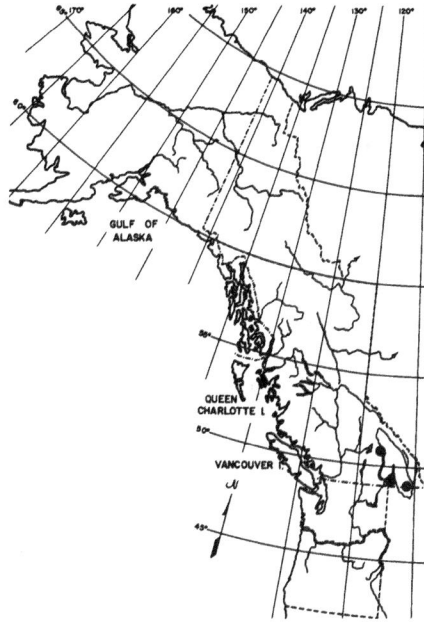

Dryopteris filix-mas (L.) Schott

Gen. Fil. *Plate 67*, 1834; Macoun, Cat. Can. Pl. pt. 5, 274, 1890; Henry, Fl. S.
British Columbia 7, 1915; Maxon [in] Abrams, Ill. Fl. Pac. States 1:16, *Fig. 31*,
1923; Frye, Ferns Northwest 128, Fig. XLIV, 1934; Peck, Man. Higher Pl. Ore. 47,
1941; Morton [in] Gleason, Ill. Fl. 1:52, *Fig.*, 1952.
Aspidium filix-mas (L.) Sw., Schrad. Bot. J. **1800**:38, 1801.
Thelypteris filix-mas (L.) Nieuwl., Am. Midl. Nat. **1**:226, 1910.

Rhizome stout, ascending, covered with long scales. Stipes stout, comparatively short, densely covered with lance-linear, long-attenuate, pale brown
scales. Fronds erect, deciduous, lanceolate to lance-oblong, slightly narrowed towards the base and gradually acuminate upwards; pinnate-pinnatifid to almost 2-pinnate. Pinnae lance-linear to ovate-lanceolate,
sharply acuminate, more or less petioled; the lower pinnae shorter, with
crenate or serrate, obtuse, oblong segments or pinnules; glandless but with
rachis and midrib bearing narrow, strongly toothed scales. Sori.midway
between the margin and the midvein, mostly confined to the upper half of
the frond; indusia glabrous, orbicular-reniform, firm, persistent, brown
to pale brown; sinus prominent. (n = 82)

HABITAT Woods and shaded talus slopes.

COMMENTS Despite the great extent of its range, this species shows
surprisingly little variation in appearance. It is sporadic in the coastal
mountains as far north as the Nass River, British Columbia, in the Selkirk

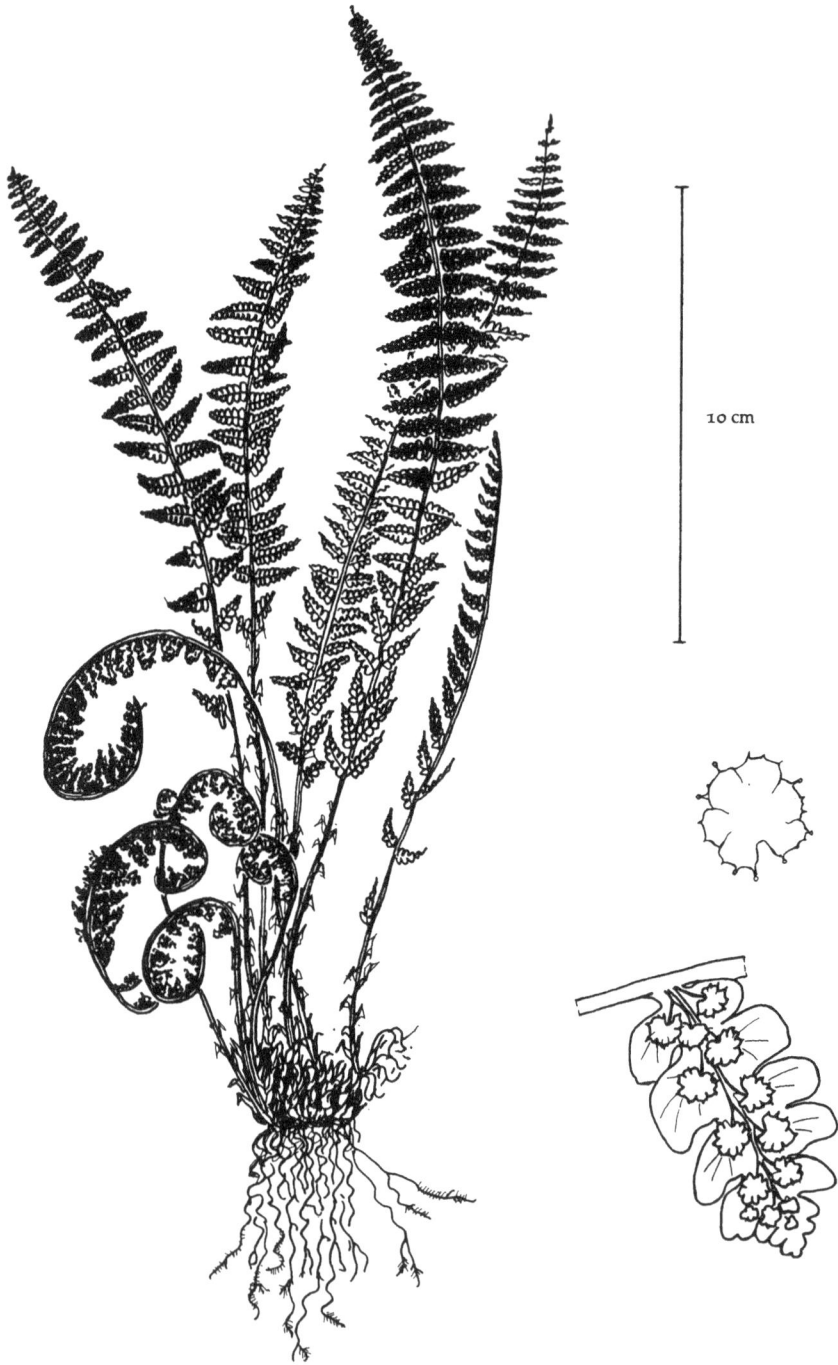

10 cm

Dryopteris fragrans

Mountains, and scattered in the Cascade Mountains in Washington and the Blue Mountains in Oregon. Ewan (1944) reports a natural hybrid between this species and *T. limbosperma* (*Thelypteris oreopteris*) from Alice Arm, British Columbia. A re-examination of the specimen by Taylor (1953) discounts this probability.

R A N G E British Columbia to Newfoundland, south to California, Texas, Oklahoma, and Vermont; Greenland; Iceland; Eurasia.

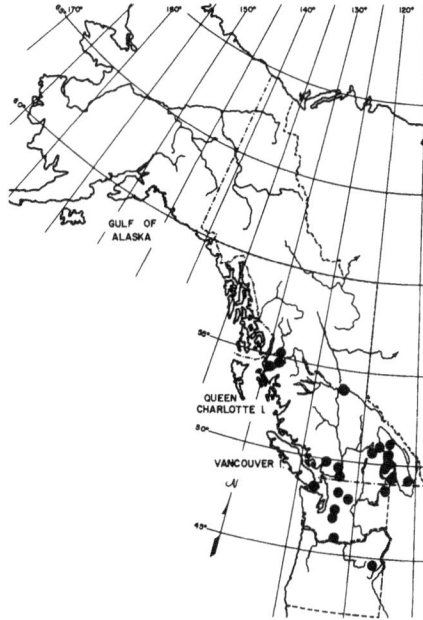

Dryopteris fragrans (L.) Schott

Gen. Fil. *Plate* 9, 1843; Macoun, Cat. Can. Pl. pt. 5, 276, 1890; Morton [in] Gleason, Ill. Fl. **1**:52, *Fig.*, 1952; Anderson, Fl. Alaska 15, 1959; Wiggins & Thomas, Fl. Alaskan Arctic Slope 44, 1962; Hultén, Fl. Alaska 56, 1968.
Aspidium fragrans (L.) Sw., Schrad. Bot. J. **1800**: 35, 1801.
Thelypteris fragrans (L.) Niewl., Am. Midl. Nat. **1**:226, 1910.

Rhizome stout and short, erect, covered by many persistent, curled, old fronds. Fronds spreading, evergreen. Stipes short, glandular, densely clothed with lustrous brown or reddish, often glandular-puberulent, ovate-lanceolate scales. Blades coriaceous, pinnate-pinnatifid; linear-lanceolate or broadly linear, tapering both ways from the middle, the lowest pinnae very much reduced. Pinnae oblong-lanceolate, obtuse, pinnately crenate or incised, not spinulose, often overlapping and inrolled, glandular, persistently scaly below. Sori medial; indusia large, whitish becoming brown with age, densely crowded, sometimes overlapping, persistent, glandular-margined. ($n = 41$)

H A B I T A T Shaded or exposed cliffs of basic, often calcareous rocks; talus slopes.

RANGE Alaska to New-
foundland, south to northern
British Columbia, Minnesota,
Wisconsin, Michigan, north-
east New York, and northern
New England; Eurasia.

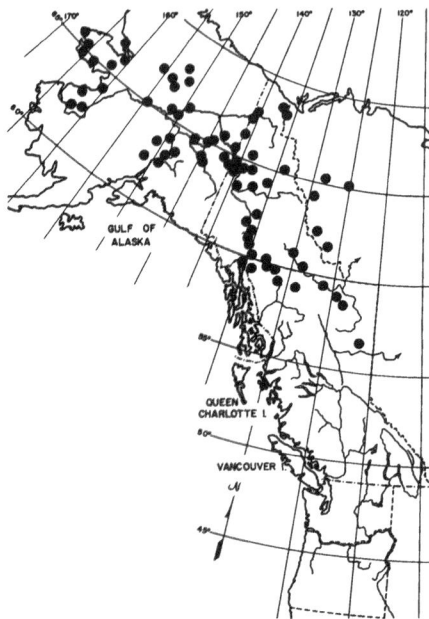

COMMENTS A common, markedly northern and arctic species. The per-
sistence of the curled-up old fronds, their obvious scaliness and sickly
sweet smell when fresh are useful identifying characteristics. In North
America the form that occurs south of the arctic has been distinguished as
var. *remotiuscula* Komarov which is less scaly beneath and with the pinnae not overlapping. R. M. Tryon (1942) described a hybrid between the
present species and *D. spinulosa* var. *intermedia* (Muhl.) Underw.

GYMNOCARPIUM Newm. Phytol. 4:371, 1851

Rhizomes long-creeping, branched, sparsely scaly with uniformly
coloured, glabrous, entire scales. Fronds scattered, arising singly from the
rhizome, membranous, thin, deciduous. Stipes straw-coloured with a
darker base, slender, longer than the blades. Blades deltoid or pentagonal,
ternately compound, the lower pinnae articulated at the base, long-petioled, inequilateral, glabrous and scaleless, sometimes glandular. Sori
essentially circular; indusium lacking. (Name from Greek *gymnos*, naked,
and *karpos*, fruit, with reference to the naked sori.)

A genus of very few species of north temperate woodlands and shaded
rocky slopes.

Rachis and blades glandless or nearly so; first pinnule on the lower side of the
basal pinnae about 1/3 as long as the main rachis or longer; pinnules acutish;
margins not recurved.   *G. dryopteris*

Rachis and blades densely glandular; first pinnule on the lower side of the basal pinnae about 1/4 as long as the main rachis or shorter; pinnules obtuse, margins often recurved. *G. robertianum*

Gymnocarpium dryopteris (L.) Newm.

Phytol. **4**, app. xxiv, 1851; Macoun, Cat. Can. Pl. pt. 5, 270, 1890; Henry, Fl. S. British Columbia 4, 1915; Piper & Beattie, Fl. Northwest Coast 3, 1915; Maxon [in] Abrams, Ill. Fl. Pac. States **1**:13, *Fig.* 25, 1923; Frye, Ferns Northwest 135, *Fig.* xlviii, 1934; Peck, Man. Higher Pl. Ore. 46, 1941; Morton [in] Gleason, Ill. Fl. **1**:50, *Fig.*, 1952; Anderson, Fl. Alaska 15, 1959; St. John, Fl. Southeast. Wash. 6, 1963; Hultén, Fl. Alaska 56, 1968; Calder & Taylor, Fl. Queen Charlotte Islands pt. 1, 150, 1968.
Phegopteris dryopteris (L.) Fée, Gen. Fil. 243, 1850–2.
Dryopteris linnaeana C. Chr., Ind. Fil. 275, 1905.
Thelypteris dryopteris (L.) Slosson [ex] Rydb., Fl. Rocky Mts. 1044, 1917.
D. disjuncta (Rupr.) Morton, Rhodora **43**:217, 1941.

Rhizome slender, wide-creeping, bearing a few thin, pale brownish scales. Fronds somewhat distant in rows, erect, up to 40 cm tall, deciduous. Stipes slender, lustrous, straw-coloured, glabrous or sometimes with a few papery scales, arising from a chaffy base. Blade broadly deltoid-pentagonal, ternately pinnate, glabrous, rarely glandular. Basal pinnae asymmetrically deltoid, at least 1/3 the length of the main rachis or longer, their petioles articulated to the rachis; ultimate segments oblong, obtuse, crenate to slightly pinnatifid; their margins not recurved. Sori small, circular, submarginal; indusia absent. ($n = 40, 80$)

HABITAT Damp woods, shaded rocky slopes and ledges, soil often subacid.

RANGE Alaska to Newfoundland, south to Oregon, mountains of eastern Arizona and New Mexico, northern Kansas, West Virginia, Pennsylvania; Greenland; Iceland; Eurasia.

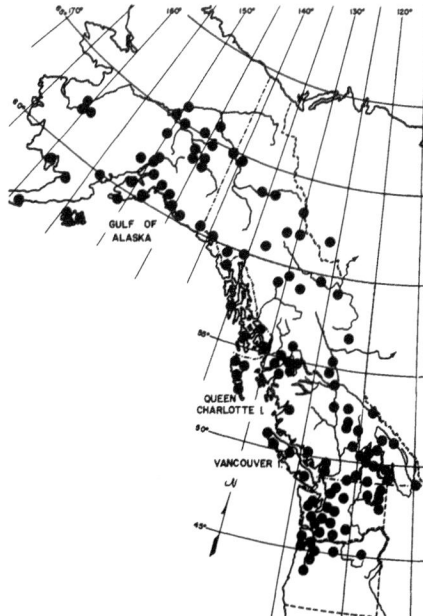

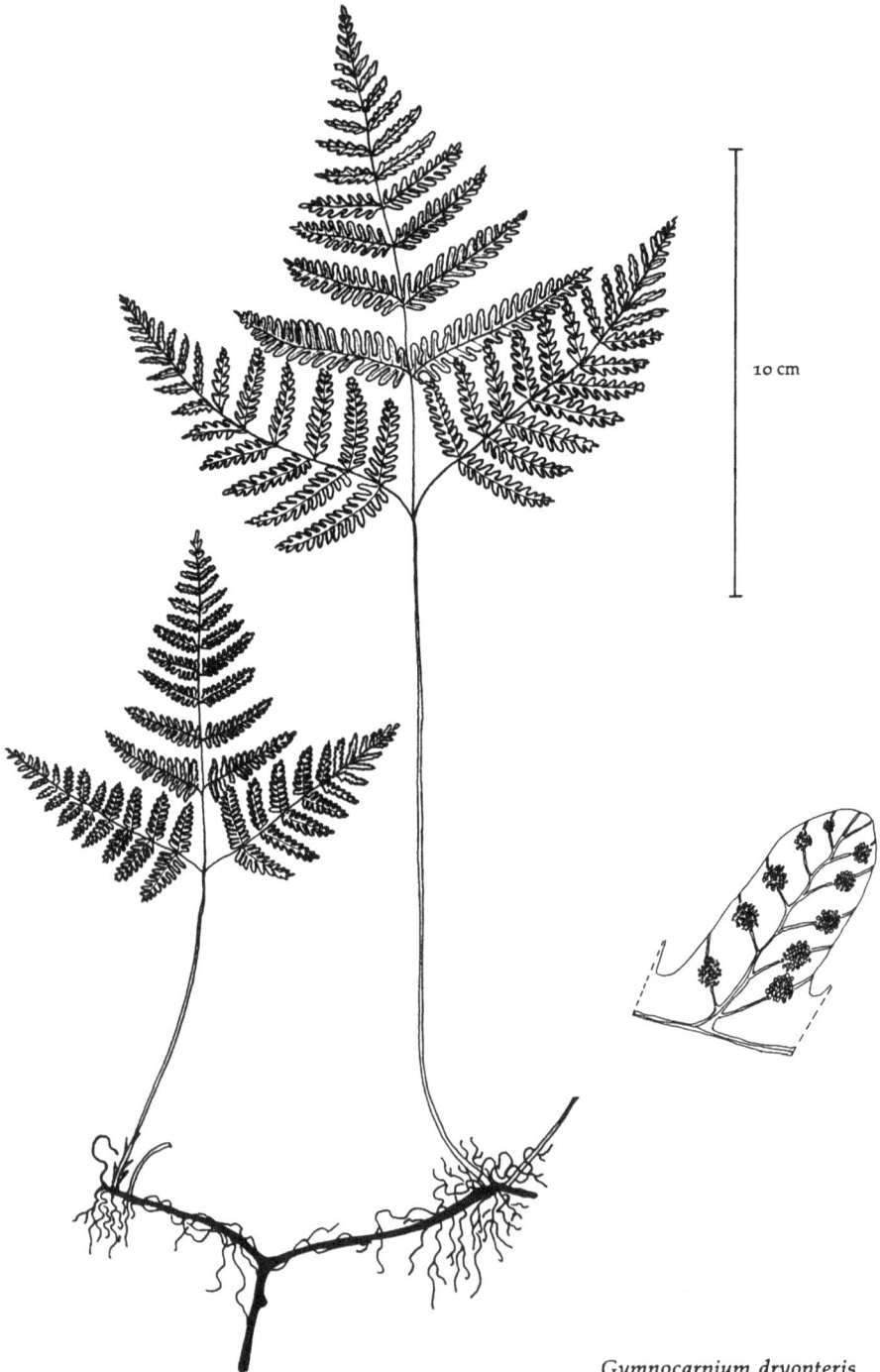

10 cm

Gymnocarpium dryopteris

COMMENTS Very common throughout in suitable habitats. The western representative with larger, frequently 3-pinnate fronds and larger, more toothed segments and smaller spores has been recognized as var. *disjunctum* (Led.) Ching. Wagner (1963) reports that this variety has only $n = 40$ as compared with $n = 80$ in the eastern var. *dryopteris*. *G. dryopteris* can be distinguished from the following species by the essential absence of glands, and by the pentagonal shape of the blade with the lowermost pinnule on the basal pinna being 1/3 the length of the main rachis. Hybrids with *G. robertianum* have been described by Root (1961). Further study has led Wagner (1966b) to conclude that these "hybrids" belong to an apomictic species which he has named *G. heterosporum* Wagner.

Gymnocarpium robertianum (Hoffm.) Newm.

Phytol. **4**: app. xxiv, 1851; Frye, Ferns Northwest 135, *Fig.* xlix, 1934; Morton [in] Gleason, Ill. Fl. **1**:50, *Fig.*, 1952; Anderson, Fl. Alaska 15, 1959; Hultén, Fl. Alaska 57, 1968.
Phegopteris robertianum (Hoffm.) A. Br. [ex] Aschers., Fl. Brand. **2**:198, 1859.
Dryopteris robertiana (Hoffm.) C. Chr., Ind. Fil. 289, 1905.
Thelypteris robertiana (Hoffm.) Slosson [ex] Rydb., Fl. Rocky Mts. 1044, 1917.

Rhizome slender, wide-creeping, somewhat scaly; scales ovate, pale brown, sparingly glandular. Fronds somewhat distant in rows, erect, up to 35 cm tall, deciduous. Stipes yellowish, dull, minutely glandular-puberulent, sparsely scaly. Blade deltoid-ovate to narrowly deltoid, minutely glandular-puberulent, 2-pinnate-pinnatifid. Basal pinnae narrowly deltoid, almost symmetrical, about 1/4 or less the length of the main rachis, their petioles articulated to the rachis; ultimate segments oblong, margins entire to somewhat crenate, often revolute. Sori small, circular, submarginal; indusia absent. ($n =$ ca 80)

HABITAT Limestone ledges and talus; cool, calcareous sites.

RANGE Alaska to Newfoundland, south to British Columbia, Idaho, Iowa, Ohio, Tennessee, and North Carolina; Eurasia.

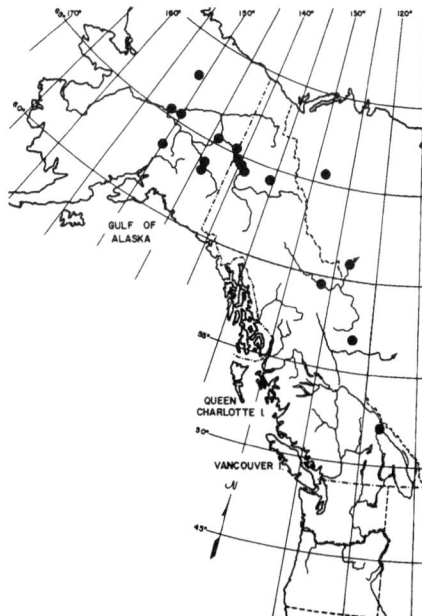

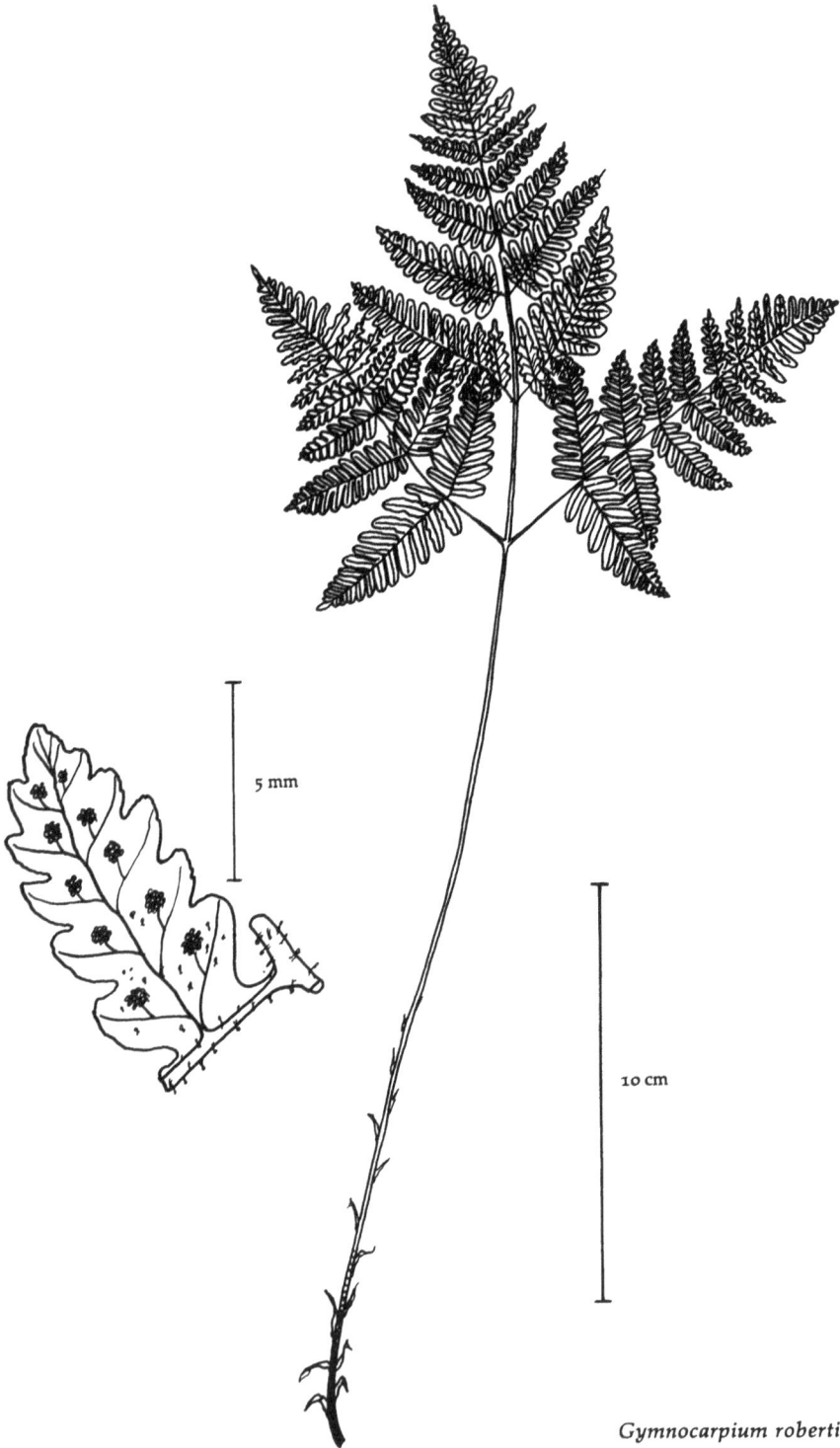

5 mm

10 cm

Gymnocarpium robertianum

COMMENTS Rare, being found in British Columbia only as far south as 52° north latitude at the Big Bend of the Columbia River. Much more common in Alaska and the Yukon. As mentioned before, this species has been reported by Root (1961) to hybridize with *G. dryopteris* (*G. heterosporum* Wagner).

MATTEUCCIA Todaro, Syn. Pl. Sicil. 30, 1866

Rhizome deeply subterranean, black, much branched, ending in an erect, very stout caudex covered with thin brown scales. Fronds markedly dimorphic; the sterile fronds herbaceous, forming a vaselike arrangement in the centre of which develop the much shorter, dark, indurated fertile fronds. Stipes much shorter than the blades. Sterile blades pinnate-pinnatifid, fertile blades simply pinnate. Veins free. Sori elongated, several to a segment, covered by the more or less scarious, revolute margin of the frond; indusium hoodlike, hyaline, soon withering. (Name commemorates Carlo Matteucci, 1800–68, an Italian physicist.)

A small genus of only two or three species of coarse but decorative ferns found only in the northern hemisphere.

Matteuccia struthiopteris (L.) Todaro

Gior. Sci. Nat. Econ. Palermo 1:235, 1866; Macoun, Cat. Can. Pl. pt. 5, 281, 1890; Henry, Fl. S. British Columbia 3, 1915; Morton [in] Gleason, Ill. Fl. 1:37, Fig., 1952; Anderson, Fl. Alaska 10, 1959; Hultén, Fl. Alaska 52, 1968. *Onoclea struthiopteris* (L.) Hoffm., Deutsch. Fl. 2:11, 1759. *Struthiopteris pensylvanica* Willd., Sp. Pl. 5:289, 1810. *Matteuccia nodulosa* (Michx.) Fern., Rhodora 17:164, 1915. *Pteretis nodulosa* (Michx.) Nieuwl., Am. Midl. Nat. 4:334, 1916. *Pteretis pensylvanica* (Willd.) Fern., Rhodora 47:123, 1945.

Caudex erect, covered by the bases of old stipes, stout, arising from a widely creeping and branching rhizome. Sterile fronds in a circle in a vaselike arrangement, up to 2 m tall, pinnate-pinnatifid. Stipe short, green to black at the base, furrowed on the upper face, chaffy when young with thin, papery, pale brown scales. Blade of sterile fronds broad, narrowly obovate-lanceolate, gradually narrowed towards the base, abruptly short-acuminate above; rachis furrowed above, narrowly wing-margined. Pinnae spreading-ascending, long-acuminate, alternate, deeply pinnatifid into blunt, oblong segments with slightly revolute margins. Fertile fronds stiffly erect inside the circle of sterile fronds and considerably shorter. Stipe of fertile fronds dark brown, about as long as the blade. Fertile blade olivaceous, finally blackish with crowded, spreading-ascending pinnae with podlike segments. ($n = 40$)

HABITAT Edge of swamps and rich bottom land, particularly in circum-neutral alluvial soil.

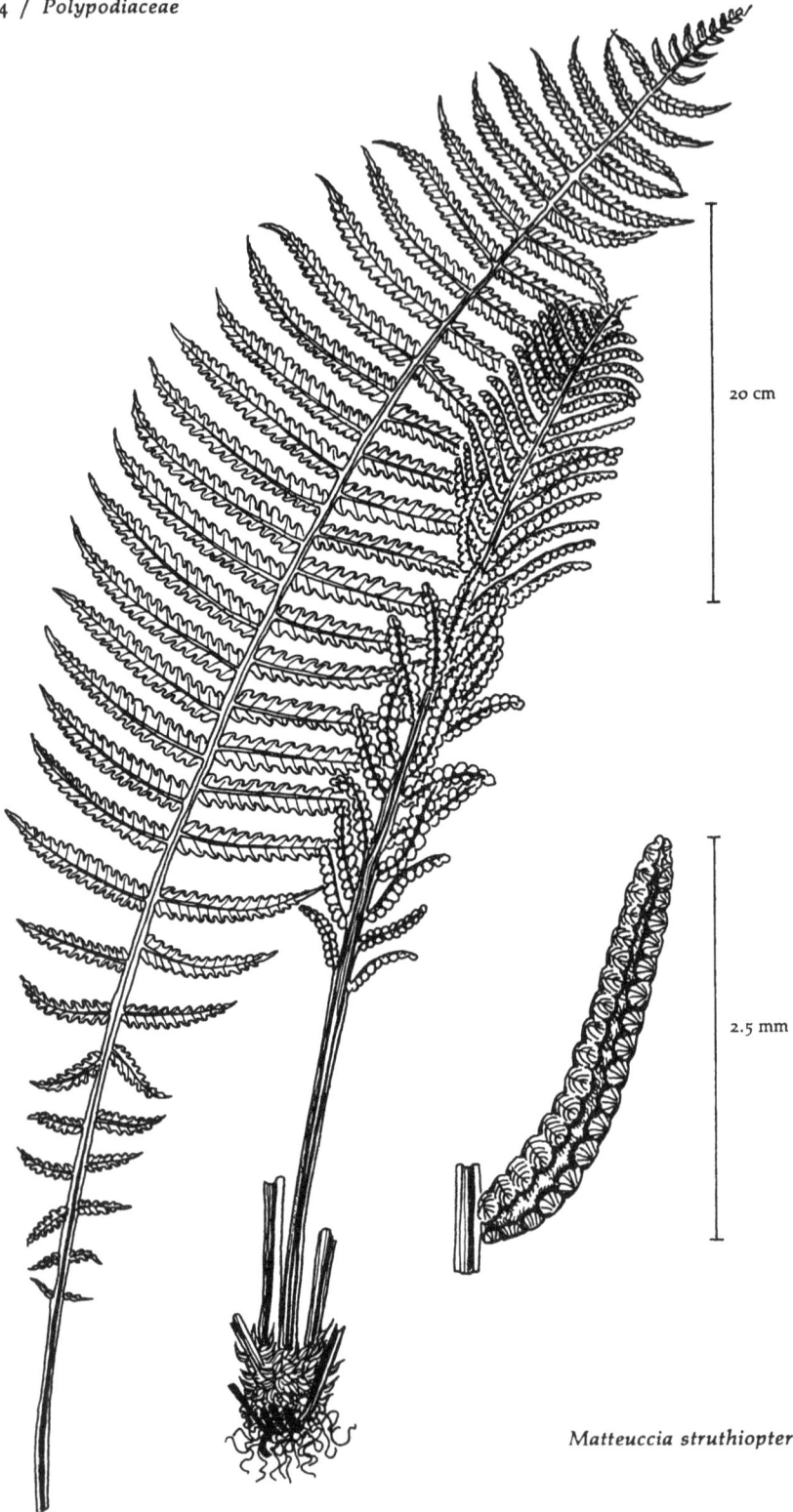

20 cm

2.5 mm

Matteuccia struthiopteris

COMMENTS Locally very common in British Columbia forming extensive colonies, particularly in association with the larger river systems; not
as yet known south of the 49th parallel in the west. The large size and
the shape of both the sterile and the fertile fronds are unmistakable. By
some botanists the North American plant is assigned to var. *pensylvanica*
(Willd.) Morton, distinguishable by the scales on the stipe-base being
uniformly coloured, lacking the dark central band found in European
specimens.

RANGE Alaska to New-
foundland, south to British
Columbia, South Dakota,
Missouri, Illinois, Indiana,
Ohio, West Virginia, north-
ern Virginia, and New
England; Europe.

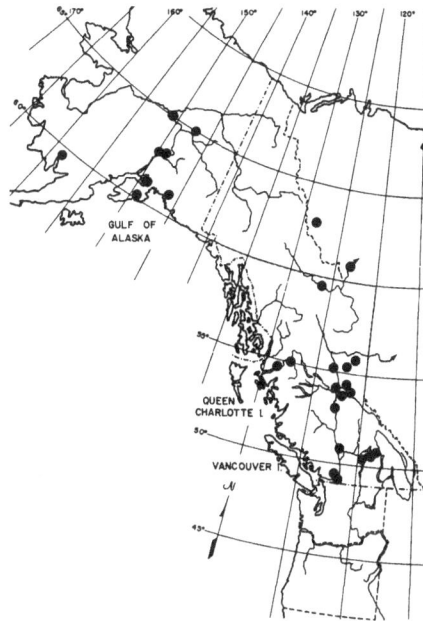

PELLAEA Link, Fil. Sp. Hort. Berol. 59, 1841

Small, stiff, rock ferns. Rhizome short and nodose or slender and creeping,
scales linear-subulate, brown, attenuate. Fronds often much bunched,
stiff, mostly erect, frequently glabrous. Stipes wiry, dark, and lustrous.
Blades 1- to 3-pinnate, coriaceous, often dimorphic, more or less glaucous.
Segments linear to round-oval, more or less distinctly articulate. Sori
without indusium, terminal or subterminal on the veinlets, usually laterally confluent and submarginal, protected by the reflexed or revolute,
indusiumlike margin. (Name from Greek *pellos*, dusky, alluding to the
dark-coloured stipe.)

About eighty species of dry temperate regions.

(The key and descriptions that follow are adapted from A. F. Tryon,
1957.)

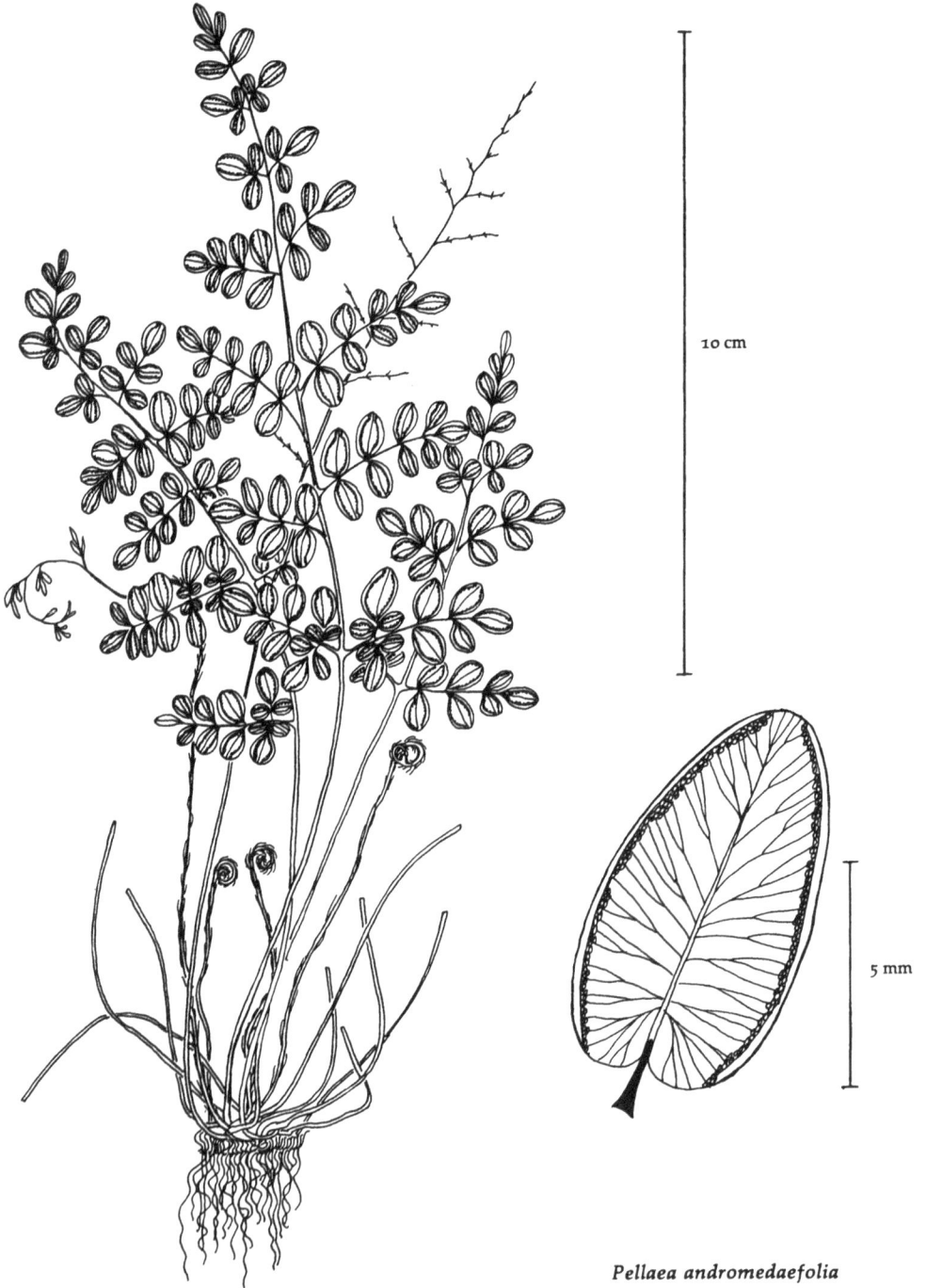

10 cm

5 mm

Pellaea andromedaefolia

1	Stipe and rachis straw-coloured to ruddy brown or darker at the base and mottled above; blades 3- to 4-pinnate. *P. andromedaefolia*
1	Stipe and rachis chestnut-coloured to black.
2	Scales of rhizome concolorous; stipe and rachis terete or elliptical.
3	Fronds monomorphic; pinnae sessile or the stalks somewhat decurrent on the rachis and arising at acute angles to it; stipe and rachis glabrous or sparsely pubescent; scales of rhizome discrete, uniform rust-brown in colour.
4	Rhizome relatively massive with numerous compressed, short, articulated stipe bases; mature fronds with rachises mostly green in terminal portion of the blade, apical segments strongly decurrent; basal pinnae without persistent stalks, often sessile or subsessile. *P. breweri*
4	Rhizome moderately stout with few stipe bases persistent, these more or less spreading; mature fronds with rachises mostly brownish in the terminal portion of the blade, the apical segments stalked or somewhat decurrent; basal pinnae with persistent stalks. *P. glabella*
3	Fronds dimorphic; pinnae stalked, not decurrent, and arising at broad angles to the rachis; stipe and rachis scurfy with appressed pubescence; scales of rhizome matted, rust-coloured with tips light tan. *P. atropurpurea*
2	Scales of rhizome bicolorous with a sclerotic central stripe or base.
5	Blades 2-pinnate; pinnules narrowly linear, margins revolute. *P. brachyptera*
5	Blades 1-pinnate; pinnae cordate-oblong to suborbicular, margins scarcely revolute. *P. bridgesii*

Pellaea andromedaefolia (Kaulf.) Fée

Gen. Fil. 129, 1850–2.
Pteris andromedaefolia Kaulf. Enum. Fil. 158, 1824.

Rhizome slender, dichotomously branched, usually long-creeping. Scales of rhizome and stipe base appressed, bicolorous, ruddy brown, elongate lanceolate-triangular, usually cordate, the sclerotic central portion lustrous, more or less straight, margins pectinate-serrulate, apex filiform. Fronds up to 60 cm long, erect, monomorphic. Stipe and rachis variously glabrous, pubescent, or scaly, straw-coloured to reddish-brown becoming grey with age, stipe breaking irregularly without lines of articulation, rachis straight or more or less flexuous. Blade about twice as long as the stipe, elongate-triangular, rhomboid or deltoid, usually tripinnate. Pinnae green, occasionally somewhat reddish, generally ascending at a broad angle to the rachis, segments usually numerous, long-stalked; segments up to 2 cm long and 1.2 cm broad, ellipsoidal or ovate, retuse, entire or ternate, veins evident, borders narrow, whitish lutescent.

HABITAT Dry, rocky ravines or ledges, sometimes along shaded stream banks.

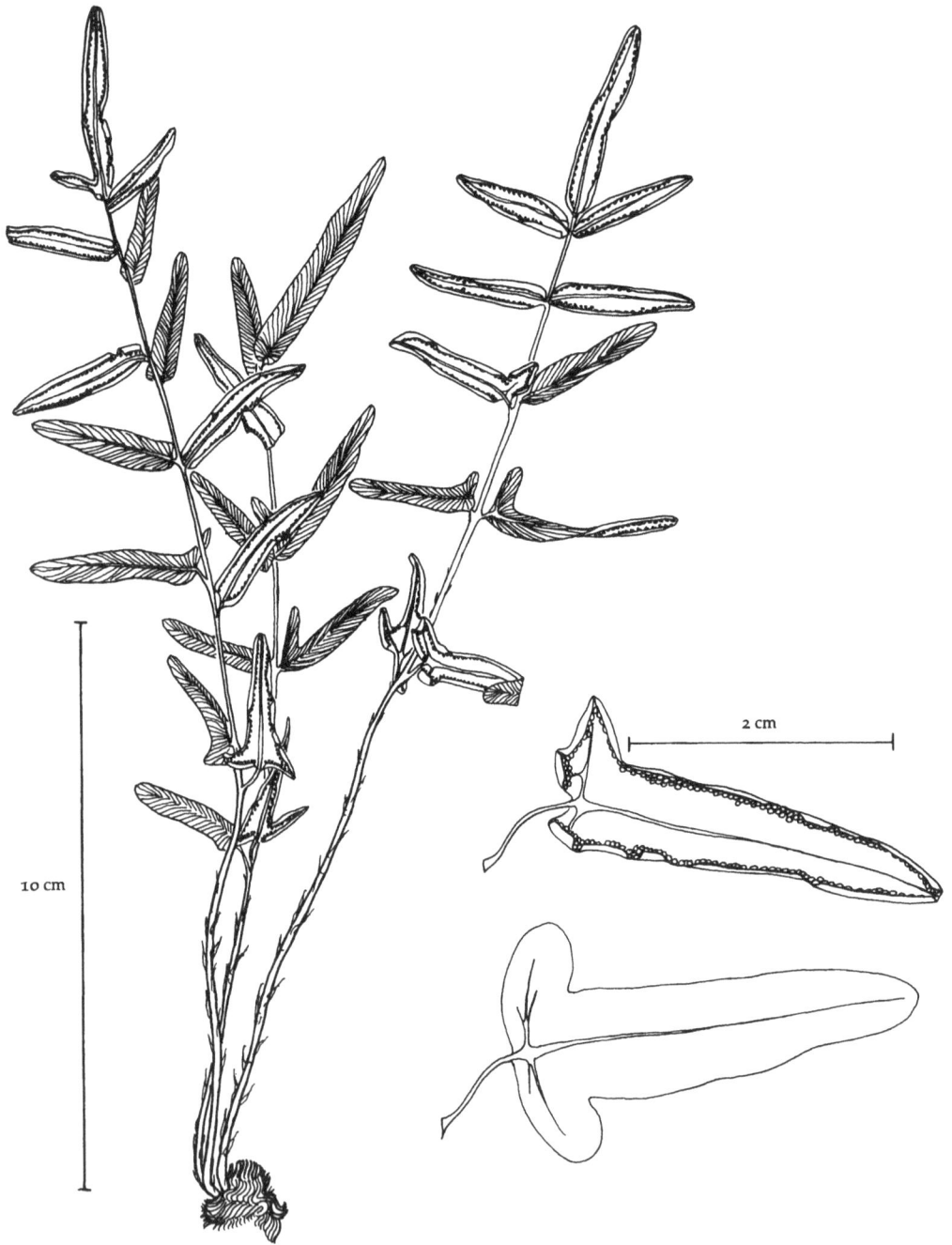

10 cm

2 cm

Pellaea atropurpurea

RANGE Southern Oregon to California and Baja California.

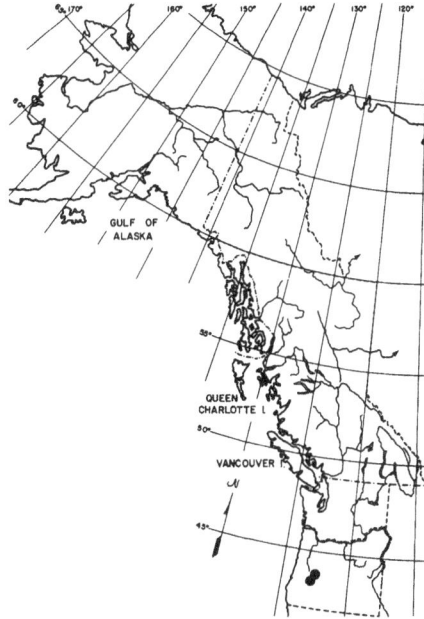

COMMENTS According to A. F. Tryon (1957) the report of the occurrence of this species in Oregon is based on a single collection by Howell made in Douglas County in 1887. In the University of Oregon herbarium, there is, however, a second specimen collected near Jasper in Lane County in 1934. The fact that the species still grows in the same area is substantiated by a collection made there as recently as 1968.

Pellaea atropurpurea (L.) Link

Fil. Sp. Hort. Berol. 59, 1841; Morton [in] Gleason, Ill. Fl. **1**:33, *Fig.*, 1952; Correll, Ferns & Fern Allies Texas 115, *Plate 22*, 1956.

Rhizome short, moderately stout, densely clothed with matted, rust-coloured, linear-subulate scales. Fronds tufted, somewhat dimorphic, up to 50 cm long, straight and stiff. Stipe and rachises dark purple to blackish, terete or elliptical, strongly pubescent to hispidulous with more or less crisp, appressed hairs. Blade narrowly deltoid-ovate, 1- to 3-pinnate, lower pinnae as a rule considerably longer than the upper ones, sterile blades usually evergreen. Lower pinnae entire but more commonly pinnate with 3–15 pinnules, stalked. Upper pinnae entire, subsessile or short-stalked, sometimes auriculate at the base on the upper side, or ternately lobed, subcoriaceous; lower surface of pinnae often pubescent along the veins. Fertile pinnules narrower than the sterile, the border broad, white or opaque, crenulate, forming a continuous indusium. ($n=87$)

HABITAT Dry limestone crevices, ledges, and talus.

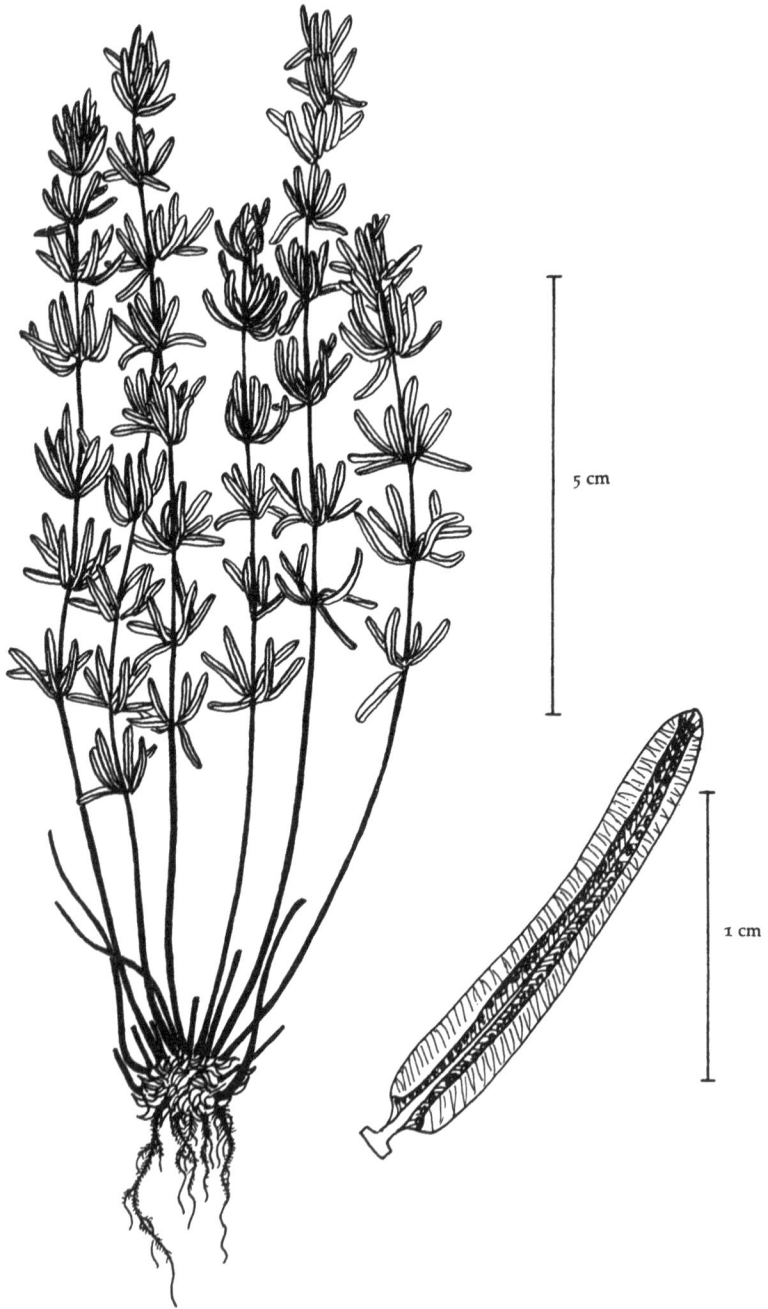

5 cm

1 cm

Pellaea brachyptera

RANGE Southern British
Columbia, South Dakota,
Saskatchewan, Ontario,
Quebec, Michigan, Vermont,
south to Florida, Texas, New
Mexico, Arizona; Mexico to
Guatemala.

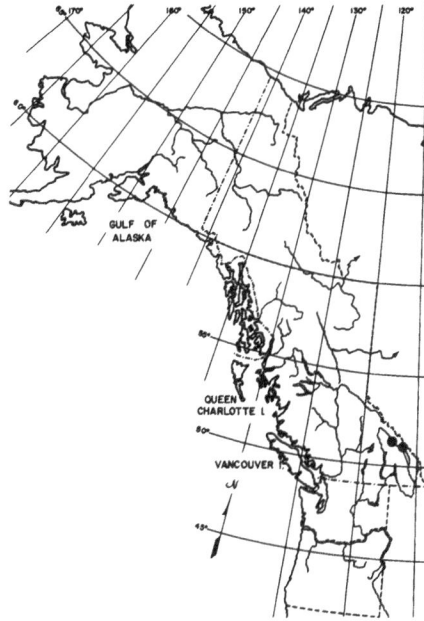

COMMENTS In our area known only from the upper Columbia River
valley in British Columbia. Most likely to be confused with *P. glabella*
from which it can be distinguished by its larger size, rough hairy stipe and
rachises, and with many of the pinnae themselves pinnate. The species
is apparently completely apogamous with only 32 spores to a sporangium.

Pellaea brachyptera (Moore) Baker

Syn. Fil. 477, 1874; Maxon [in] Abrams, Ill. Fl. Pacific States 1:31, *Fig. 63*, 1923;
Frye, Ferns Northwest 93, *Fig. xxii*, 1934; Peck, Man. Higher Pl. Ore. 50, 1941.

Rhizome moderately stout, horizontal and elongate, nodose. Scales of
rhizome and base of stipes appressed, ruddy-brown with a narrow darker
brown stripe, the borders thinner and pale, strongly dentate, apex filiform, twisted. Fronds up to 40 cm long, stiff, monomorphic. Stipe and
rachis furrowed longitudinally, essentially glabrous, glaucous, castaneous,
stipe wiry, as long as the blade or longer. Blade linear; 2-pinnate or,
rarely, 3-pinnate; greyish-green. Pinnae nearly equal in length, markedly
oblique to the rachis, pinnate with 5–13 pinnules, subsessile or short-
stalked. Pinnules linear, subcoriaceous, margins strongly revolute, crenate, mucronate.

HABITAT Talus slopes, crevices, and outcrops of basic rocks.

COMMENTS In our region limited to a few stations in southwestern
Oregon. Distinguishing features are its semicircular pinnae with long
linear segments. This is a sexual diploid species with 64 spores per
sporangium.

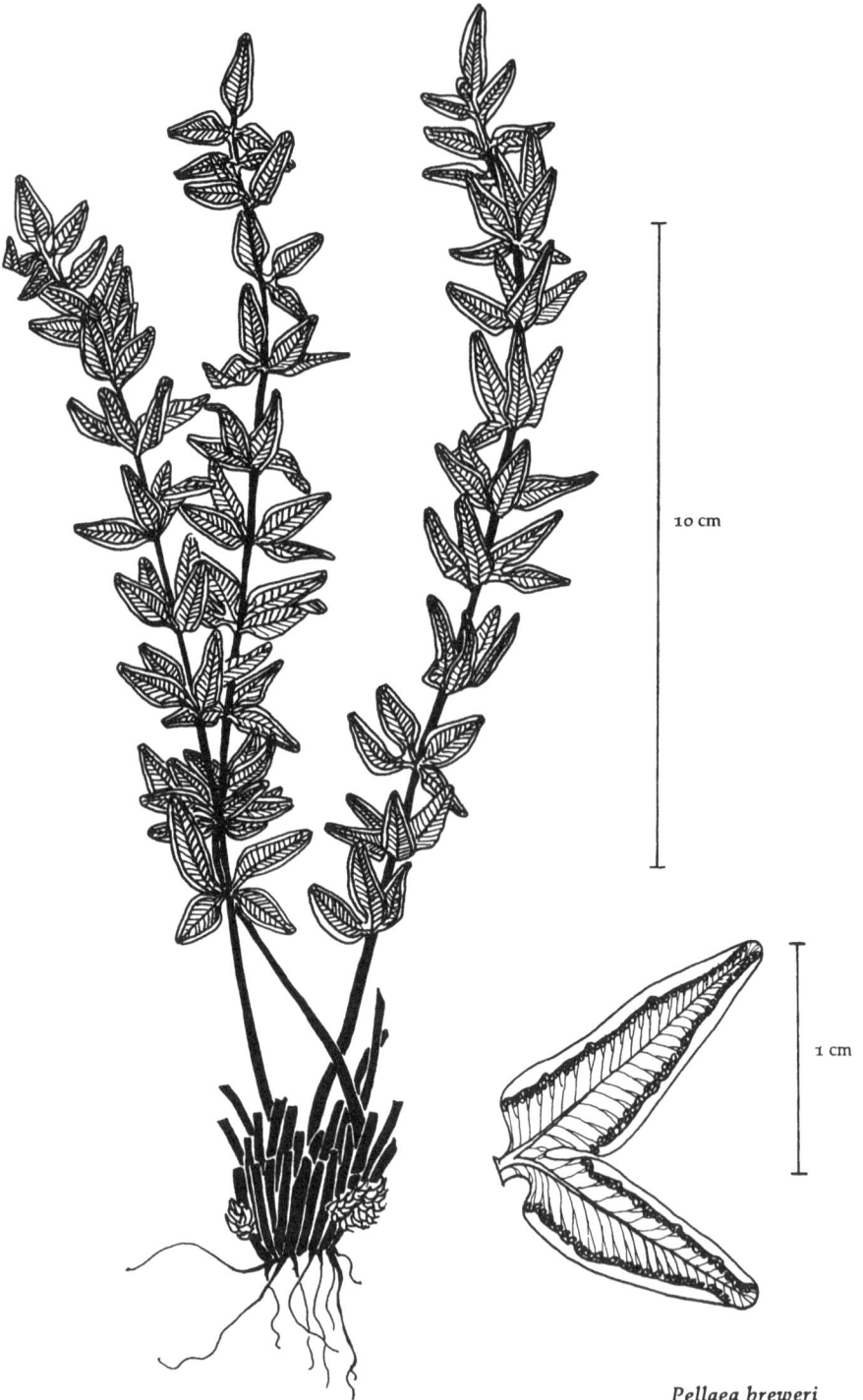

Pellaea breweri

RANGE Southwest Oregon
and adjacent northwest
California.

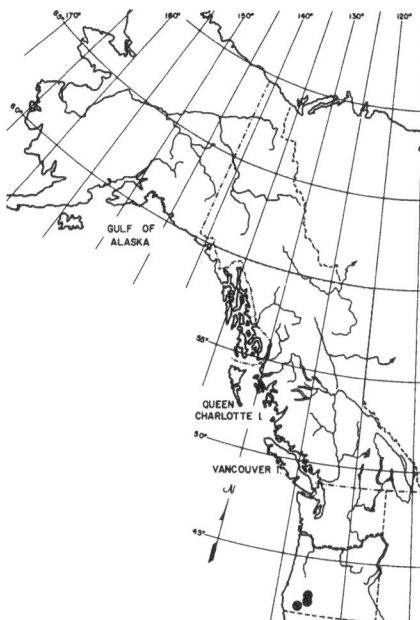

Pellaea breweri D. C. Eaton

Proc. Am. Acad. **6**:555, 1865; Maxon [in] Abrams, Ill. Fl. Pac. States **1**:29, *Fig. 58*, 1923; Frye, Ferns Northwest 93, *Fig.* xxII, 1934; Peck, Man. Higher Pl. Ore. 50, 1941.

Rhizome compact, somewhat ascending, massive, crowded with erect compressed bases of old fronds. Rhizome scales matted, uniform rusty-brown as are those of the stipe bases, lustrous, acicular, flexuous, attenuate at the apex. Fronds variable up to 21 cm tall, erect from an arcuate base, monomorphic. Stipe and rachis terete, often flat when dry; stipes stout with prominent lines of articulation, lustrous, castaneous, sparsely pubescent; rachis green at apex. Blade linear-oblong, acute, pinnate or pinnate-pinnatifid, glabrous, green but somewhat glaucous. Pinnae all about the same length, inserted at acute angles on the rachis, upper pinnae entire, sessile, usually strongly decurrent on the rachis; lower pinnae clearly 2-lobed, the upper lobe usually the larger, subsessile or with short stalks, decurrent. Segments linear ovate to deltoid, usually somewhat lobed, apex blunt, whitish border narrow, crenulate. Sporangia almost concealed by the reflexed margin of the segment. ($n = 29$)

HABITAT Usually on granitic but sometimes on basic rocks, at high altitudes.

COMMENTS Resembles *P. glabella* somewhat, but mature fronds may be recognized by the constricted bands on the stipe. This species is dip-

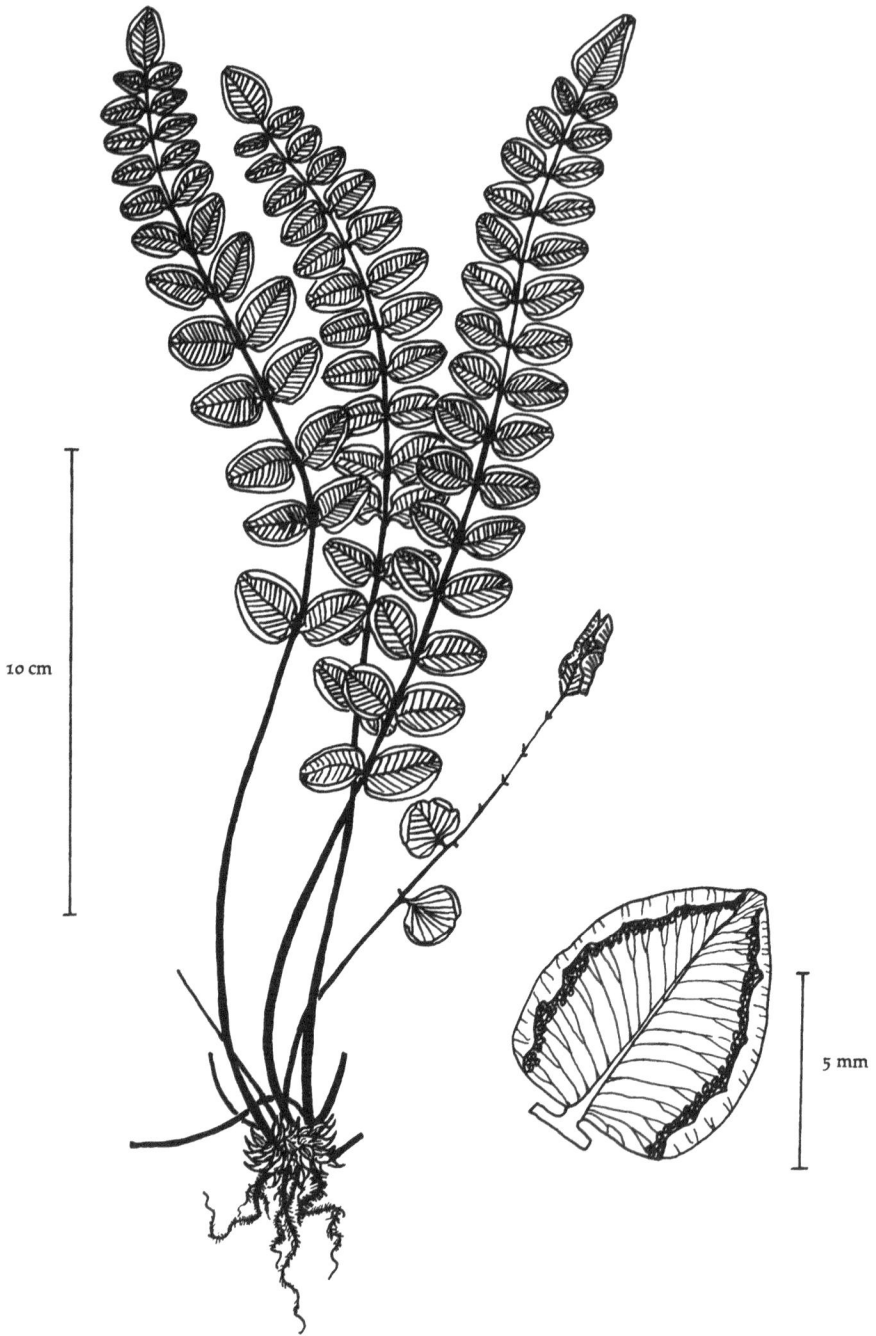

10 cm

5 mm

Pellaea bridgesii

loid and reproduces sexually. Found only in the Cascades of Washington and Oregon with eastward extensions from the latter.

RANGE Washington to California, east to Wyoming, Utah, and Nevada.

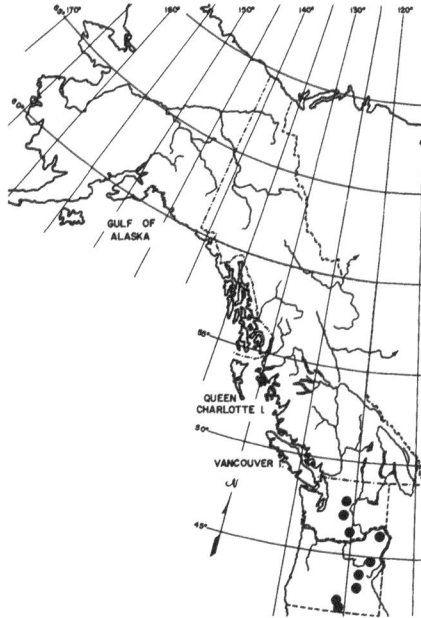

Pellaea bridgesii Hook.

Sp. Fil. **2**:238, *Plate* 142, 1858; Maxon [in] Abrams, Ill. Fl. Pac. States **1**:30, *Fig. 59*, 1923; Frye, Ferns Northwest 92, *Fig.* xxii, 1934; Peck, Man. Higher Pl. Ore. 50, 1941.

Rhizome short-creeping, the divisions short and close; scales conspicuous in dark-brown tufts, narrowly linear-attenuate, mostly with a narrow, somewhat thick, blackish stripe, margins paler, denticulate. Fronds closely tufted, numerous, erect, up to 35 cm tall. Stipes dark chestnut, shining, long persistent, nearly as long as the blades. Blades linear to linear-oblong, glabrous, coriaceous, grey-green, 1-pinnate. Sterile pinnae suborbicular, usually flat. Fertile pinnae broadly oval to cordate-oblong, nearly sessile, opposite, usually conduplicate and falcate. Revolute margins narrow, whitish, soon reflexed exposing the broad intramarginal sori.

HABITAT Granite slopes at high altitudes.

COMMENTS The systematic position of this well-defined species is doubtful. A. F. Tryon (1957) excludes it from *Pellaea* and considers that it has affinities with *Notholaena*. Pending further study it is, however, retained in *Pellaea* because of superficial resemblances to species of this genus in our area. In our area it is extremely local, being known only from Baker County, Oregon.

1 cm

10 cm

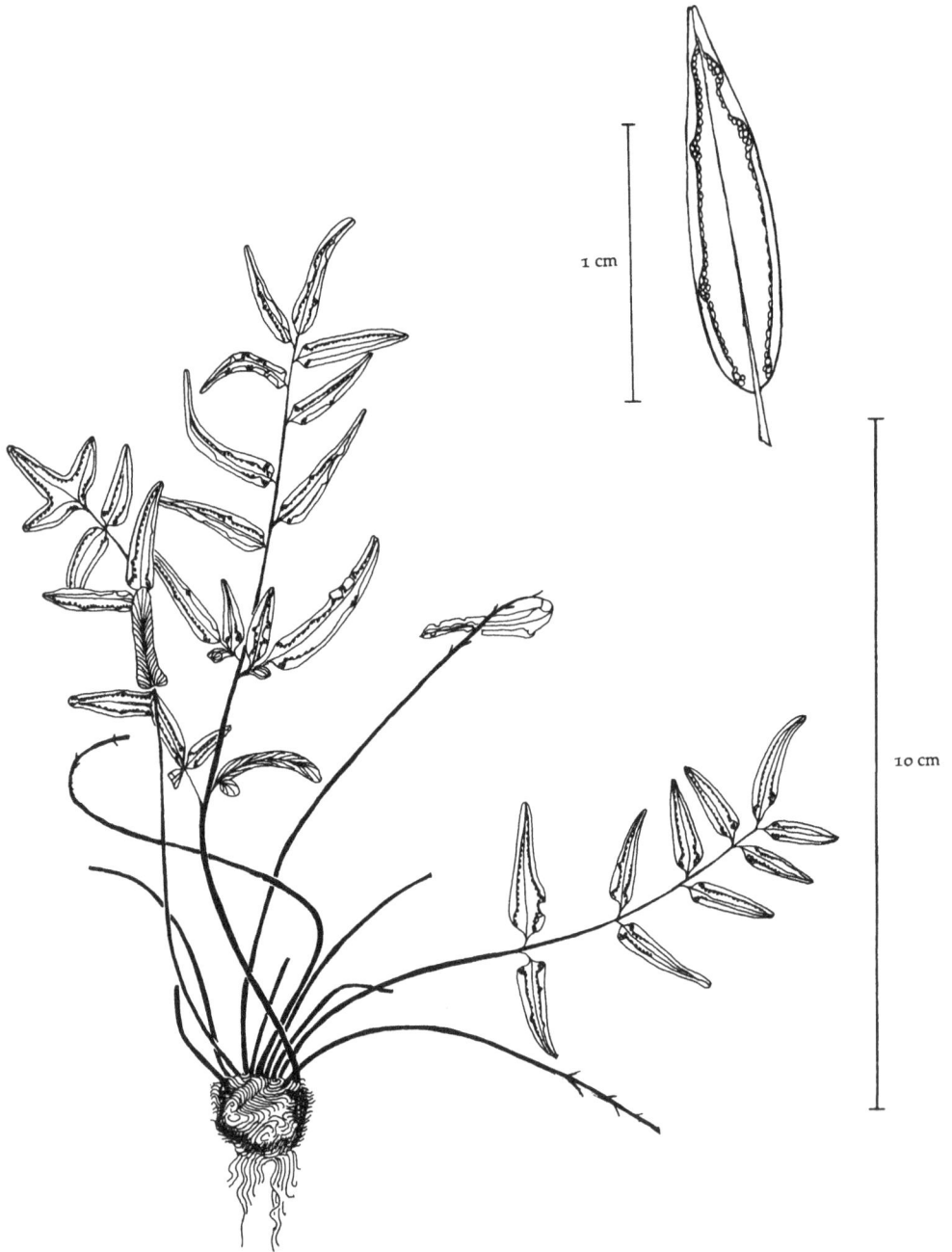

Pellaea glabella

RANGE Northeastern Oregon and adjacent Idaho, south in the Sierras to central California.

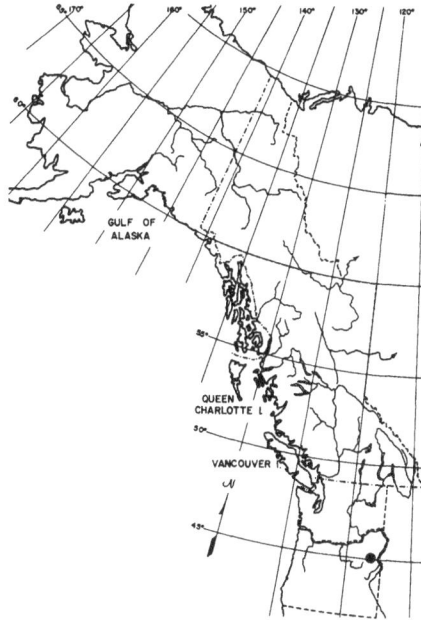

Pellaea glabella Mett.

[Ex] Kuhn, Linnaea 36:87, 1869; Henry, Fl. S. British Columbia 5, 1915; Maxon [in] Abrams, Ill. Fl. Pac. States 1:30, *Fig.* 60, 1923; Frye, Ferns Northwest 91, *Fig.* XXVII, 1934.
P. occidentalis (E. Nelson) Rydb., Mem. N.Y. Bot. Gard. 1:4, 1900.
P. suksdorfiana Butters, Am. Fern J. 11:40, 1921.
P. atropurpurea var. *simplex* (Butters) Morton, Leafl. West. Bot. 6:156, 1951.

Rhizome short, stout, ascending. Scales of the rhizome and base of stipes sinuate, irregularly dentate, bright reddish-brown. Fronds up to 20 cm long, densely cespitose. Stipes dark reddish-brown, glabrous or with a few long, flaccid, jointed hairs. Blade oblong to lanceolate, usually 1-pinnate or occasionally 2-pinnate below, evergreen, coriaceous, blue-green. Segments oblong-lanceolate, 3–5-lobed, frequently with 2 or more small auricles, segments of basal pinnae withering early leaving persistent petiolules; apex blunt or submucronate. Revolute margins membranous, white, almost entire. ($n = 29, 116$)

HABITAT Dry limestone ledges and crevices, or sandstone cliffs.

COMMENTS According to A. F. Tryon (1957) and confirmed by A. F. Tryon and Britton (1958) this species exists in three distinct races. Two of these (var. *simplex* Butters in the West and var. *glabella* in the East) are apogamous tetraploids with n and $2n = 116$ and 32 spores in each sporangium (Wagner, Farrar and Chen (1965) later reported an additional sexual diploid race of var. *glabella* in Missouri); the third (var. *occidentalis*

10 cm

4 mm

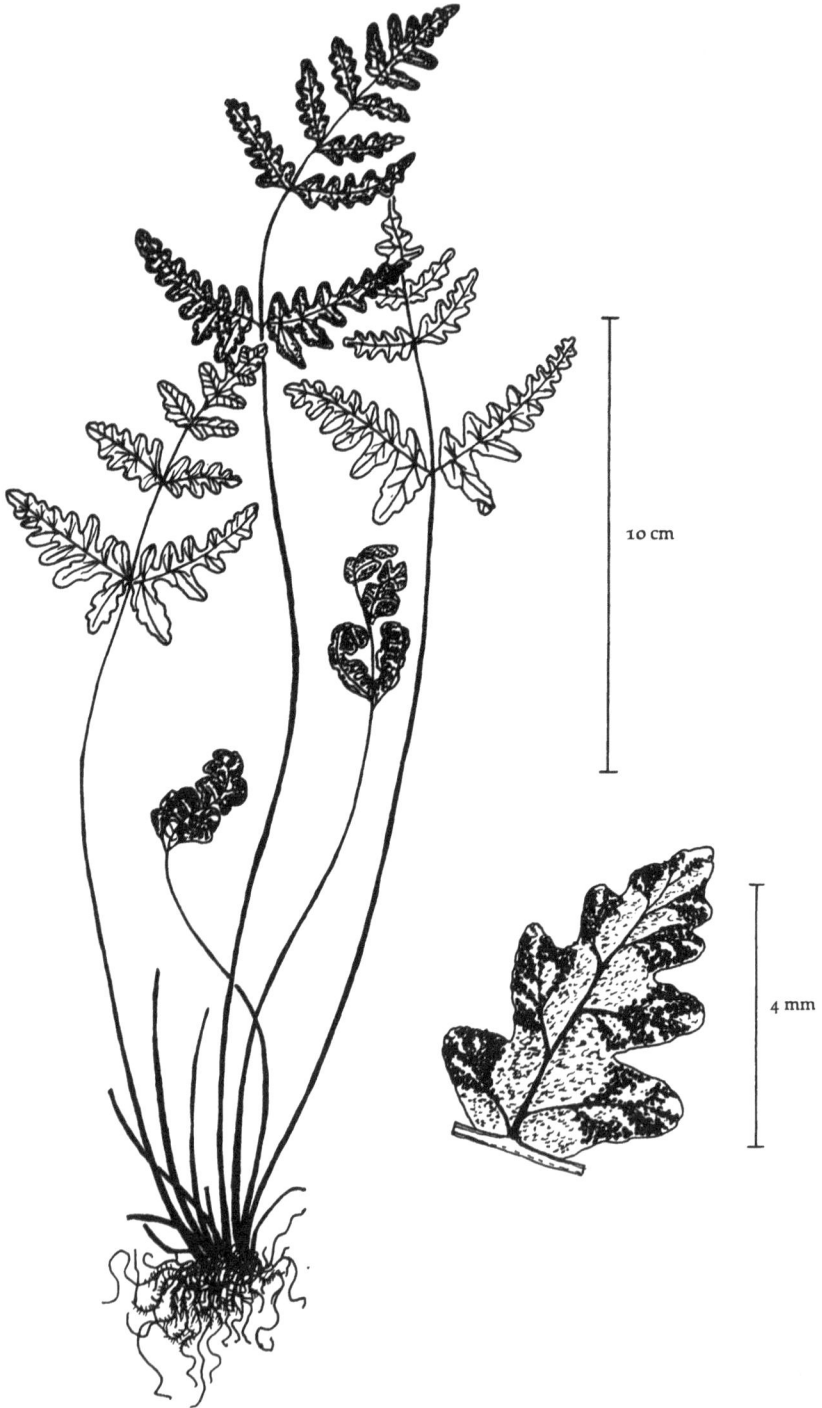

Pityrogramma triangularis

(Nelson) Butters in the Midwest) is a sexual diploid with $n = 29$ and the normal number of 64 spores per sporangium. The only form in our area is var. *simplex* which is known from a very few limestone localities in southern British Columbia and from central Washington. In the past it has been confused with *P. atropurpurea* from which it differs in its smaller size, monomorphic fronds and in the stipe and rachis being glabrous or at most sparsely pubescent.

RANGE British Columbia to Ontario and Vermont, south to Arizona and Texas.

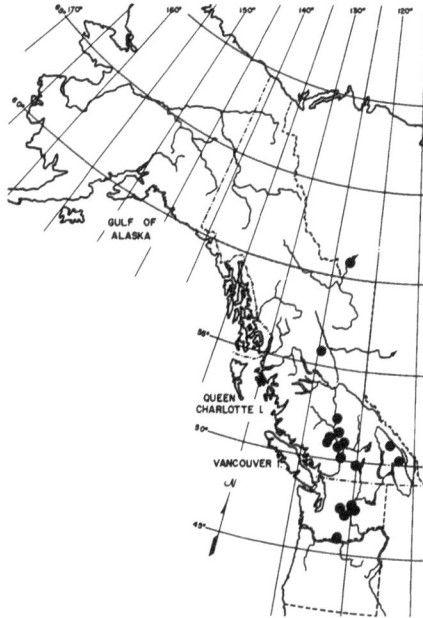

PITYROGRAMMA Link, Handb. d. Gewachs 3:19, 1833

Smallish ferns of dry hillsides, rock ledges, and crevices. Rhizome short-creeping, covered with stiff dark scales. Fronds tufted, erect or drooping, all alike. Stipes not articulated to the rhizome, rather wiry, dark, and glossy. Blades 1- to 3-pinnate, linear to deltoid-pentagonal in outline, usually without scales, sometimes glandular above, covered beneath by a conspicuous white to golden-yellow powder. Sori following the veinlets, often confluent, indusium absent. (Name from Greek *pityron*, bran, and *gramma*, letter or line, referring to the scaly, linear sori.)

A genus of about fifteen species, mainly of tropical regions, only one of which occurs in our region.

Pityrogramma triangularis (Kaulf.) Maxon

Contr. U.S. Nat. Herb. **17**:173, 1913; Macoun, Cat. Can. Pl. pt. 5, 258, 1890; Henry, Fl. S. British Columbia 3, 1915; Piper & Beattie, Fl. Northwest Coast 2,

1915; Maxon [in] Abrams, Ill. Fl. Pac. States **1**:20, *Fig.* 39, 1923; Frye, Ferns Northwest 102, *Fig.* xxvii, 1934; Peck, Man. Higher Pl. Ore. 49, 1941; St. John, Fl. Southeast. Wash. 7, 1963.
Gymnogramma triangularis Kaulf., Enum. Fil. 73, 1824.
Ceropteris triangularis (Kaulf.) Underw., Bull. Torrey Bot. Club **29**:630, 1902.

Rhizome thickish, short-creeping or somewhat ascending; scales narrow, stiff, brown, often black-keeled. Fronds erect, numerous, tufted, up to 35 cm tall. Stipes stiff and wiry, very dark brown, lustrous, glabrous except at the base, about twice as long as the blade. Blades deltoid-pentagonal, pinnate, two basal pinnae relatively large, pinnate or pinnately divided, deltoid but markedly asymmetrical about the midvein, the largest pinnules on the lower side. Other pinnae linear-oblong, pinnately lobed, segments rounded-obtuse, mostly decurrent, coriaceous, covered with white or yellowish waxy powder below, essentially glabrous above. Sori along branched veins, confluent with age. ($n = 30, 60$)

H A B I T A T Rocky slopes and crevices, in the open or half shade.

C O M M E N T S In dry weather the fronds curl up showing their characteristic light-coloured lower surface covered with whitish or yellowish powder. This is essentially a coastal species with interesting outliers in the Columbian gorge and in Whitman County, Washington. According to Alt and Grant (1960) the tetraploid race apparently occurs scattered throughout the area of the diploid population.

R A N G E Southwestern British Columbia to northern Baja California, east to Idaho, Nevada and Arizona.

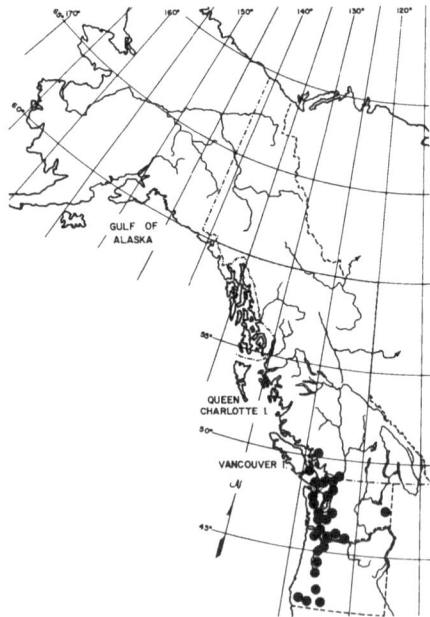

POLYPODIUM L., Sp. Pl. 1082, 1753

Rhizomes not stout, often wide-creeping mostly quite scaly. Stipes short, distinctly articulated to the rhizome. Fronds evergreen or withering in dry seasons, all alike, scattered, pinnatifid to pinnate (in our species), segments alternate to subopposite, not reduced towards the base. Veins all 1- or 2-forked, sometimes anastomosing, not reaching the margin. Sori round or somewhat oval, largish, each borne (in our species) at the end of a free veinlet; indusium always lacking. (Named from Greek *polys*, many, and *pous*, foot, alluding to the numerous knob-like prominences on the rhizome.)

A very large genus of several hundred species of terrestrial and epiphytic ferns, mainly of tropical and subtropical regions.

1   Veins of blade anastomosing; segments of blade stiffly coriaceous, rounded at the apex, midrib scaly below at first.   *P. scouleri*

1   Veins often branching but free; segments herbaceous to membranous, not rounded at the apex, not scaly on midribs.

   2   Segments usually more than 30 mm long, ratio of length of segment to its width 4:1 or more; tips acute to attenuate; rhizome with licorice taste.   *P. glycyrrhiza*

   2   Frond segments usually less than 30 mm long, ratio of length of segment to its width 3:1 or less; tips obtuse to acute; rhizome sweet or acrid but not licorice to taste.

     3   Sori round, marginal; rhizome scales usually with a darker central stripe, margins coarsely toothed; rhizome acrid.

       4   Frond segments narrowly ovate, tips acute; glandular paraphyses usual.   *P. virginianum*

       4   Frond segments oblong to obovate, tips obtuse; glandular paraphyses very rare.   *P. montense*

     3   Sori oval, medial; rhizome scales concolorous, margins not coarsely toothed; rhizome sweet.   *P. hesperium*

Polypodium glycyrrhiza D. C. Eaton

Am. J. Sci. II. **22**:138, 1856; Macoun, Cat. Can. Pl. pt. 5, 258, 1890; Henry, Fl. S. British Columbia 3, 1915; Piper & Beattie, Fl. Northwest Coast 3, 1915; Maxon [in] Abrams, Ill. Fl. Pac. States **1**:8, *Fig.* 15, 1923; Frye, Ferns Northwest 150, *Figs.* LVIII, 5, & 6, 1934; Peck, Man. Higher Pl. Ore. 52, 1941; Hultén, Fl. Alaska 58, 1968; Calder & Taylor, Fl. Queen Charlotte Islands pt. 1, 153, 1968.
P. vulgare L. var. *occidentale* Hook., Fl. Bor. Am. **2**:258, 1840.
P. falcatum Kellogg, Proc. Calif. Acad. Sci **1**:20, 1854, not L.f. 1781.
P. vulgare L. var. *commune* Milde, Fil. Eur. Atlan 18, 1867.
P. occidentale (Hook.) Maxon, Fern Bull. **12**:102, 1904.

Rhizome with a strong licorice taste, comparatively thick, creeping, scaly; scales pale-brown to straw-coloured, deeply cordate to peltate, narrow-

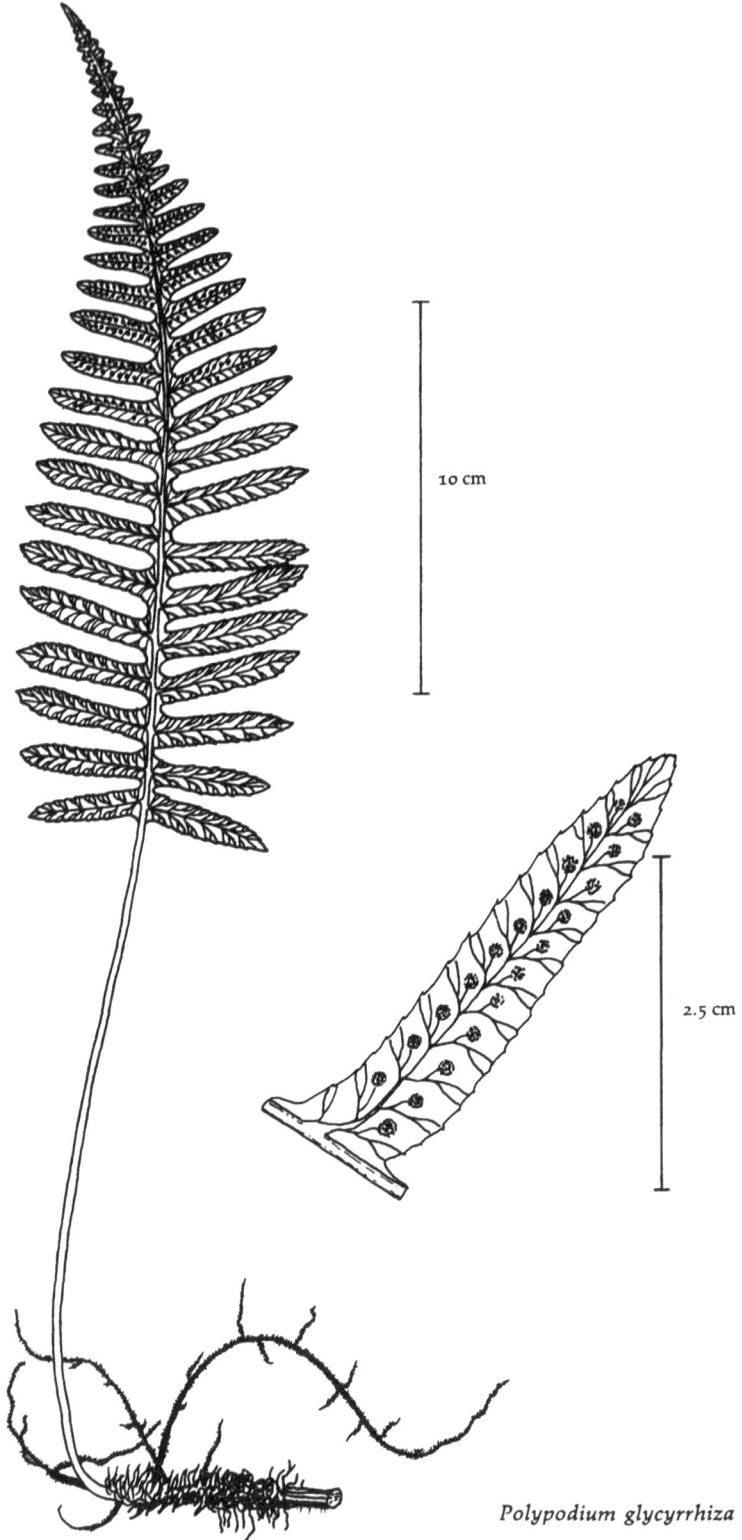

10 cm

2.5 cm

Polypodium glycyrrhiza

ovate to ovate, more or less entire, sometimes with a prolonged capillary tip. Fronds up to 60 cm long; stipe stout and about 1/2 as long as the blade. Blade oblong to ovate sometimes subdeltoid; segments narrowly oblong-attenuate, more or less falcate, tips acute to acuminate, margins finely serrate with appressed to spreading teeth. Sori usually round, occasionally slightly oval, midway between margin and midrib of segments; paraphyses lacking. ($n = 37$)

HABITAT Trees or rocks mostly near the coast at low elevations.

RANGE Alaska to Central California.

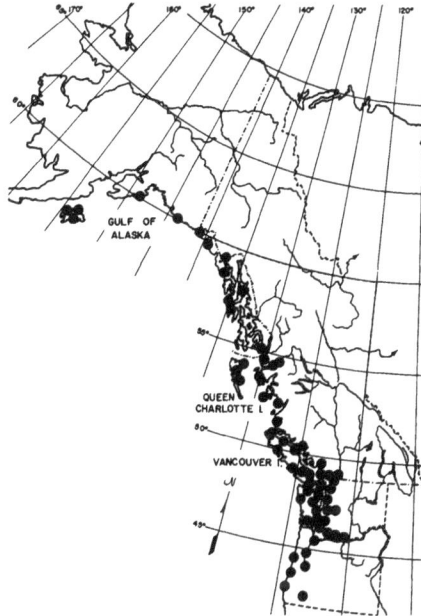

COMMENTS In the southern half of its range this species extends eastward only to the Cascade Mountains where it may be found growing in the valleys of the major rivers. It is readily distinguishable by its tapering, somewhat falcate blade segments, and in the field by its licorice taste. It commonly grows as an epiphyte on the big leaf maple (*Acer macrophyllum* Pursh). Hybrids between the present species and *P. hesperium* are known; they are intermediate in their morphology and have aborted spores.

Polypodium hesperium Maxon

Proc. Biol. Soc. Wash. **13**:200, 1900; Macoun, Cat. Can. Pl. pt. 5, 257, 1890; Henry, Fl. S. British Columbia 3, 1915; Piper & Beattie, Fl. Northwest Coast 3, 1915; Maxon [in] Abrams, Ill. Fl. Pac. States 1:8, *Fig. 14*, 1923; Frye, Ferns Northwest 150, *Fig.* LVII, 1934; Peck, Man. Higher Pl. Ore. 52, 1941; St. John, Fl. Southeast. Wash. 7, 1963.

5 cm

1 cm

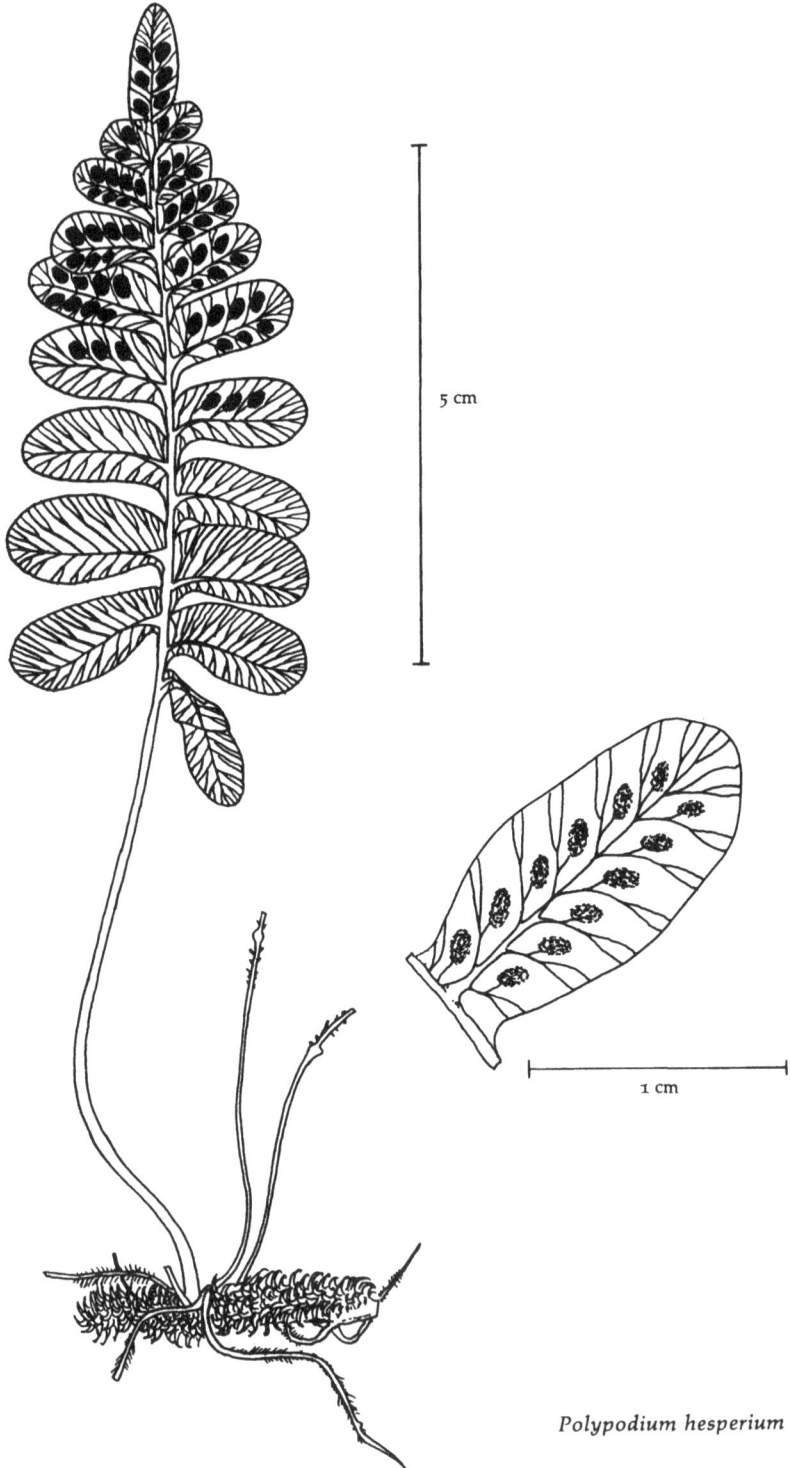

Polypodium hesperium

P. vulgare L. var. *columbianum* Gilbert, List N. Am. Pterid. 19, 38, 1901.
P. vulgare L. var. *perpusillum* Clute, Fern Bull. **18**:98, 1910.
P. prolongilobum Clute, Fern Bull. **18**:97, 1910.
P. amorphum Suksdorf, Werenda **1**:16, 1927.

Rhizome comparatively thick, sweet, creeping, scaly; scales tan to chestnut-brown, lance-ovate, margins somewhat crenate-serrate. Frond up to 37.5 cm long; stipe stout, about 1/2 as long as the blade. Blade oblong; segments oblong, tip obtuse to acute, margins entire to serrate. Sori oval, midway between margin and midrib of the segments; glandular paraphyses very rare. ($n = 74$)

HABITAT Crevices and rocky slopes.

RANGE Alaska to North Dakota, south to Baja California, Arizona and New Mexico.

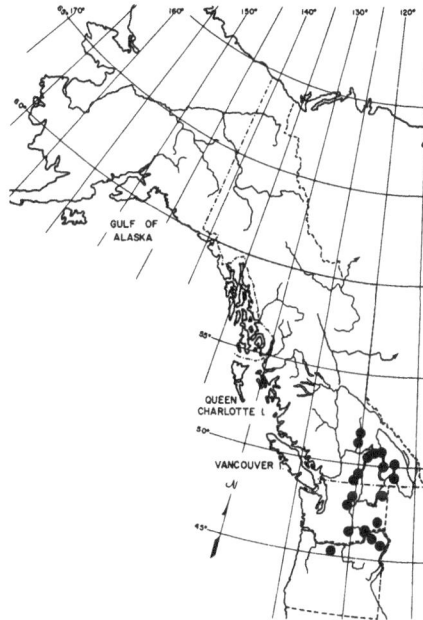

COMMENTS This is essentially a low montane species that never quite reaches the coast. It can usually be recognized by its relatively small frond segments with blunt, rounded tips. Sterile triploid hybrids between this species and both *P. glycyrrhiza* and *P. montense* have been found by Lang.

Polypodium montense F. A. Lang

Madroño **20**: 57, 1969.

Rhizome comparatively thin, acrid, often pruinose, scaly; scales dark brown to chestnut, frequently with a darker central stripe, narrow-ovate to ovate, often constricted above the point of attachment, usually with a prolonged capillary tip, margin coarsely toothed. Frond up to 30 cm long;

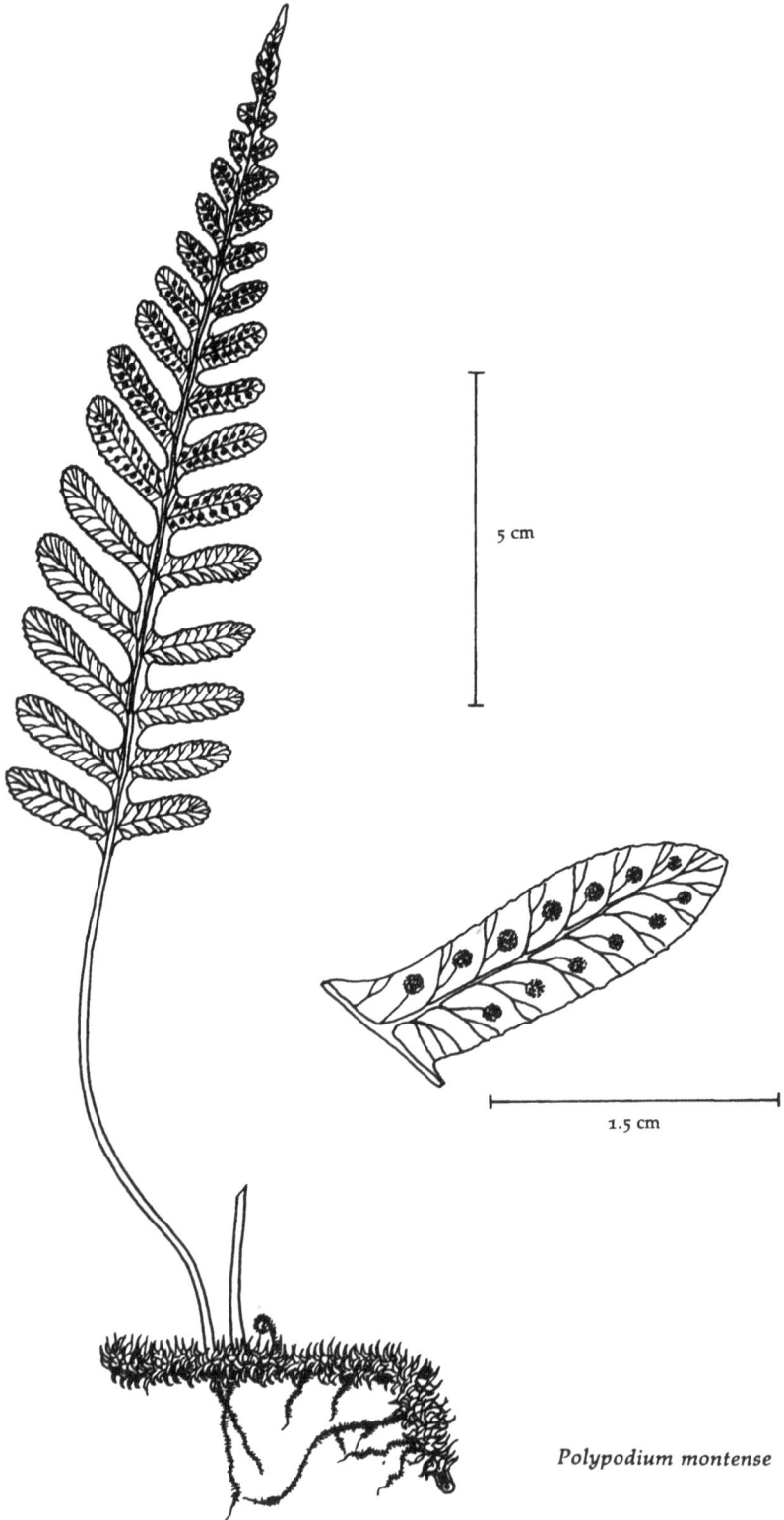

5 cm

1.5 cm

Polypodium montense

stipe slender, about 2/3 as long as the blade. Blade oblong; segments oblong to obovate, tips obtuse, rarely acute, margins entire to crenulate. Sori round, submarginal, glandular paraphyses very rare. $(n = 37)$

HABITAT Rock crevices in the mountains.

COMMENTS This recently recognized species in the *P. vulgare* complex occurs typically as a montane rock fern of comparatively high elevations, although it may descend into adjacent valleys. It is similar in its general features to the tetraploid *P. hesperium* and Lang has found the sterile triploid hybrid between them. The best field characteristics for distinguishing the two species are the round, rather than oval, sori of *P. montense* and their submarginal rather than medial, location.

The "*P. vulgare*" complex in our area has long been in a state of taxonomic confusion which has been somewhat clarified recently by the studies of Lang. He considers that in addition to *P. virginianum* three other species of the complex occur in our area, viz. *P. glycyrrhiza* $(n = 37)$, *P. hesperium* $(n = 74)$ and a previously unrecognized species that he has named *P. montense* $(n = 37)$. He has found sterile hybrids between *P. glycyrrhiza* and both the diploid species. This situation to some extent parallels the findings of Shivas (1961) who reports that the *P. vulgare* complex in Europe consists of three species, viz. *P. australe* Fée $(n = 37)$, *P. vulgare* L. $(n = 74)$, and *P. interjectum* Shivas $(n = 111)$ with hybrids between them.

RANGE British Columbia to Wyoming, south to California and Colorado.

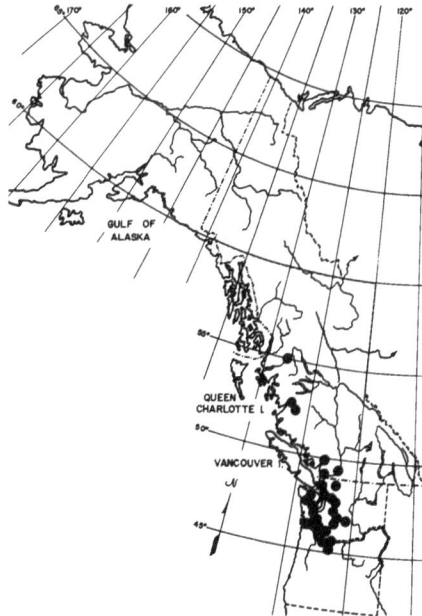

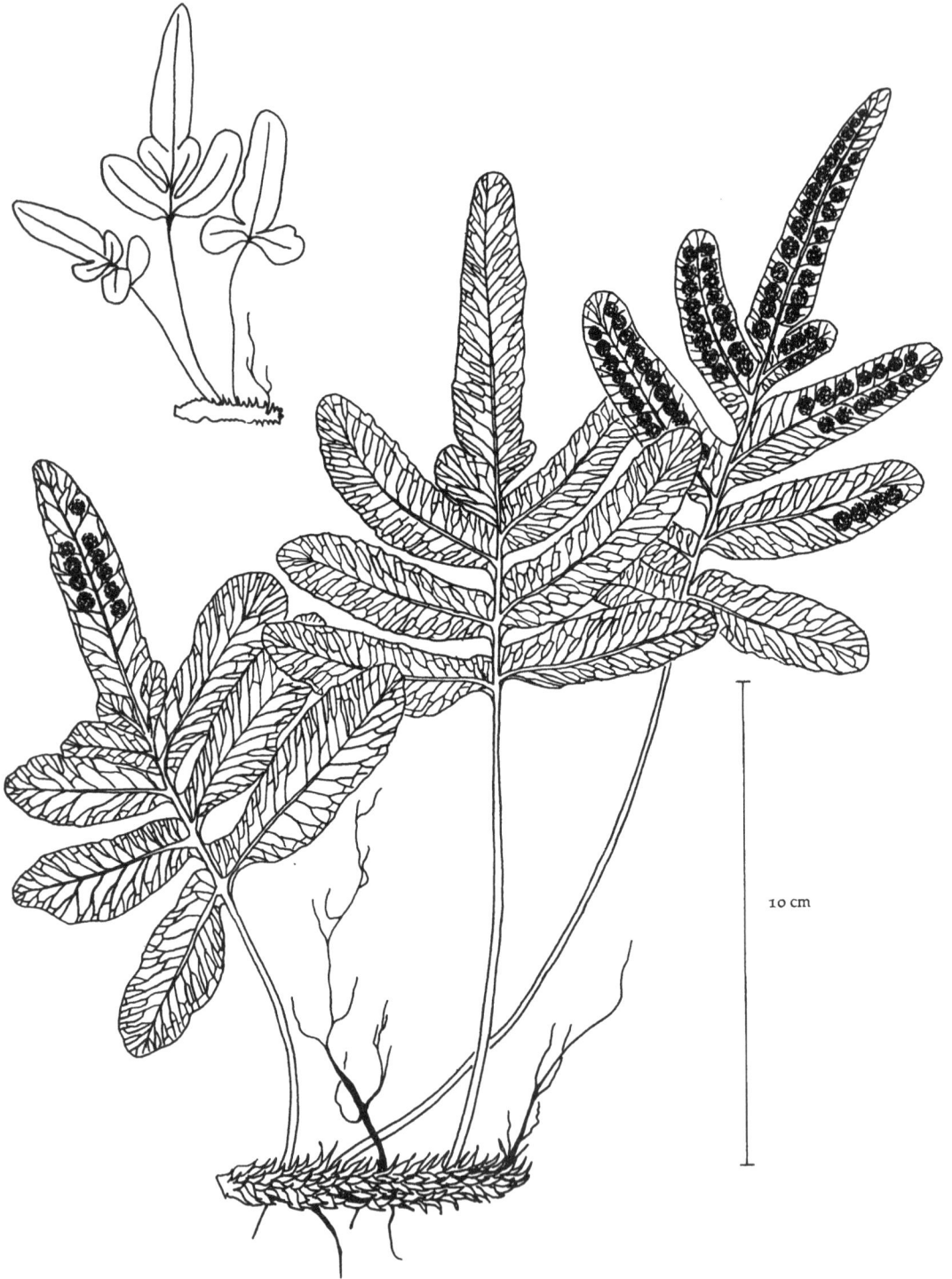

Polypodium scouleri

Polypodium scouleri Hook & Grev.

Icon. Fil. **1**:56, 1829; Macoun, Cat. Can. Pl. pt. 5, 258, 1890; Henry, Fl. S. British Columbia 3, 1915; Piper & Beattie, Fl. Northwest Coast 3, 1915; Maxon [in] Abrams, Ill. Fl. Pac. States **1**:7, *Fig.* 13, 1923; Frye, Ferns Northwest 153, *Fig.* LVII, 1934; Peck, Man. Higher Pl. Ore. 52, 1941; Calder & Taylor, Fl. Queen Charlotte Islands pt. 1, 152, 1968.

Rhizome stout, sparsely scaly, naked in age; scales broadly ovate, dark brown. Fronds few, up to 50 cm in length, usually somewhat scattered; stipes glabrous, stout, stiff, shorter than the blade. Blades deltoid-ovate, acute, pinnate to pinnatifid. Segments linear to linear-oblong, stiffly coriaceous, rounded at the apex with a crenate to obscurely crenate-serrulate cartilaginous border, terminal pinnae distinct, much longer than the upper lateral ones. Midvein prominent, deciduously scaly beneath, lateral veins fusing to form a single series of areoles. Sori very large, round, crowded against the midvein, mostly confined to the upper pinnae. $(n = 37; 2n = 111)$

HABITAT Trees or rocks near the coast.

RANGE British Columbia to Baja California.

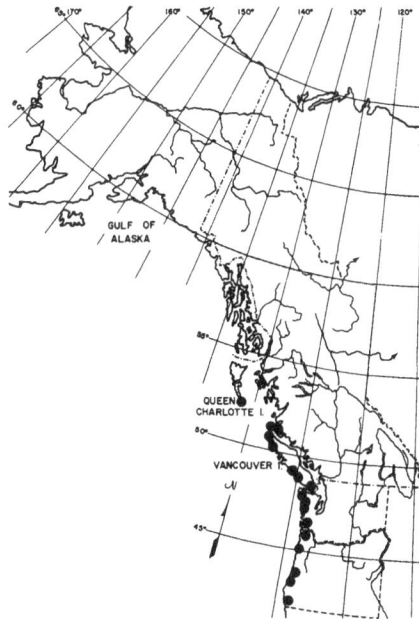

COMMENTS This species is readily identified by its firm, leathery fronds with broad, blunt segments. It is confined to the outer coast from the southern end of the Queen Charlotte Islands, British Columbia, south to Baja California. Manton's report (1951) of a triploid plant ($2n = 111$) from California is of special interest as no tetraploid has apparently been found as yet.

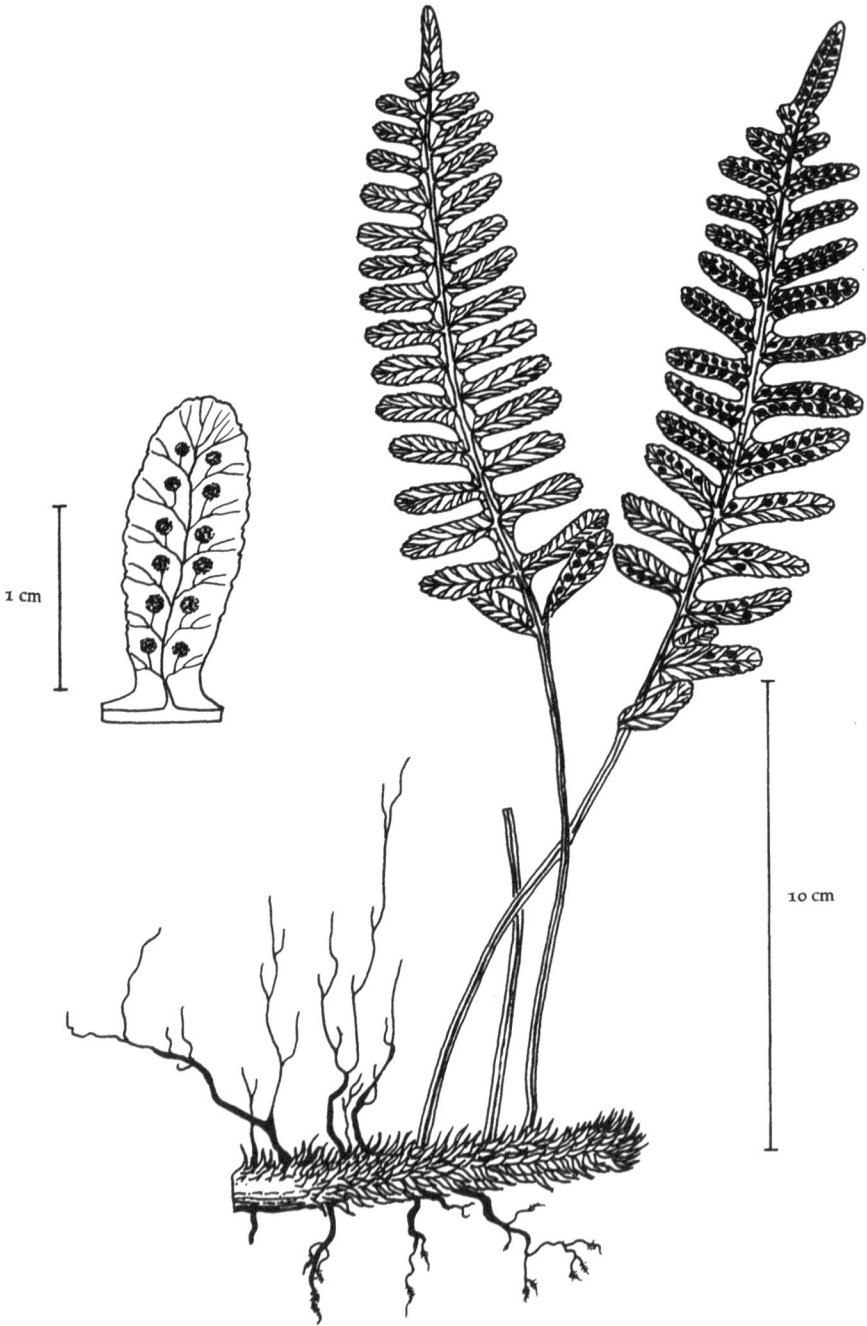

1 cm

10 cm

Polypodium virginianum

Polypodium virginianum L.

Sp. Pl. 1085, 1753; Eaton, Ferns N. Am. **1**: *Plate* 41, 1879; Fernald, Rhodora **24**: 141, 1922; Morton [in] Gleason, Ill. Fl. **1**:34, *Fig.*, 1952.
P. vulgare many eastern Am. auth., not L.

Rhizome spongy, creeping, somewhat acrid to taste, scaly; scales light brown, concolorous but often with a darker central stripe, deeply cordate with sinus usually closed; attenuate to a capillary apex, entire. Fronds up to 35 cm long; stipe slender, variable in proportion to the length of the blade. Blades oblong-linear to oblong-lanceolate, acuminate; segments linear-oblong to lanceolate, mostly entire or inconspicuously denticulate, obtuse at apex. Sori round, nearly marginal, sporangia mixed with long-stalked glandular paraphyses. ($n = 37, 74$)

HABITAT Mossy rocks, rocky slopes in shady woodland, rarely epiphytic on the bases of tree trunks.

COMMENTS Very local in our area being known only from east of the Rocky Mountains in northeastern British Columbia and the Yukon. Can best be distinguished from *P. montense* by its narrowly ovate blade segments with acute tips. In eastern North America this species occurs in both diploid and tetraploid races with the sterile hybrid between them also being known; cytological information is not available from our area.

RANGE Yukon; northeastern British Columbia to Newfoundland; south to Alberta; Arkansas, Alabama, Georgia, and New England.

POLYSTICHUM Roth, Fl. Germ. 3:69, 1799

Rather coarse, evergreen ferns of woods and rocky places. Rhizomes erect to decumbent, copiously chaffy. Fronds stiffly upright or recurved, sometimes more or less spreading, usually borne in a crown, in our species pinnately or 2-pinnately compound. Blades more or less scaly and leathery in texture; pinnae often unequal at the base, auriculate at the base of the upper margin, margins usually serrate, teeth sharp to spinose, veins free. Sori circular; indusia round, centrally peltate. (From Greek *polus*, many, and *stichos*, a row, referring to the regular rows of sori.)
   A large genus mainly of boreal and temperate regions.

1   Blades simply pinnate, the pinnae variously incised but never pinnately lobed or pinnatifid.

 2   Pinnae mostly oblong-lanceolate, the lower ones about as long as wide, progressively reduced; teeth conspicuously spreading.
                                                                P. lonchitis
 2   Pinnae linear-attenuate, the basal ones at least 2–3 times as long as wide, not reduced below; teeth incurved.   *P. munitum*

1   Blades 2-pinnate, or at least pinnately lobed or divided at base.

  3   Pinnae pinnately lobed to pinnatifid at base; segments few, lobes large with pungent teeth.

   4   Stipe short or lacking; about 6 pairs of lateral veins and teeth on median pinnae; pinnae more or less triangular; teeth spinulose, spreading, rarely incurved.   *P. kruckebergii*

   4   Stipe 1/6 to 1/4 the length of the blade; about 12 pairs of lateral veins and teeth on median pinnae; pinnae more or less oblong; teeth mucronate, ascending or incurved.   *P. scopulinum*

  3   Pinnae nearly or quite pinnate, at least at base.

   5   Segments crenate-dentate, cuneate at base.   *P. mohriodes*

   5   Segments spinulose-aristate, in general broadest at base but not cuneate.

    6   Rachis with proliferous buds; basal pinnule on the upper side of lower pinnae conspicuously longer than the next.   *P. andersonii*

    6   Rachis without proliferous buds; basal pinnule on upper side not conspicuously longer than the next.   *P. braunii*

Polystichum andersonii Hopkins

Am. Fern J. 3:116, *Plate 9*, 1913; Henry, Fl. S. British Columbia 6, 1915; Maxon [in] Abrams, Ill. Fl. Pac. States 1:12, *Fig. 23*, 1923; Frye, Ferns Northwest 121, 1934; Peck, Man. Higher Pl. Ore. 48, 1941; Anderson, Fl. Alaska 14, 1959; Hultén, Fl. Alaska 55, 1968; Calder & Taylor, Fl. Queen Charlotte Islands pt. 1, 155, 1968.

Rhizome very stout, ascending, sometimes assurgent, with numerous, thin, pale chestnut, linear to ovate, denticulate scales. Fronds up to 10 dm tall, in a crown at the apex of the rhizome. Stipes stout, grooved above, chaffy. Blades pinnate-pinnatifid to bipinnate in part, dark green above, densely chaffy on the rachis below, narrowly lance-oblong to lance-eliptic, long acuminate; rachis grooved above, characteristically bearing one or more chaffy buds near the tip. Pinnae alternate, narrowly triangular, attenuate, finely scaly below, glabrate above, pinnately lobed or divided, pinnate below. Pinnules oblique, elliptical, decurrent, strongly serrate, the teeth long-awned; basal pinnule on the upper side of the lower pinnae longer than the one next to it. Sori large, nearly midway between the midvein and the margin; indusia erose-dentate, teeth somewhat gland tipped. ($n = 82$)

HABITAT Cool, moist woods and shaded rocky slopes.

RANGE Alaska to northern Oregon; Idaho and Montana.

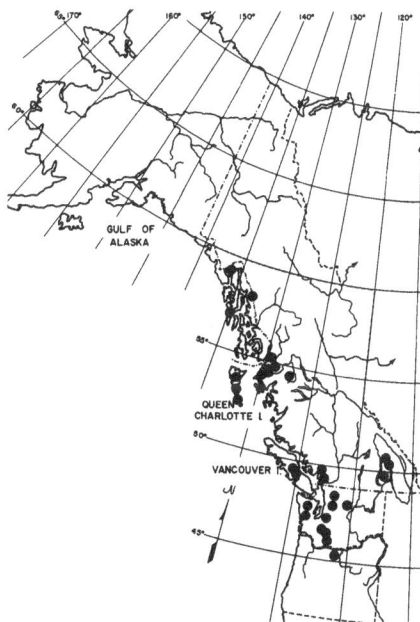

COMMENTS This is one of a large group of closely related species of wide distribution. It may ultimately prove to be only a geographical variant of *P. braunii* (Spenner) Fée. It can best be distinguished from the latter by the buds on the fronds and by the fact that the lowest pinnule on the upper side of the lower pinnae is longer than the one next to it. In Oregon and Washington it occurs west from the Cascade Mountains but in British Columbia and northward it is generally more coastal. Wagner has found a sterile natural triploid hybrid between this species and *P. munitum.*

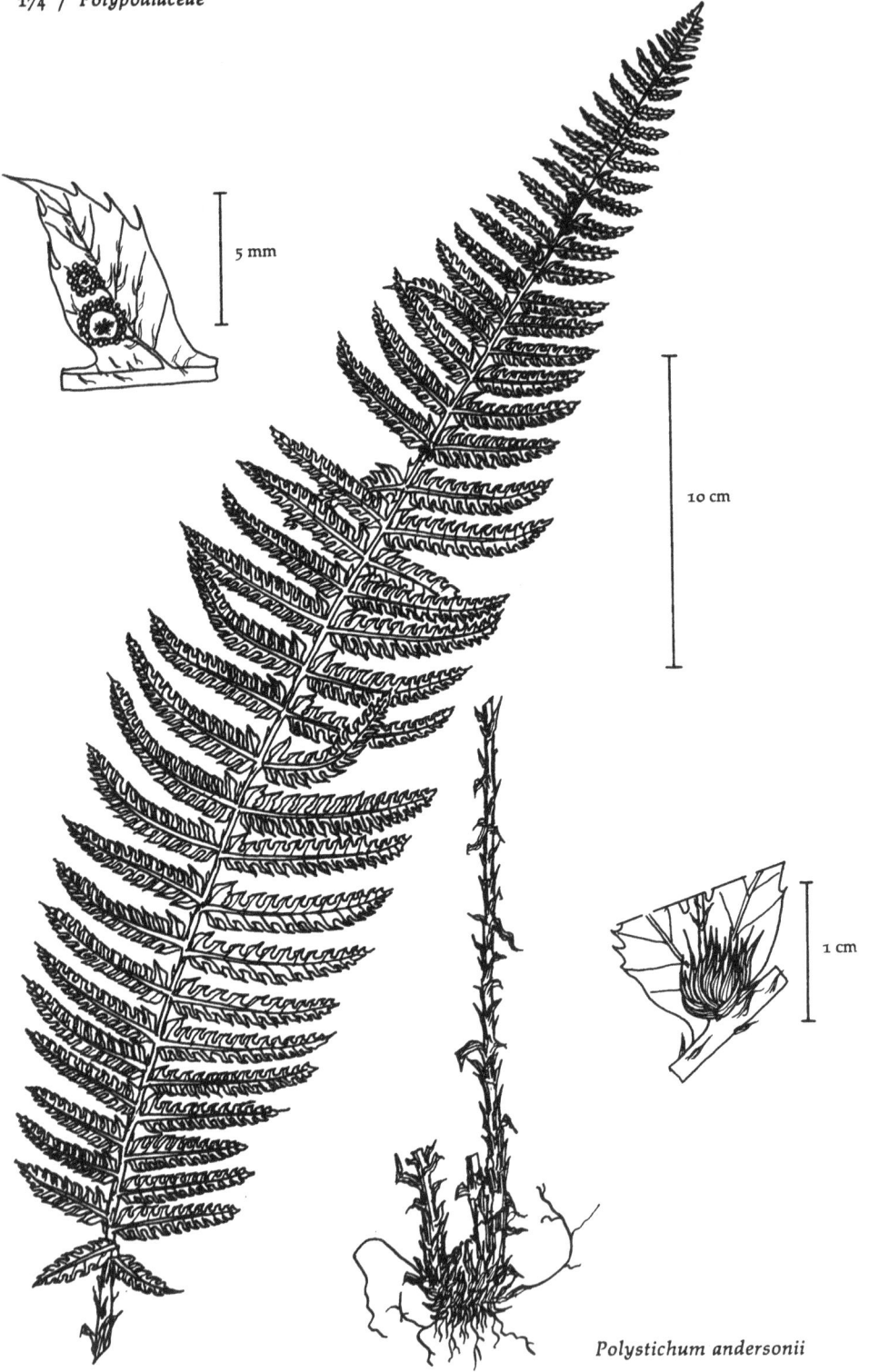

5 mm

10 cm

1 cm

Polystichum andersonii

Polystichum braunii (Spenner) Fée

Mém. Fam. Foug. 5:278, 1852; Macoun, Cat. Can. Pl. pt. 5, 278, 1890; Henry,
Fl. S. British Columbia 6, 1915; Morton [in] Gleason, Ill. Fl. 1:56, *Fig.*, 1952;
Anderson, Fl. Alaska 14, 1959; Hultén, Fl. Alaska 54, 1968; Calder & Taylor, Fl.
Queen Charlotte Islands pt. 1, 155, 1968.
Aspidium braunii Spenner, Fl. Frib. 1:9, *Plate 2*, 1825.
P. alaskense Maxon, Am. Fern J. 8:35, 1918.

Rhizome very stout, erect. Fronds numerous, ascending in a vaselike
crown up to 12 dm tall. Stipe stout, flattened and furrowed above, densely
chaffy with pale brown scales, the larger one long-attenuate. Blades subcoriaceous, lustrous green above, more or less deciduous, elliptic-lanceolate, conspicuously tapering to base and tip, 2-pinnate. Pinnae lanceolate
to narrowly oblong, obtuse to acute, densely chaffy on the rachis below,
subopposite below, alternate above. Pinnules narrowly ovate to trapezoid-
oblong, obtuse, nearly rectangular at base, slightly auricled on the upper
side, sharply serrate with incurved, bristle-tipped teeth. Sori in 2 rows
near the midvein; indusia often with erose margins. ($n = 82$)

HABITAT Cool, shaded talus slopes and rich woods.

RANGE Southern coast of
Alaska and adjacent British
Columbia; Great Lakes basin
to Newfoundland; south to
Wisconsin and Pennsylvania;
Eurasia.

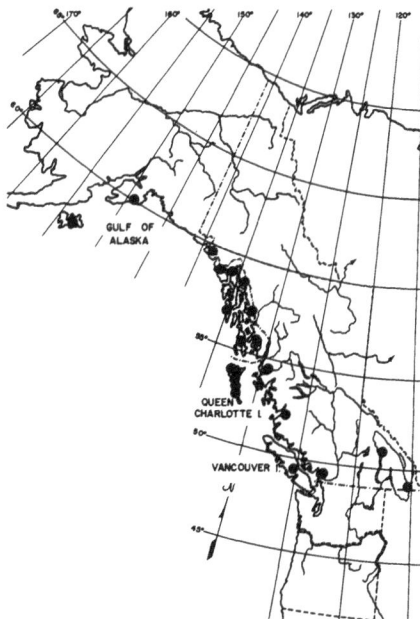

COMMENTS This is a more northern species than *P. andersonii* and so
far has not been found south of British Columbia. The absence of proliferous buds on the fronds and the relatively uniform size of the basal pinnules on the lower pinnae help to distinguish this species from *P. andersonii.* Meyer (1959a) has found a sterile natural hybrid of this species

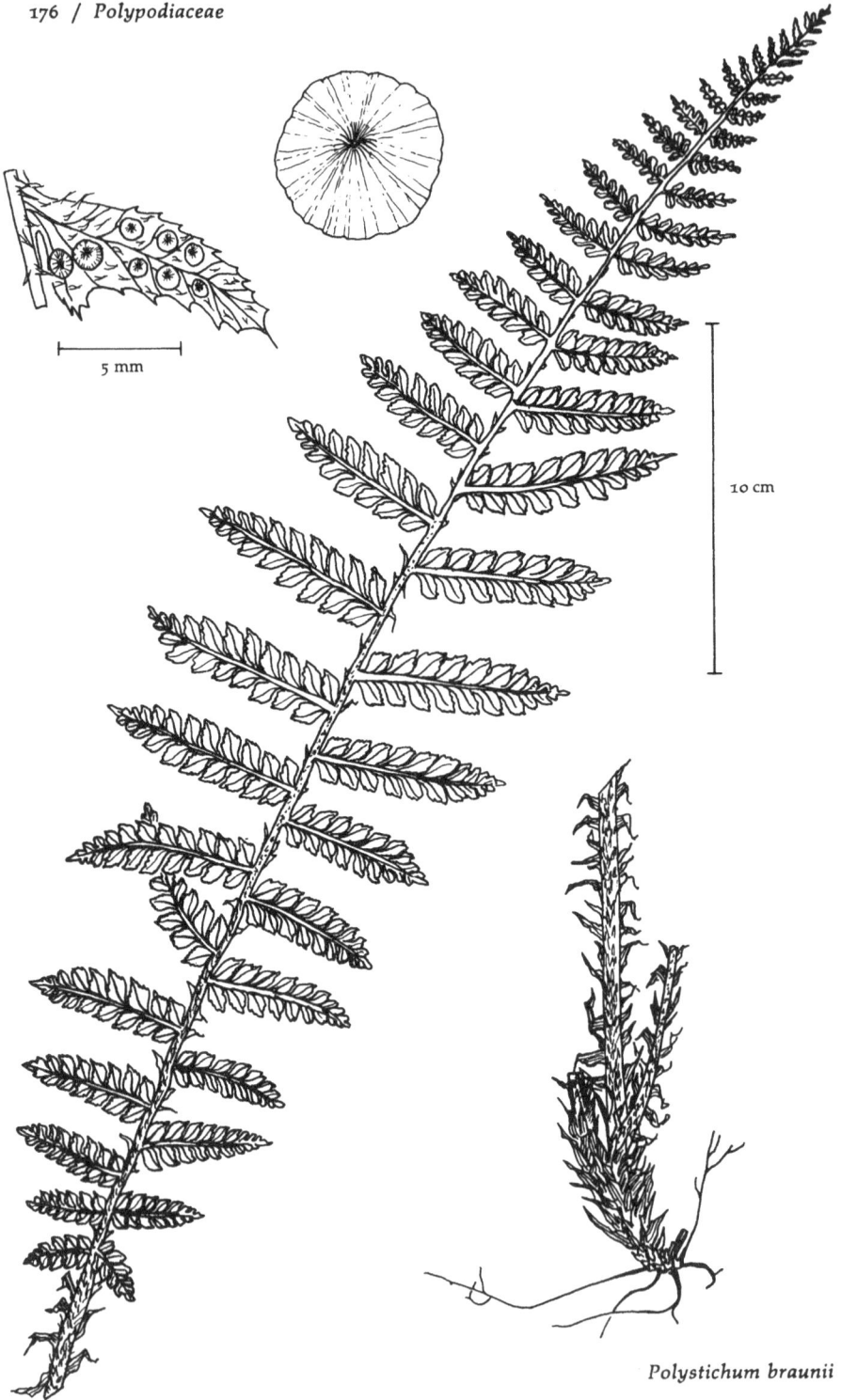

5 mm

10 cm

Polystichum braunii

with *P. lonchitis*; the same hybrid had previously been presumed by Ewan (1944) on the basis of an herbarium specimen from Fort Simpson, British Columbia.

Polystichum kruckebergii Wagner

Am. Fern J. 56:4, 1966.

Rhizome erect, small but stout, giving rise to tufts of about 6 fronds and stipe bases. Fronds linear-lanceolate, about 20 cm long by 2.5 cm broad; blade coriaceous and rigid, dark green and somewhat glossy above, slightly paler and duller below, sparsely glandular; stipe pale green or straw-coloured, short or nearly lacking, except in shade forms where it may be up to 10 cm long, bases with numerous, pale tan, concolorous scales that diminish in size up the rachis. Rachis at first with reddish to pale whitish scales, later more or less glabrous. Pinnae overlapping except in shade forms, mostly ovate-triangular, deeply and sharply toothed, acute with conspicuous anterior auricles, larger pinnae usually with one pair of basal pinnules, lowermost progressively smaller, more or less triangular, margins typically with 4–8 conspicuous, spreading, hard, cartilaginous teeth, major veinlets usually about 6 pairs from the costa; sori, except in shade forms, becoming confluent at maturity; indusia peltate with wavy margins. ($n = 82$)

HABITAT Crevices in rock bluffs or talus in alpine to subalpine zones.

RANGE Erratically from southern British Columbia south to Utah and California.

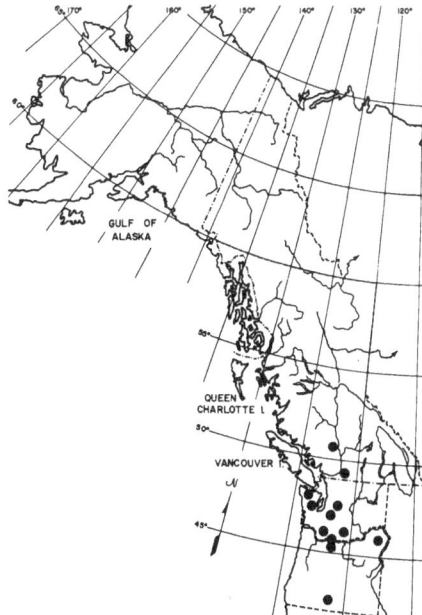

COMMENTS Rarer than *P. scopulinum* which is a larger plant with oblong, rather than more or less triangular, pinnae that may be as much as 4 cm long. *P. kruckebergii* usually has 6 marginal teeth and lateral veins per side of the median pinnae while *P. scopulinum* has twice as many. Wagner (1966a) considers that this species is an allotetraploid hybrid derived from a cross between *P. lonchitis* and *P. mohrioides.*

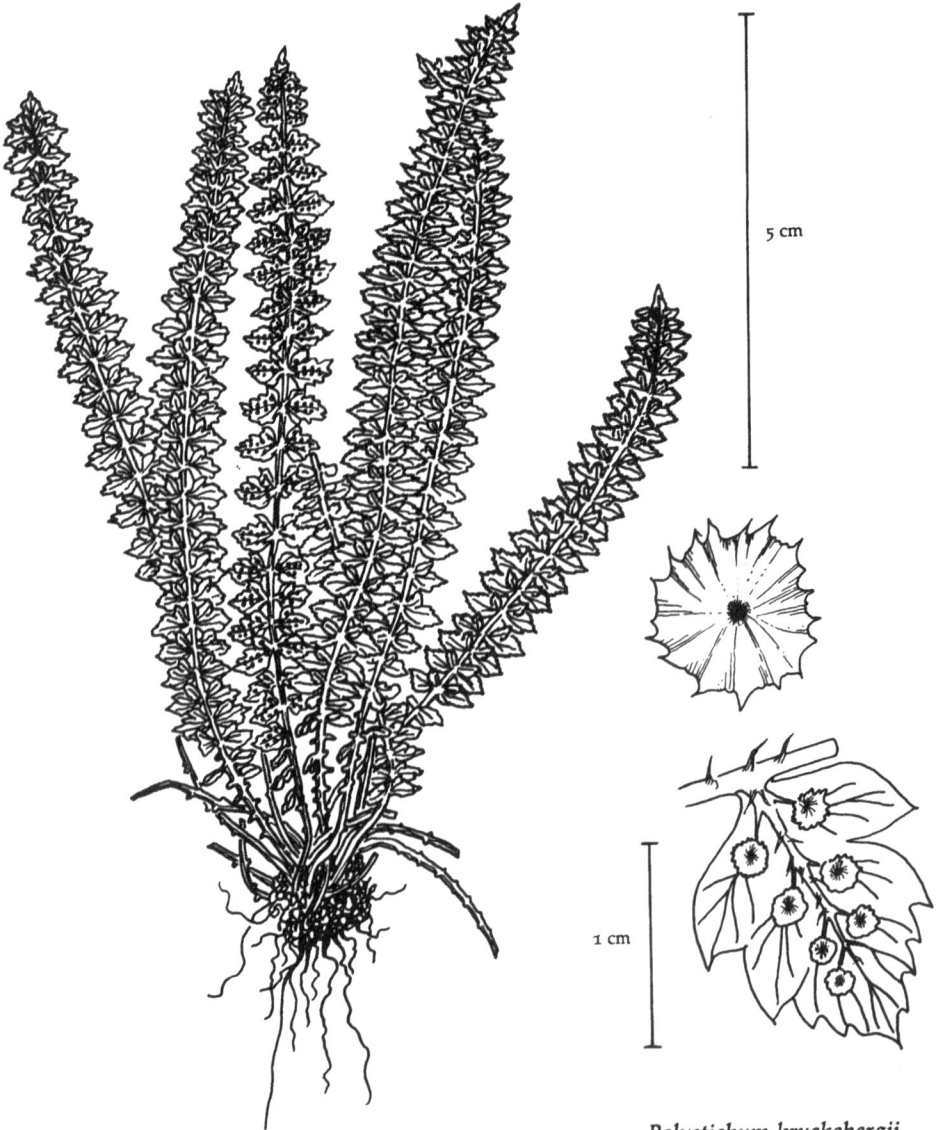

5 cm

1 cm

Polystichum kruckebergii

Polystichum lonchitis (L.) Roth

Rom. Arch. Bot. **2**:106, 1799; Macoun, Cat. Can. Pl. pt. 5, 277, 1890; Henry, Fl. S. British Columbia 6, 1915; Piper & Beattie, Fl. Northwest Coast 7, 1915; Maxon [in] Abrams, Ill. Fl. Pac. States **1**:9, *Fig.* 17, 1923; Frye, Ferns Northwest 113, *Fig.* xxxvi, 1934; Peck, Man. Higher Pl. Ore. 47, 1941; Morton [in] Gleason, Ill. Fl. **1**:56, *Fig.*, 1952; Anderson, Fl. Alaska 14, 1959; St. John, Fl. Southeast. Wash. 4, 1963; Hultén, Fl. Alaska 53, 1968; Calder & Taylor, Fl. Queen Charlotte Islands pt. 1, 158, 1968.

Rhizome stout, erect or ascending, chaffy, and covered with old stipe bases. Fronds several, ascending in a close crown, up to 50 cm tall. Stipes short, about 1/6 the length of the blade or less, straw-coloured, densely chaffy with lustrous, rusty-brown, mostly large, ovate, denticulate scales. Blades evergreen, coriaceous, linear to narrowly linear-oblanceolate, acuminate, tapering gradually to the base, narrowed at the apex, simply pinnate, the rachis stout with deciduous scales. Pinnae numerous, alternate, crowded, spreading at right angles, the basal deltoid, sometimes very small, equilateral, the middle and upper ones oblong-lanceolate, falcate, their bases auriculate above and cuneate below, unevenly serrate-dentate, the teeth conspicuously spreading-spinulose. Sori large, touching, usually in two rows, nearly midway between the midvein and the margin, mostly confined to the upper part of the frond. ($n = 41$)

HABITAT Crevices of rocks, often basic; cool, shaded talus and rocky slopes.

COMMENTS General at higher elevations, particularly common in the subalpine zone. The hard, rather stiff, leathery fronds with unlobed pinnae and conspicuous spreading teeth are distinguishing features.

RANGE Alaska to Newfoundland, south to northern California, Arizona, New Mexico and Colorado; upper Great Lakes; southern Quebec; Greenland; Iceland; Eurasia.

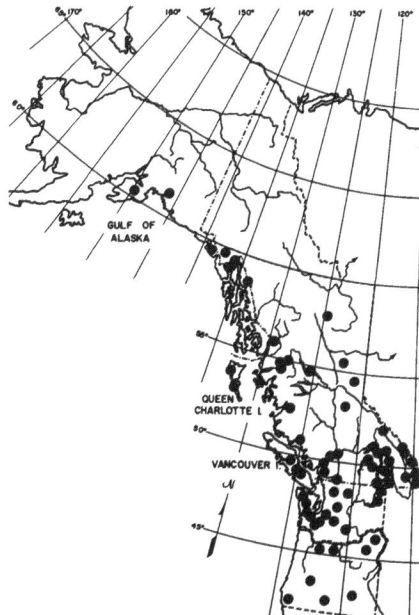

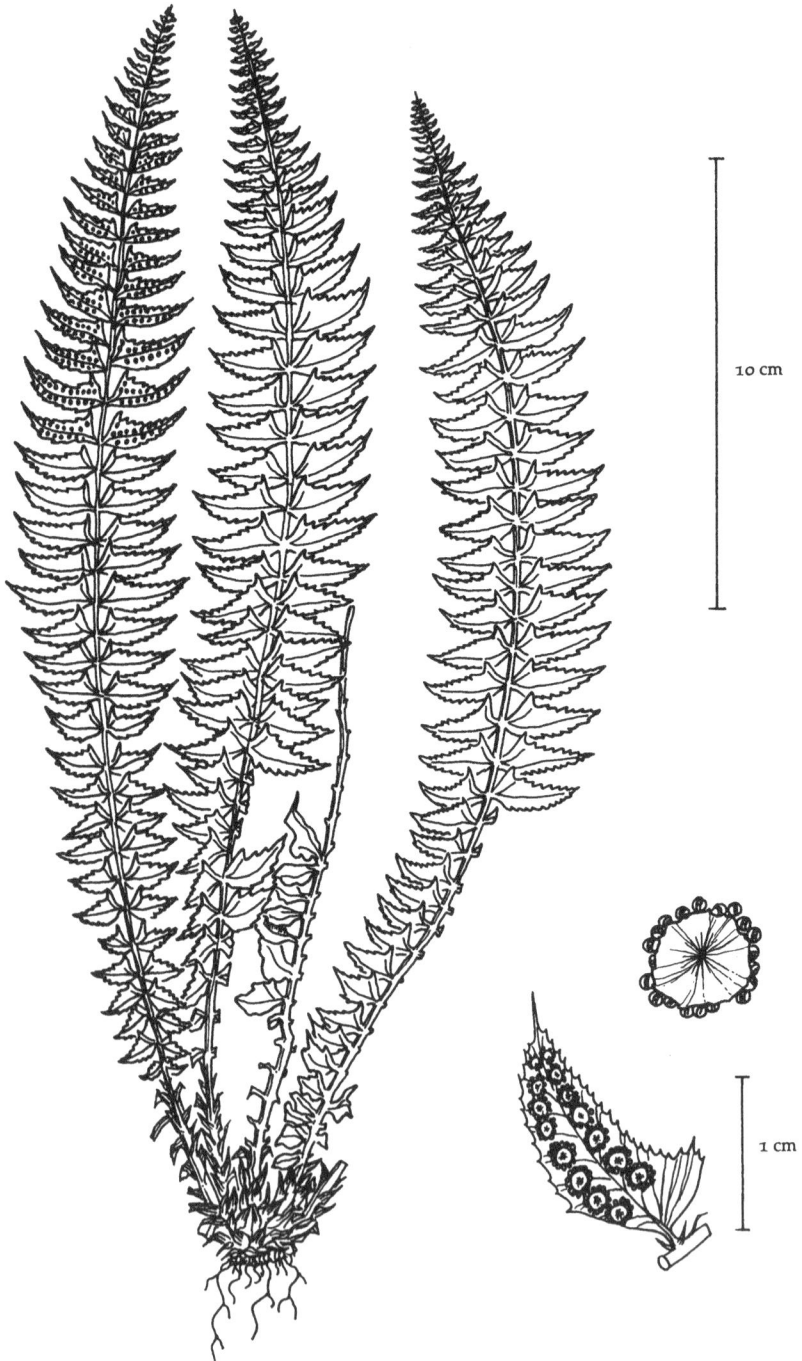

Polystichum lonchitis

Polystichum mohrioides (Bory) Presl

Tent. Pterid. 83, 1836; Henry, Fl. S. British Columbia 6, 1915; Maxon [in] Abrams, Ill. Fl. Pac. States 1:10, *Fig. 20*, 1923; D. C. Eaton, Ferns N. Am. 2:251, *Plate* LXXX, 1880; Fern. Rhodora 26:90–5, 1924; Frye, Ferns Northwest 118, *Figs.* XXXVI, XL, 1934; Peck, Man. Higher Pl. Ore. 48, 1941.
P. lemmonii Underw., Nat. Ferns, 6th ed., 116, 1900.

Rhizome short, ascending. Fronds densely clustered, rather fleshy, up to 40 cm tall. Stipes frequently glandular and finely puberulent, straw-coloured from a very chaffy base; scales pale cinnamon-brown, numerous, of two types – lanceolate scales mixed with larger scales that are ovate to acuminate; attenuate, remotely denticulate, more or less scattered above. Blades linear to narrowly lance-oblong, somewhat acute, deeply 2-pinnatifid or 2-pinnate in the lower third. Pinnae numerous, usually somewhat lobed above, deeply pinnately divided at base, crowded, ovate to deltoid-ovate or deltoid-oblong, rounded at the ends. Segments oval, obtuse, crenate or crenately lobed, not armed. Sori borne on the apical pinnae, one or two to each segment. Indusia large, thin, erose-toothed. ($n = 41$)

HABITAT Rocky alpine slopes.

RANGE Washington, south along the Andes to Tierra del Fuego and subantarctic islands.

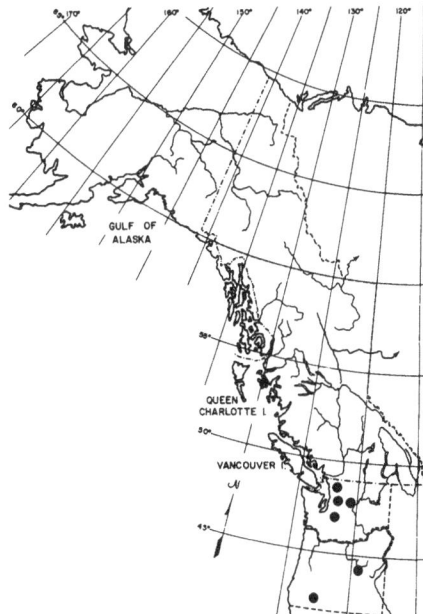

COMMENTS May easily be confused with *P. scopulinum* from which it can be differentiated by the upper segments of the frond being merely crenate or crenate-dentate, lacking the stiff, sharp-pointed teeth of the latter species. Wagner has found the triploid hybrid *P. mohrioides* × *scopulinum.*

10 cm

5 mm

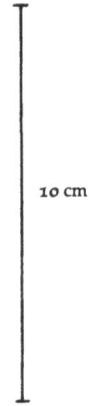

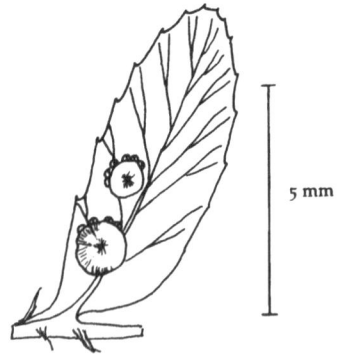

Polystichum mohrioides

Polystichum munitum (Kaulf.) Presl

Tent. Pterid. 83, 1836; Macoun, Cat. Can. Pl. pt. 5, 278, 1890; Henry, Fl.
S. British Columbia 6, 1915; Piper & Beattie, Fl. Northwest Coast 7, 1915;
Maxon [in] Abrams, Ill. Fl. Pac. States 1:10, *Fig.* 18, 1923; Frye, Ferns North-
west 115, *Figs.* xxxvi, xxxvii, 1934; Peck, Man. Higher Pl. Ore. 47, 1941;
Anderson, Fl. Alaska 14, 1959; St. John, Fl. Southeast. Wash. 5, 1963; Hultén,
Fl. Alaska 54, 1968; Calder & Taylor, Fl. Queen Charlotte Islands pt. 1, 159,
1968.

Rhizome stout, woody, ascending, covered with reddish-brown scales.
Fronds stiffly erect, forming a crown, up to 15 dm tall. Stipes densely
chaffy; scales bright glossy brown, often dark-centred, ovate-acuminate,
mixed with laciniate-ciliate smaller ones. Blades evergreen, dark shiny
green above, paler beneath, linear-lanceolate, short acuminate, simply
pinnate, scarcely contracted at base. Pinnae alternate, numerous, often
very scaly beneath, spreading, subfalcate, linear-attenuate, prominently
auriculate at the base above, cuneate below, sharply 2-serrate to incised,
teeth incurved, rigidly long aculeate. Sori large, nearly equidistant
between the midvein and the margin when in one row, sometimes
crowded in several rows; indusia papillose-dentate to long-fringed.
($n = 41$)

H A B I T A T Damp rich woods and shaded slopes.

R A N G E Alaska to California, northern Idaho, and
Montana.

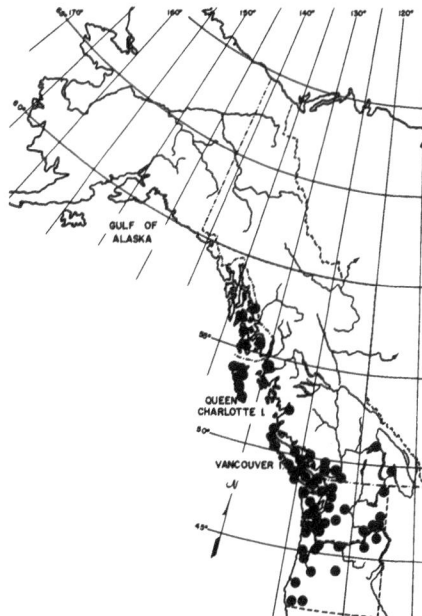

C O M M E N T S Widespread and very common particularly west of the
coastal mountains. Quite a variable species both in stature and outline of
the fronds, the variability in great measure being related to differences

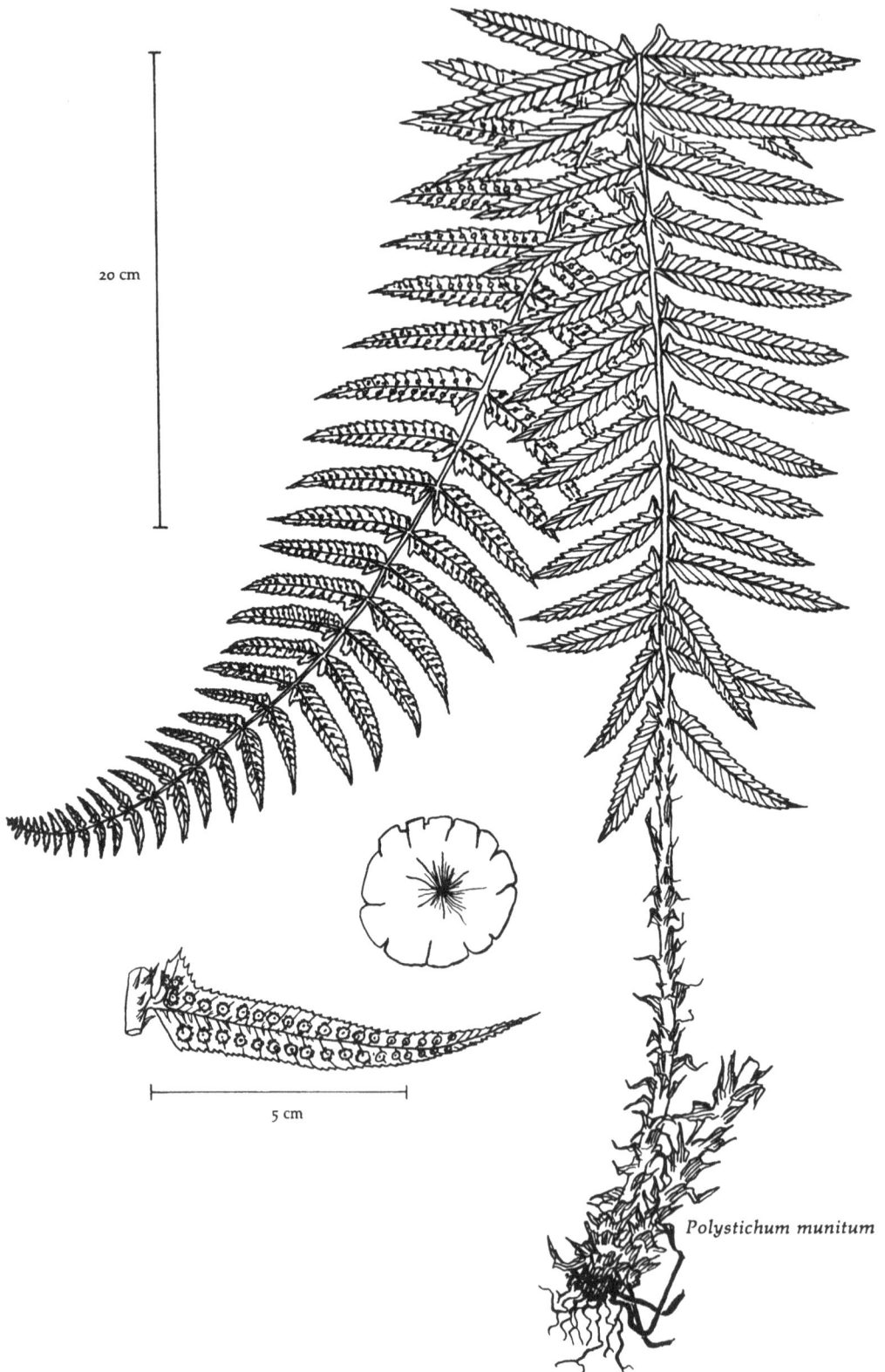

20 cm

5 cm

Polystichum munitum

in soil conditions and exposure. What has been described as an extreme response to full sun and dry conditions is shown by forma *imbricans* (D. C. Eaton) Clute, which has a narrow linear blade and closely crowded, obliquely overlapping pinnae. This form in cultivation under mesic conditions for several years, however, has retained its morphology unchanged. This suggests that the differences between it and the typical *P. munitum* may prove to be quite fundamental. Wagner has found a natural hybrid between this species and *P. scopulinum*; it showed 41 pairs plus 41 single chromosomes at meiosis.

Polystichum scopulinum (D.C. Eaton) Maxon

Fern Bull. **8**:29, 1900; Henry, Fl. S. British Columbia 6, 1915; Piper & Beattie, Fl. Northwest Coast 6, 1915; Maxon [in] Abrams, Ill. Fl. Pac. States **1**:11, *Fig.* 21, 1923; Fernald, Rhodora **26**:90–5, 1923; Frye, Ferns Northwest 118, *Figs.* xxxv, xxxvi, 1934; Peck, Man. Higher Pl. Ore. 48, 1941; Morton [in] Gleason, Ill. Fl. **1**:56, *Fig.*, 1952.
P. mohrioides (Bory) Presl var. *scopulinum* (Eaton) Fern., Rhodora **26**:92, 1924.

Rhizome relatively stout, erect or decumbent, conspicuously covered with linear, light-brown scales. Fronds mostly spreading or ascending, evergreen, up to 30 cm long. Stipes, 1/6 to 1/4 the length of the frond, grooved above, the upper parts straw-coloured, basal parts dark chestnut, densely chaffy; scales narrow- to ovate-lanceolate, remotely dentate, long attenuate, cinnamon- or light brown, considerably intermixed with numerous, smaller, narrower scales. Blades leathery to somewhat fleshy, pinnate, narrowly lanceolate, somewhat tapered to apex and base, pale green. Pinnae subopposite, essentially sessile, deltoid-ovate to deltoid-oblong, obtuse, pinnately lobed or divided, particularly towards the base, teeth 10–12 per side of a median pinna, sharp but not bristlelike except sometimes below, cartilaginous at tip, oblique; lower pinnae more widely separated. Sori conspicuous in 2 median rows, often fusing with age; indusia thin, pale, more or less erose. ($n = 82$)

HABITAT Cliffs and dry crevices of basic rocks.

COMMENTS Very local in the Cascade Mountains of British Columbia and scattered southward. Similar to *P. kruckebergii* and sometimes growing with it. (For distinguishing features see under the latter species.) Some botanists (Fernald, 1924) prefer to regard this as a variety of *P. mohrioides* but the recent discovery by Wagner that Washington plants are tetraploid ($n = 82$) gives added justification for its recognition as a distinct species. He has also found natural triploid hybrids between it and both *P. munitum* and *P. mohrioides*.

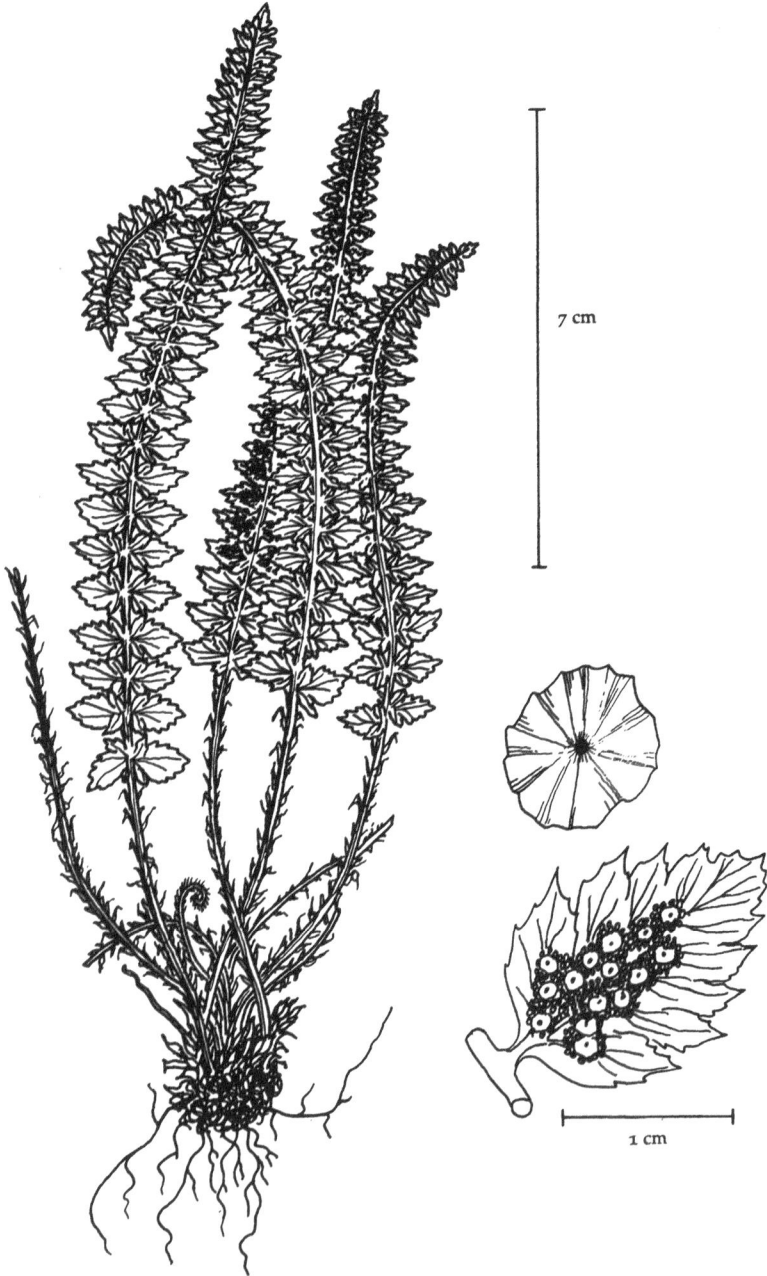

7 cm

1 cm

Polystichum scopulinum

RANGE Southern British Columbia, south to California, east to Idaho, Utah, and Arizona; Gaspé Peninsula, Quebec.

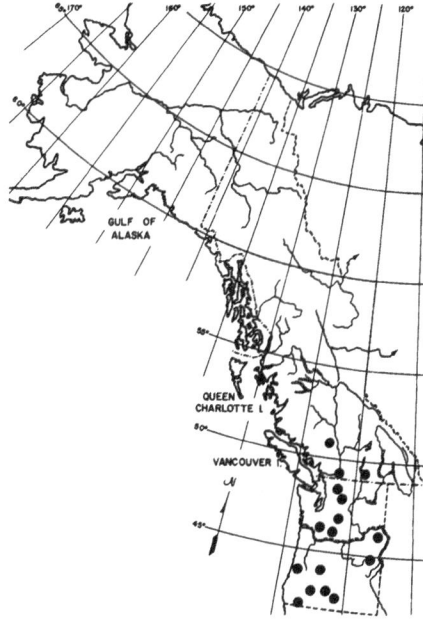

PTERIDIUM Gled. [ex] Scop., Fl. Carn. 169, 1760

Large, coarse ferns of open situations in acid soil. Fronds up to 5 m tall. Stipes about as long as the blade, covered by a feltlike mass at the darker base. Blades large, deltoid to broadly triangular, pinnately compound. Ultimate segments mostly entire, sometimes toothed or lobed. Sori continuous and marginal. (Name is a diminutive of *pteris*, the classical Greek name for a fern.)

Consists of a single world-wide species.

Pteridium aquilinum (L.) Kuhn

[In] v.d. Decken, Reisen in Ost-Afrika 3:11, 1879; Macoun, Cat. Can. Pl. pt. 5, 262, 1890; Henry, Fl. S. British Columbia 4, 1915; Piper & Beattie, Fl. Northwest Coast 4, 1915; Maxon [in] Abrams, Ill. Fl. Pac. States 1:23, *Fig.* 43, 1923; Frye, Ferns Northwest 78, *Figs.* XIX, XX, 1934; Peck, Man. Higher Pl. Ore. 51, 1941; R. M. Tryon, Jr., Rhodora 43:1–31, 37–67, 1941; Morton [in] Gleason, Ill. Fl. 1:28, *Fig.*, 1952; Anderson, Fl. Alaska 11, 1959; St. John, Fl. Southeast. Wash. 8, 1963; Hultén, Fl. Alaska 43, 1968; Calder & Taylor, Fl. Queen Charlotte Islands pt. 1, 159, 1968.
Pteris aquilina L., Sp. Pl. 1075, 1753.

Rhizome subterranean, much branched, wide-spreading, clothed with hairs but not scales. Fronds very large, deciduous, scattered along the rhizome. Stipes stout, relatively long, somewhat straw-coloured. Blade coarse, usually 3-pinnately compound. Pinnae subopposite; pinnules

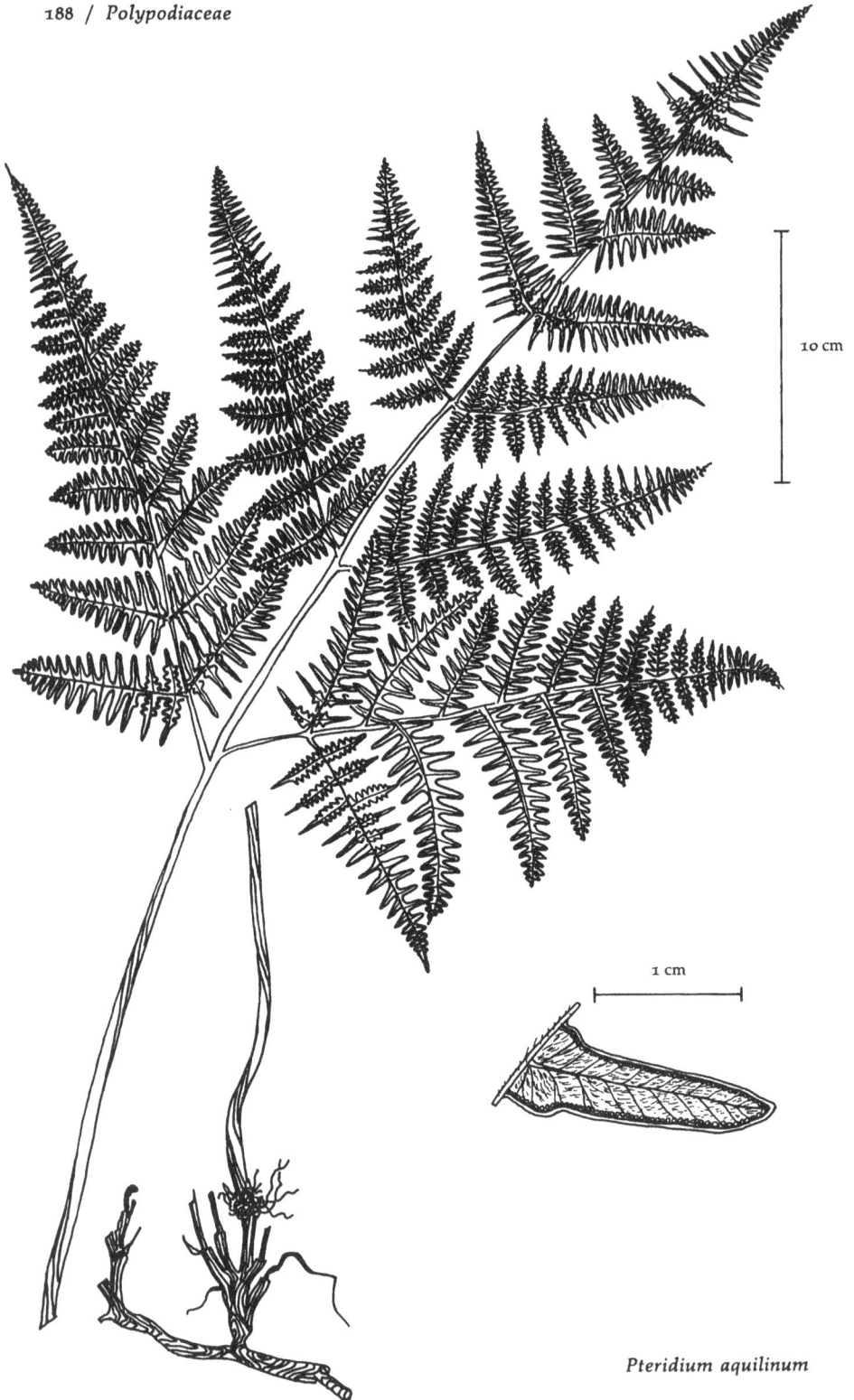

10 cm

1 cm

Pteridium aquilinum

alternate, ultimate divisions very numerous, oblong to linear with revolute margins. Sori marginal, mostly continuous, covered by a false outer indusium formed by the revolute margin and a minute, often nearly obsolete, hyaline inner indusium. ($n = 52$)

HABITAT Open places, generally in subacid soil.

RANGE Cosmopolitan in temperate to tropical regions of both hemispheres.

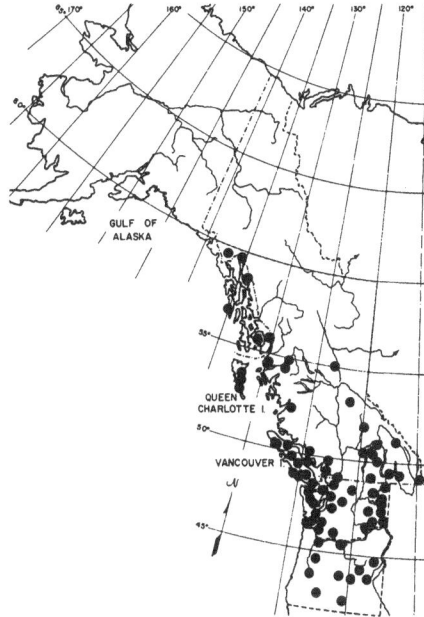

COMMENTS Found generally throughout except at high altitudes. Our geographic form is var. *pubescens* Underw. (var. *lanuginosa* Bong.), characterized by the indusia being ciliate and pubescent with the ultimate divisions of the fronds woolly beneath.

THELYPTERIS Schmidel, Icon. Pl. (ed. Keller) 45, *Plate 11*, 1762

Rhizomes slender, wide-creeping, very dark, sparsely scaly. Scales entire, fibrous, ciliate as a rule. Stipes slender, straw-coloured. Blades thin, all alike or sometimes subdimorphic, pinnate-pinnatifid or 2-pinnate; divisions entire or nearly so, at most sparingly scaly or not at all, pubescent with sharp pointed, unicellular hairs at least on the rachis and main veins above; ultimate divisions ciliate. Veins few, simple or once forked, reaching the margins and running into the teeth (if present). Sori dorsal on the veins, medial or supramedial. Indusia small or sometimes absent, kidney-shaped, mostly ciliate or glandular. (Named from Greek *thelus,* female, and *pteris,* fern, perhaps from their rather slender appearance in contrast to the male fern.)

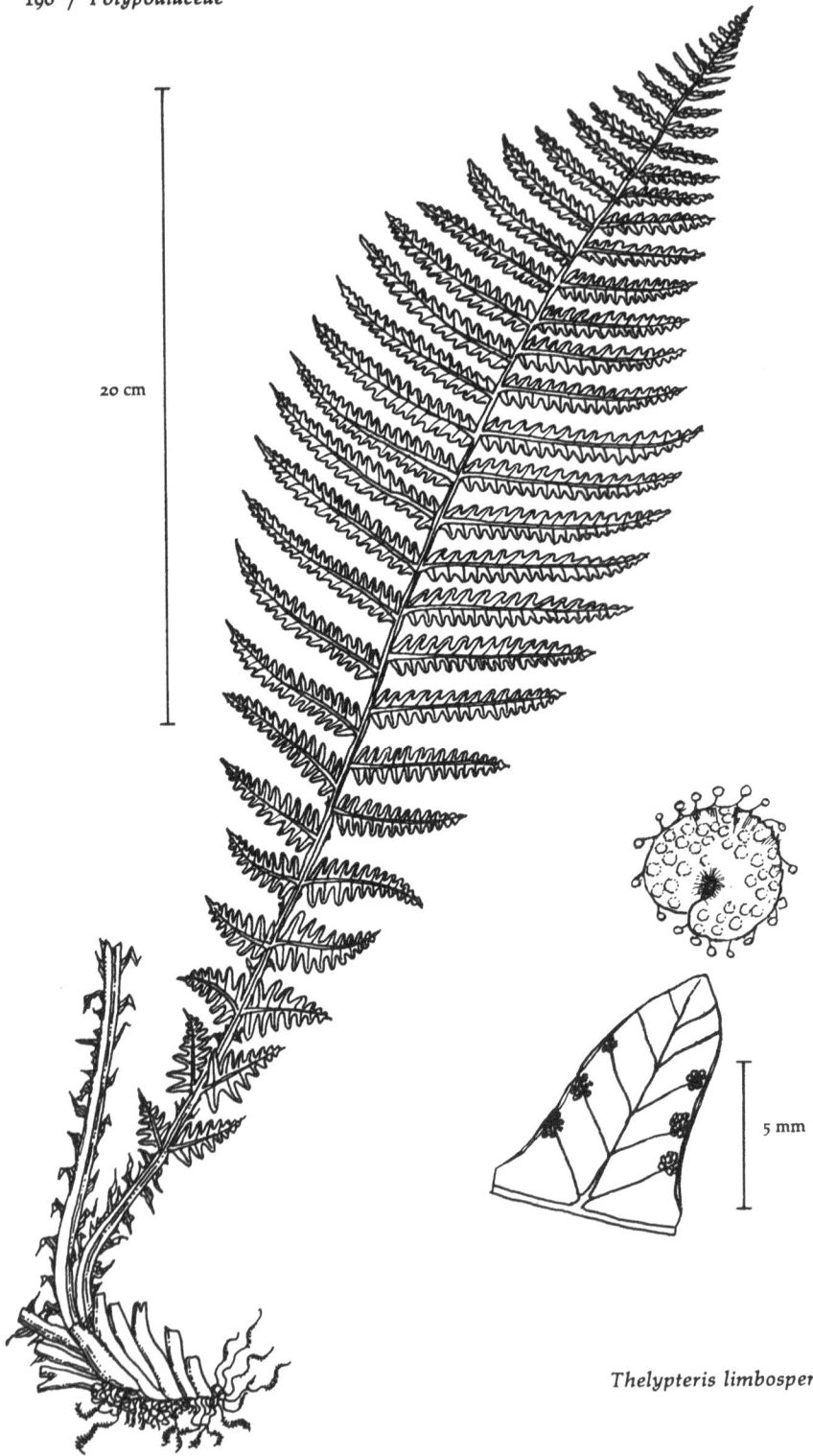

20 cm

5 mm

Thelypteris limbosperma

A very large cosmopolitan genus of temperate and subtropical regions.

1   Indusium lacking; blades narrowly deltoid; rhizome wide-spreading, so fronds scattered.   *T. phegopteris*

1   Indusium present; blades lanceolate or oblanceolate, distinctly narrowed below; rhizome much less spreading, so fronds somewhat clustered.

   2   Ultimate divisions 4–5 mm broad, margins distinctly hyaline-papillose; veins often forked.   *T. limbosperma*

   2   Ultimate divisions about 2 mm broad, margins only slightly hyaline; veins mostly simple.   *T. nevadensis*

Thelypteris limbosperma (All.) H. P. Fuchs

Am. Fern J. **48**:144, 1958; Macoun, Cat. Can. Pl. pt. 5, 272, 1890; Henry, Fl. S. British Columbia 6, 1915; Piper & Beattie, Fl. Northwest Coast 7, 1915; Maxon [in] Abrams, Ill. Fl. Pac. States **1**:15, *Fig. 28*, 1923; Frye, Ferns Northwest 126, *Fig. XLIII*, 1934; Anderson, Fl. Alaska 16, 1959; Hultén, Fl. Alaska 45, 1968; Calder & Taylor, Fl. Queen Charlotte Islands pt. 1, 160, 1968.
Aspidium oreopteris (Ehrh.) Sw., Schrad. Bot. J. **1800**:35, 1801.
Dryopteris montana (Vogler) Ktze., Rev. Gen. Pl. **2**:813, 1891.
Dryopteris oreopteris (Ehrh.) Maxon, Proc. U. S. Nat. Mus. 23:638, 1901.
Thelypteris oreopteris (Ehrh.) Slosson [in] Rydb., Fl. Rocky Mts. 1043, 1917.

Rhizome short, rather stout, more or less ascending, with lanceolate, yellowish-brown scales. Fronds few, grouped at the end of the rhizome, up to 10 dm tall, deciduous. Stipes relatively stout, grooved above, straw-coloured from a dark scaly base, about 1/4 as long as the blade. Blades lanceolate or oblanceolate, abruptly acuminate, tapering to a narrow base, firm, yellowish-green, aromatic when crushed, lower surface with numerous brownish-yellow glands, pinnate-pinnatifid. The longest pinnae (about the middle of the blade) linear-lanceolate, narrowed at the base but not stalked, shorter below; the lower distant, deltoid, quite short; all deeply pinnatifid, main vein soft, white, hairy below, at least while young. Ultimate divisions oblong, wide at the base, obtuse or subacute, sinuate-crenate or subentire, membranous margins often recurved, finely hyaline-papillose. Veins simple or once-forked. Sori small, close to the margin; indusia small, thin, early deciduous, irregularly toothed, glandular. ($n = 34$)

H A B I T A T Open rocky woods, especially mountain sides and subalpine meadows in acid soil.

C O M M E N T S Coastal except in Washington where it has been found occasionally in the Cascade Mountains. Readily distinguishable by the size and shape of the frond. It is a much taller and coarser species than *T. nevadensis*. Our form is apparently identical with that of Eurasia.

15 cm

2 mm

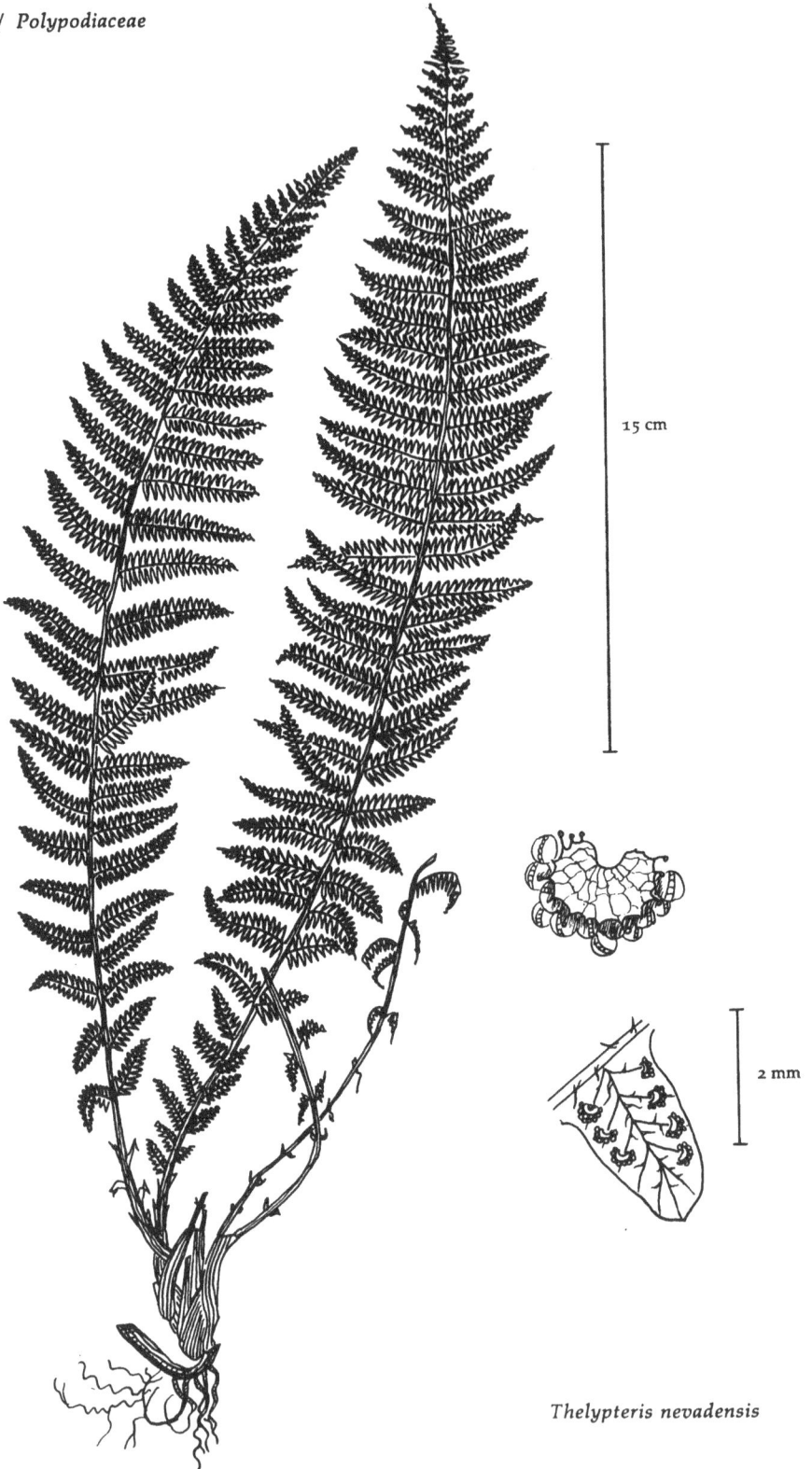

Thelypteris nevadensis

RANGE Alaska south to
northern Washington;
Eurasia.

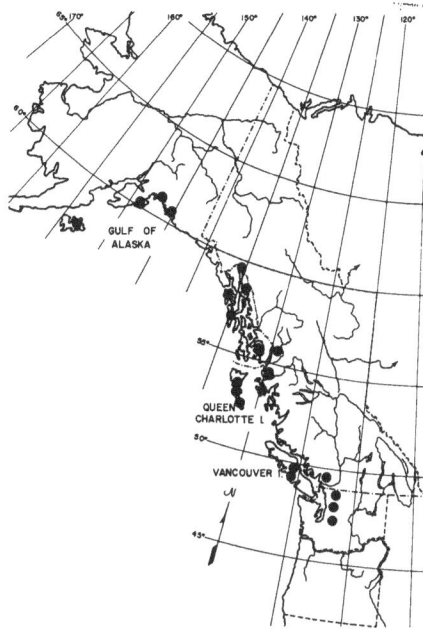

Thelypteris nevadensis (D. C. Eaton) Clute

Am. Fern J. **48**:139, 1958; Maxon [in] Abrams, Ill. Fl. Pac. States **1**:15, *Fig. 29*, 1923; Frye, Ferns Northwest 126, 1934; Peck, Man. Higher Pl. Ore. 47, 1941.
Aspidium nevadense D. C. Eaton, Ferns N. Am. **1**:73, 1878.
Dryopteris nevadensis (Eaton) Underw., Nat. Ferns, 4th ed., 113, 1893.
D. oregana C. Chr., Ind. Fil. 281, 1905.
T. oregana (C. Chr.) St. John, Proc. Biol. Soc. Wash. **41**:192, 1928.

Rhizome slender, horizontal and nearly superficial, covered by the persistent old stipe bases, scales yellowish-brown, ovate, somewhat toothed. Fronds erect in a compact tuft, membranous, deciduous. Stipes slender, very short, straw-coloured, almost glabrous. Blades pinnate-pinnatifid, narrowly elliptic-lanceolate, up to 60 cm long, acuminate, long attenuate, very narrow below. Pinnae linear to linear-lanceolate, horizontal, sessile, acuminate to almost caudate, the larger upper ones close together, the lower distant and very much reduced. Ultimate divisions oblong, somewhat oblique, close together, usually obtuse, entire or slightly crenate, minutely resinous-glandular and with a few early deciduous cilia. Veins mostly unbranched, slightly hairy beneath. Sori close to the margin, small; indusia very small, long ciliate, glandular. ($n = 26$–27)

HABITAT Wooded hillsides, damp meadows, and edges of streams.

COMMENTS Apparently not common, known from a single station in British Columbia and sparingly in the Cascades of Washington and Oregon.

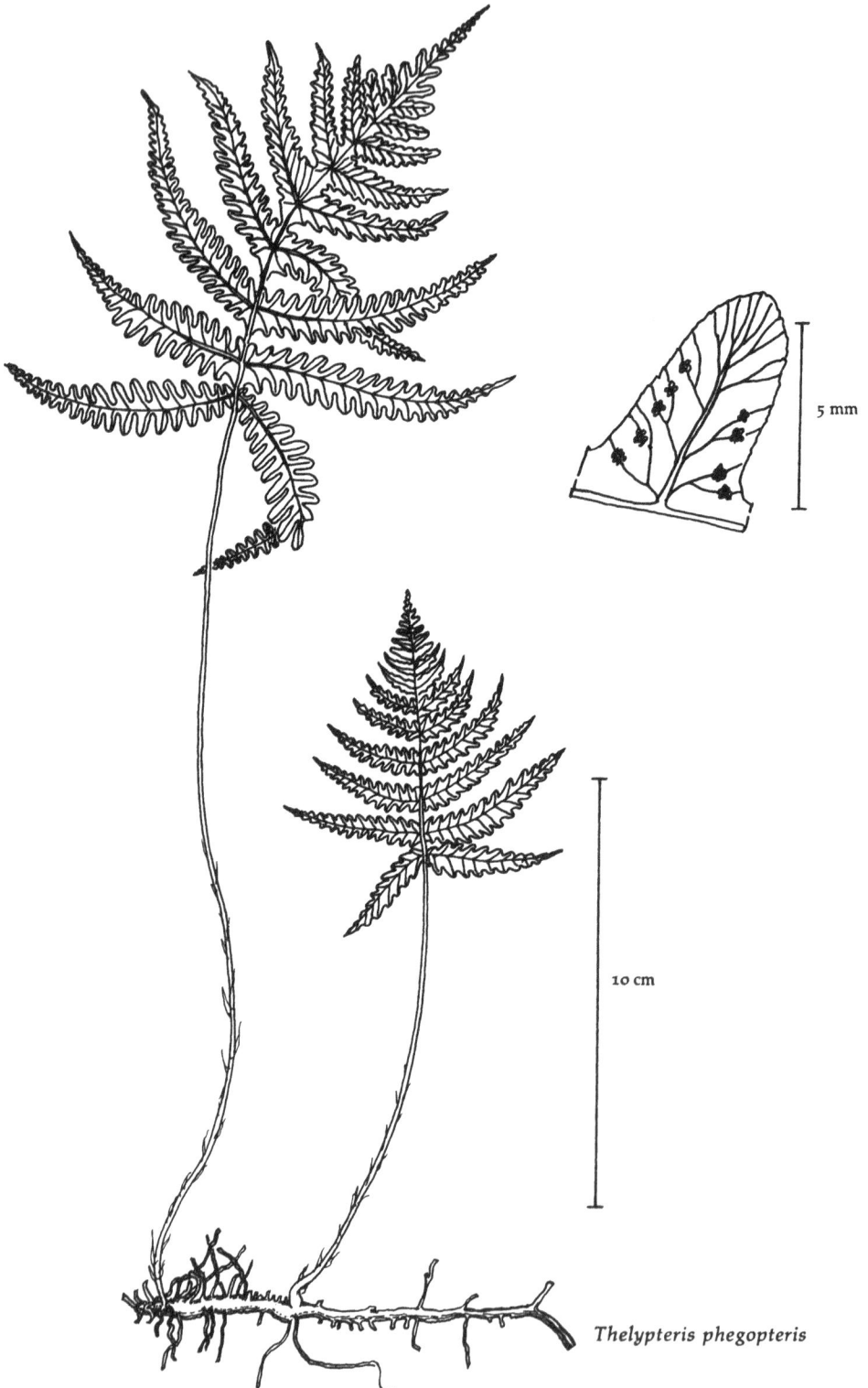

Thelypteris phegopteris

RANGE Southern British
Columbia to California.

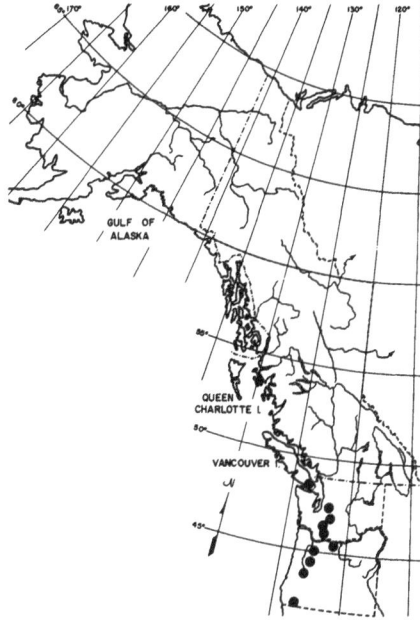

Thelypteris phegopteris (L.) Slosson

[In] Rydb., Fl. Rocky Mts. 1043, 1917; Macoun, Cat. Can. Pl. pt. 5, 269, 1890;
Henry, Fl. S. British Columbia 4, 1915; Piper & Beattie, Fl. Northwest Coast 3,
1915; Maxon [in] Abrams, Ill. Fl. Pac. States 1:14, *Fig.* 26, 1923; Frye, Ferns
Northwest 132, *Fig.* XLVIII, 1934; Peck, Man. Higher Pl. Ore. 46, 1941; Morton
[in] Gleason, Ill. Fl. 1:48, *Fig.*, 1952; Anderson, Fl. Alaska 15, 1959; Hultén, Fl.
Alaska 46, 1968; Calder & Taylor, Fl. Queen Charlotte Islands pt. 1, 161, 1968.
Aspidium phegopteris (L.) Baumg., Enum. Stirp. Transsylv. 4:28, 1846.
Phegopteris polypodioides Fée, Gen. Fil. 243, 1852.
P. phegopteris (L.) Keyserl, Polyp. Cyath. Herb. Bung. 50, 1873.
Dryopteris phegopteris (L.) C. Chr., Ind. Fil. 284, 1905.

Rhizome slender, long-creeping just below the surface of the ground with
brown, lanceolate, ciliate scales. Fronds solitary, rather widely separated,
up to 40 cm tall, deciduous. Stipes pale straw-coloured, hairy and scaly
throughout, longer than the blades. Blades triangular-ovate to deltoid,
bent nearly at right angles to the stipe, tapering to the tip, pinnate-pinnatifid. Pinnae more or less hairy on both surfaces, lanceolate, the basal
pair inclined forward away from the others, longer or slightly shorter
than the pair above, the remainder decreasing rapidly in length, all (except the lowermost) attached to the rachis by a broad, decurrent base.
Ultimate divisions herbaceous, oblong, obtuse, entire or crenate. Sori
small, submarginal; indusia lacking. (*n* and 2*n* = 90, apogamous)

HABITAT Open damp woods, wet mossy rocks, and crevices, usually
in half shade.

RANGE Alaska to New-
foundland, south to Oregon,
Iowa, Michigan, Ohio,
Tennessee, and North
Carolina; Eurasia.

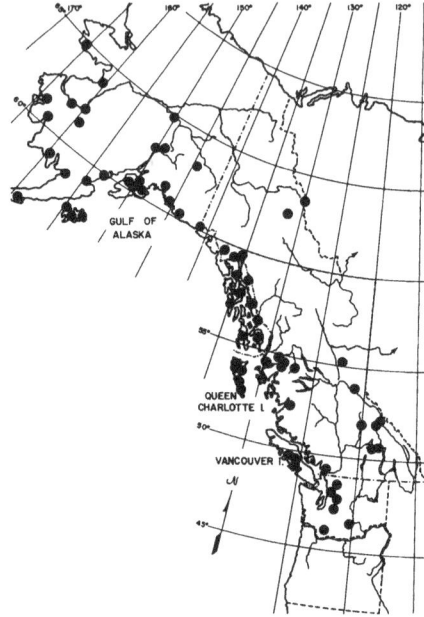

COMMENTS This attractive circumboreal species is found in cool woods
throughout. The characteristic shape of the fronds, particularly the features of the lowest pinnae serve to distinguish it. The species has been
found to be triploid (n and $2n = 90$) and apogamous both in Europe and
North America.

WOODSIA R. Br., Prodr. Fl. Nov. Holl. **1**:158, 1810

Rhizomes short-creeping or ascending with ovate-lanceolate to linear-
lanceolate brown scales. Fronds numerous, usually densely tufted, erect
or spreading. Stipes relatively stout to slender, much shorter than the
blades, in some species articulated at the base and in such cases the stipe
bases are persistent. Blades monomorphic, linear to lance-ovate, pinnate-
pinnatifid or pinnate-2-pinnatifid, glabrous or variously scaly or hairy,
sometimes glandular. Sori on the back of veinlets, round, separated at
first but later often confluent. Indusia thin, arising from below the sporangia and enclosing them with few to many hairlike or narrow or broad
lobes. (Named in honour of Joseph Woods, 1776–1864, an English
botanist.)

A small genus of rock ferns mostly of temperate and boreal regions.

1   Stipes articulated near the base, the joint appearing as a slightly
        thickened, darker ring.

2   Fronds glabrous; stipe very short, scaly at the base; pinnae as broad
     as long.   *W. glabella*

2   Fronds with hairs and scales; stipe longer.

  3   Fronds with only hairs on the lower surface, rarely with a few lanceo-
       late scales at the base of pinnae; middle pinnae 2–3-lobed on a side.
                                                                    W. alpina
  3   Fronds with both hairs and scales on the lower surface; middle pinnae
       3–6-lobed on a side.   *W. ilvensis*

1   Stipes not articulated; rachis not scaly but may be hairy.

  4   Fronds and stipes glabrous or somewhat glandular but not hairy.
                                                                   W. oregana
  4   Fronds with conspicuous white hairs mixed with glands. *W. scopulina*

Woodsia alpina (Bolton) S. F. Gray

Nat. Arr. Brit. Pl. **2**:17, 1821; Anderson, Fl. Alaska 12, 1959; Morton [in]
Gleason, Ill. Fl. **1**:44, *Fig.*, 1952; Hultén, Fl. Alaska 51, 1968.
W. hyperborea (Liljeb.) R. Br., Prodr. Fl. Nov. Holl. **1**:158, 1810.

Rhizome short, ascending, tufted, covered with old stipe bases and
numerous, somewhat broadly lanceolate, dark, papery scales. Fronds up
to 15 cm tall, commonly less than 10 cm, tufted, erect. Stipes 1/3 to 1/2
the length of the blade, straw-coloured to pale brown, dull, with narrow,
chaffy scales, particularly below; distinctly jointed near the base. Blades
somewhat oblong-lanceolate, hairy but not chaffy; rachis with a few hairs
and slender, subulate, rarely lanceolate, scales; pinnate. Pinnae narrowly
triangular-ovate, lobed into 5–7 rounded, nearly entire segments. Sori
few, usually confluent; indusia deeply cut into many very delicate hairlike
lobes longer than the sporangia. ($n = 82$)

H A B I T A T Rock crevices and fine talus, often of calciferous rocks but
not exclusively so.

C O M M E N T S In our area very rare and local, not known from south of
about 59° north latitude. Although it is generally rather similar to *W.
glabella*, it can be distinguished by the proportionally longer stipes, the
hairiness of the stipes and rachis, and the conspicuously long, old stipe
bases. The sterile hybrid between this species and *W. ilvensis* has been
reported from Europe. According to R. M. Tryon (1948) it apparently also
occurs not infrequently on the northern shore of Lake Superior. Brown
(1964) is of the opinion that this species is an allotetraploid derived from
W. ilvensis × *glabella*.

1 mm

5 cm

Woodsia alpina

RANGE Alaska to New-
foundland, south to British
Columbia; Great Lakes
region; northern New York
and northern Vermont;
Greenland; Eurasia.

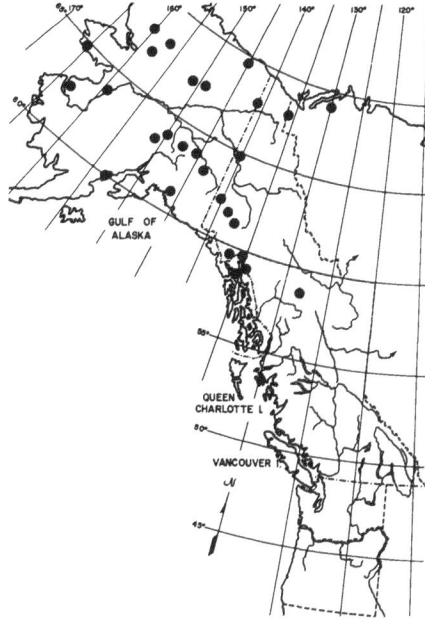

Woodsia glabella R. Br.

Franklin J., App. 754, 1823; Morton [in] Gleason, Ill. Fl. **1**:44, *Fig.*, 1952; Anderson, Fl. Alaska 12, 1959; Wiggins & Thomas, Fl. Alaskan Arctic Slope 41, 1962; Hultén, Fl. Alaska 52, 1968.

Rhizome short, ascending, clustered with numerous, very thin, ovate, chestnut-coloured scales. Fronds usually not more than 10 cm tall, tufted. Stipe very slender, yellowish-green with a few chestnut-coloured scales in the lower part, distinctly jointed at, or near, the base. Blades narrowly linear-lanceolate, pale green with lighter rachis, completely glabrous, pinnate. Pinnae obtuse, crenately lobed into 3–7 short, obtuse lobes. Sori few but frequently confluent; indusia deeply cleft into a few, very delicate hairlike lobes that conspicuously overtop the sporangia. ($n = 39$)

HABITAT Damp, shaded, mossy crevices and ledges, usually of limestone.

COMMENTS This delicate little fern has been found only in the northern part of our area; its southern limit is about 54° north latitude in British Columbia. It is apparently limited to calcareous, or at least basic rocks. It grows under conditions similar to those required by *Asplenium viride* and can easily be confused with that species. Sterile specimens, however, can be recognized quite readily by their erect, tufted habit of growth and by their stipes with distinctly jointed bases.

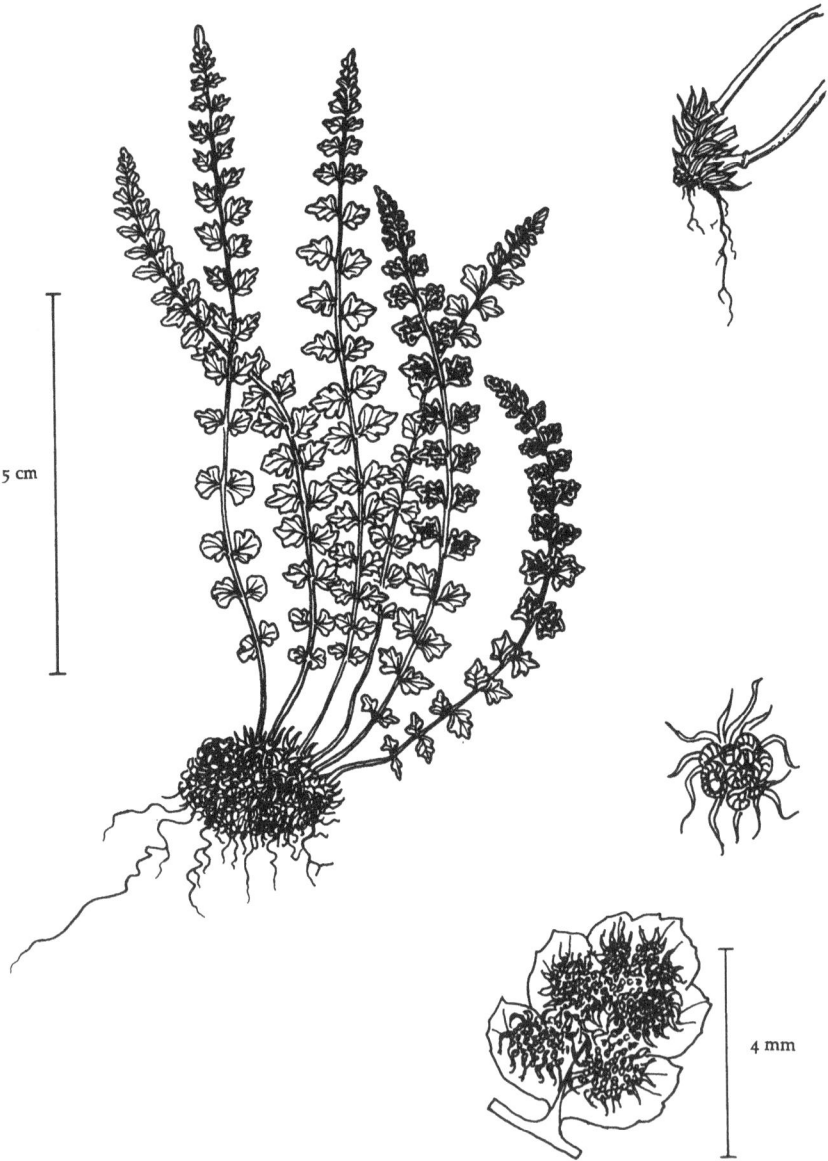

5 cm

4 mm

Woodsia glabella

RANGE Alaska to Newfoundland, south to British Columbia, Great Lakes region, northern New York, and northern New England; Greenland; arctic and alpine Eurasia.

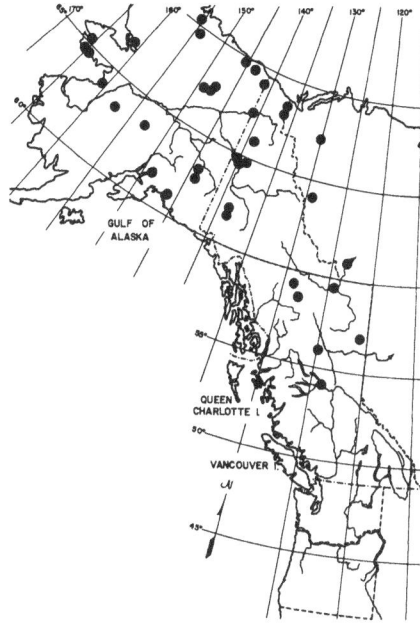

GULF OF ALASKA

QUEEN CHARLOTTE I.

VANCOUVER I.

Woodsia ilvensis (L.) R. Br.

Prodr. Fl. Nov. Holl. 1:158, 1810; Henry, Fl. S. British Columbia 7, 1915; Morton [in] Gleason, Ill. Fl. 1:44, *Fig.*, 1952; Anderson, Fl. Alaska 12, 1959; Wiggins & Thomas, Fl. Alaskan Arctic Slope 42, 1962; Hultén, Fl. Alaska 51, 1968.

Rhizome ascending, covered with persistent old stipe bases and very numerous, brown, lanceolate-attenuate scales. Fronds up to 25 cm tall in dense tufts, erect or somewhat spreading. Stipes stout, shining, dark brown below, paler above, scaly and hirsute, distinctly articulated below the middle, bases persistent densely brown-scaly. Blades lanceolate, dark green, pinnate-pinnatifid, somewhat reduced at the base, acute above, usually chaffy and pilose on the lower surface. Pinnae oblong-ovate to oblong-lanceolate, pinnatifid. Ultimate divisions oblong, obtuse, the larger crenately lobed, margins revolute, sparsely ciliate. Sori numerous, frequently confluent; indusia with numerous, long, hairlike segments conspicuously overtopping the sporangia. ($n = 41$)

HABITAT Dry rock crevices and exposed talus slopes, mostly in acid soil.

COMMENTS In the western part of its range this species comes only as far south as central British Columbia. Its distribution is probably not as scattered as the rather limited number of collections would suggest. Good recognition marks in the field are its abundantly scaly and hairy fronds with a distinct joint in the lower half of the stipe. R. M. Tryon, Jr., reports

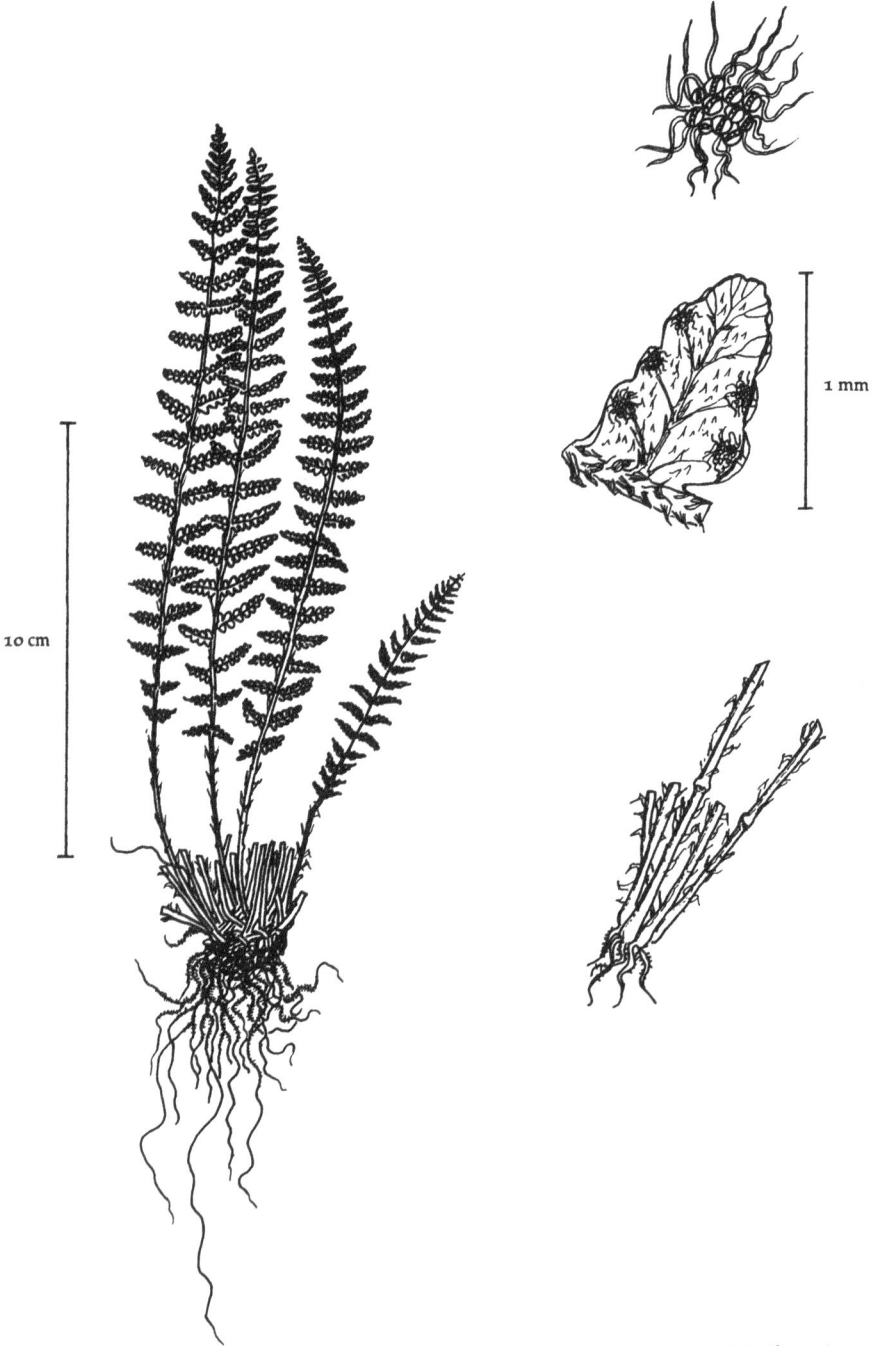

10 cm

1 mm

Woodsia ilvensis

(1948) that in the Lake Superior region *W. ilvensis* hybridizes not uncommonly with other species present. Three hybrids, all sterile, are known, viz., *W. ilvensis* × *alpina*, *W. ilvensis* × *glabella*, and *W. ilvensis* × *oregana* (var. *cathcartiana*).

RANGE Alaska to Newfoundland, south to British Columbia, Alberta, northern Iowa, northern Illinois, Michigan, Pennsylvania, and North Carolina; Greenland; Eurasia.

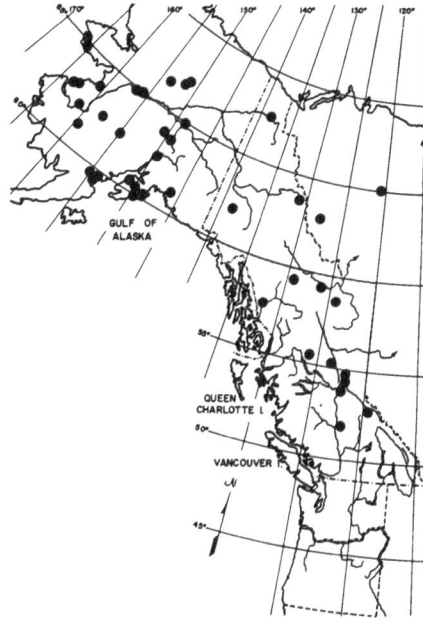

Woodsia oregana D. C. Eaton

Can. Nat. II. **2**:90, 1865; Macoun, Cat. Can. Pl. pt. 5, 284, 1890; Henry, Fl. S. British Columbia 7, 1915; Piper & Beattie, Fl. Northwest Coast 8, 1915; Maxon [in] Abrams, Ill. Fl. Pac. States **1**:6, *Fig.* 11, 1923; Frye, Ferns Northwest 109, *Fig.* XXXI, 1934; Peck, Man. Higher Pl. Ore. 46, 1941; Morton [in] Gleason, Ill. Fl. **1**:46, *Fig.*, 1952; St. John, Fl. Southeast. Wash. 9, 1963.

Rhizome short-creeping, densely covered with pale brown, linear-lanceolate scales. Fronds numerous, erect, tufted, up to 30 cm tall. Stipes stiff but rather slender, straw-coloured from a darker base, glabrous, scaly only towards the base. Blades bright green, lance-linear to lanceolate, acuminate, slightly contracted at the base, pinnate-pinnatifid or pinnate-2-pinnatifid, glabrous, sparsely or sometimes quite glandular beneath. Pinnae subopposite, triangular-oblong, blunt, rather distant, sessile. Ultimate segments of the principal pinnae close together, blunt, often pinnatifid, decurrent, margin crenate-serrulate, usually reflexed. Sori often partly concealed by the reflexed margin; indusia minute, divided nearly to the base into several short, narrow, almost threadlike segments quite concealed by the sporangia. ($n = 38$)

HABITAT Dry crevices and talus slopes.

5 cm

2.5 mm

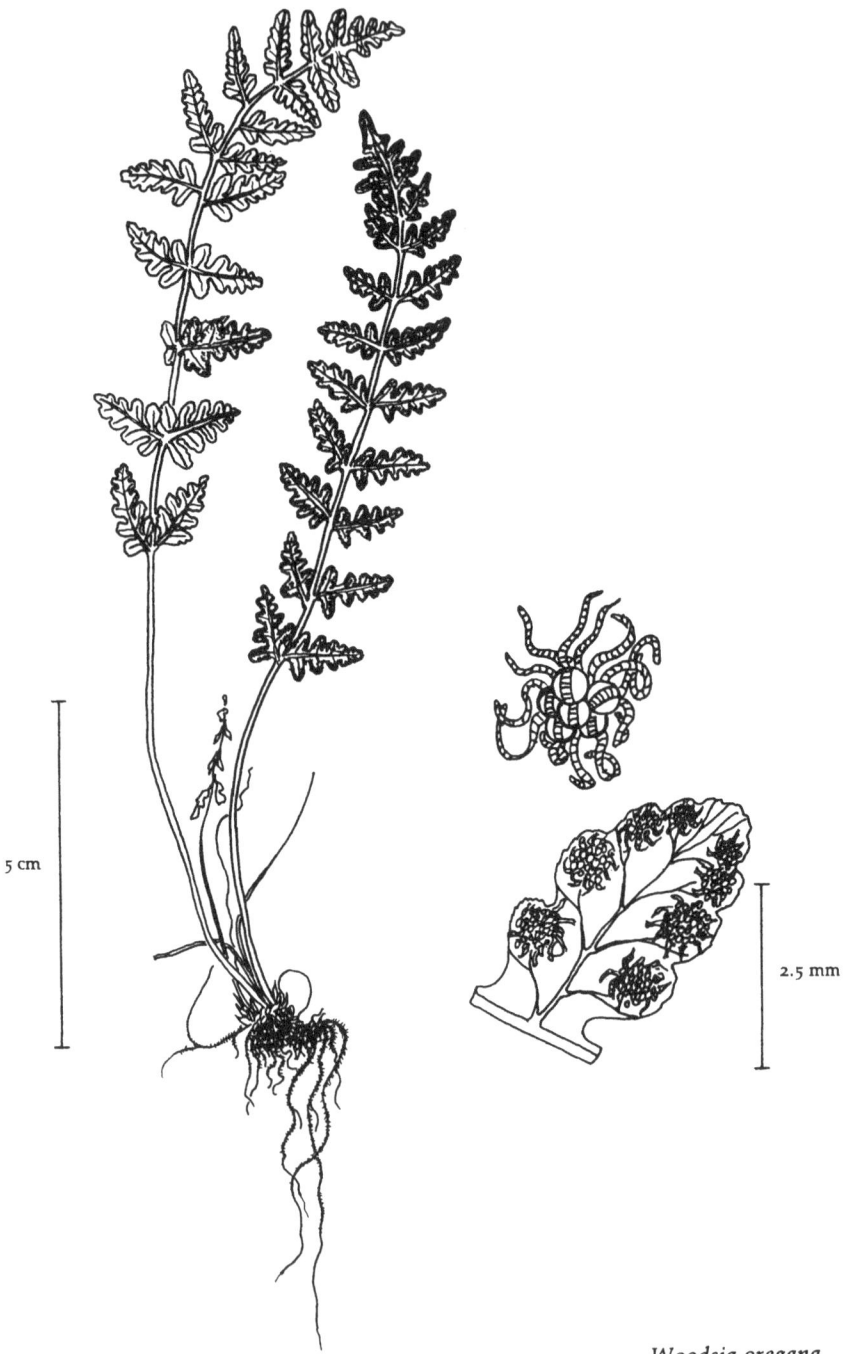

Woodsia oregana

RANGE British Columbia
and Alberta, south to Baja
California, Arizona, New
Mexico, western Oklahoma,
Nebraska, northern Iowa,
northern Wisconsin, upper
Great Lakes region; Rimouski
County, Quebec.

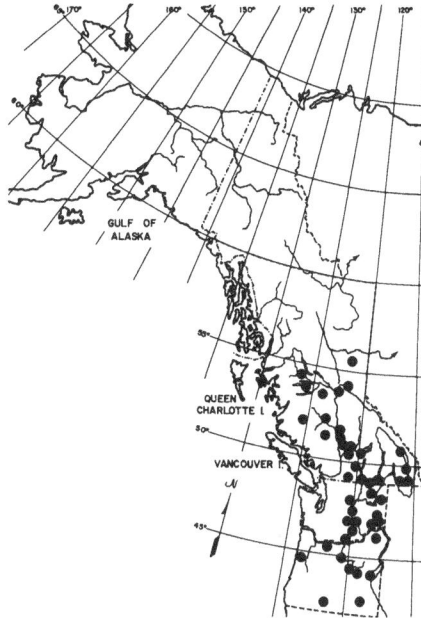

COMMENTS Quite generally distributed in dry situations as far north as about 56° north latitude. It can be distinguished from *W. scopulina*, which grows in similar habitats, by its smaller size and narrower glabrous fronds lacking flat, white, jointed hairs. R. M. Tryon, Jr. (1948) has described from the Lake Superior region *W. × maxoni* which he considers to be the hybrid *W. oregana* (var. *cathcartiana*) × *scopulina*.

Woodsia scopulina D. C. Eaton

Can. Nat. II 2:91, 1865; Macoun, Cat. Can. Pl. pt. 5, 284, 1890; Henry, Fl. S. British Columbia 7, 1915; Piper & Beattie, Fl. Northwest Coast 8, 1915; Maxon [in] Abrams, Ill. Fl. Pac. States 1:6, *Fig.* 10, 1923; Frye, Ferns Northwest 109, *Fig.* XXXI, 1934; Peck, Man. Higher Pl. Ore. 46, 1941; Morton [in] Gleason, Ill. Fl. 1:46, *Fig.*, 1952; Anderson, Fl. Alaska 12, 1959; St. John, Fl. Southeast. Wash. 9, 1963; Hultén, Fl. Alaska 50, 1968.

Rhizome stoutish, short-creeping, densely covered with lance-ovate, hair-tipped, pale brown scales often with a dark central stripe. Fronds numerous in quite large tufts, erect, up to 40 cm tall. Stipes firm, light straw-coloured above, chestnut below, pilose, only scaly at the base, not jointed. Blades linear to oblong-lanceolate, short acuminate, somewhat narrowed towards the base, 2-pinnate or pinnate-pinnatifid, pilose with few to many flat, whitish hairs mixed with glands, not scaly. Pinnae oblong-lanceolate to oblong-ovate, acuminate, rather distant, sessile. Ultimate divisions oblong-obtuse, crenately lobed or pinnatifid, rather close together. Sori submarginal; indusia deeply cut into a number of flat, lanceolate, linear-attenuate segments concealed by the sporangia. ($n = 38$)

H A B I T A T Dry crevices and talus slopes.

C O M M E N T S Widely spread and locally quite common, particularly in the dry interior regions. It can be distinguished from *W. oregana* by its viscid, very brittle, pilose fronds. Its natural hybrid with *W. oregana* var. *cathcartiana* has already been mentioned. So far this hybrid has not been reported from the West; it should be looked for wherever the two species are found growing together.

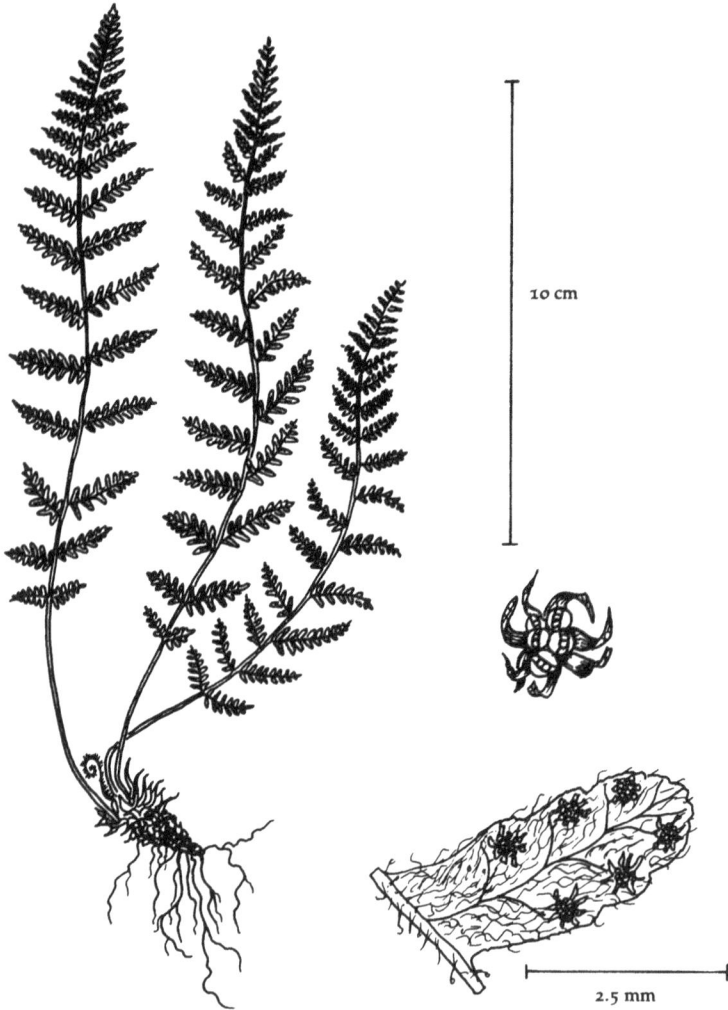

10 cm

2.5 mm

Woodsia scopulina

RANGE Alaska to Saskatchewan, south to California, New Mexico, and South Dakota; upper Great Lakes region and Gaspé, Quebec.

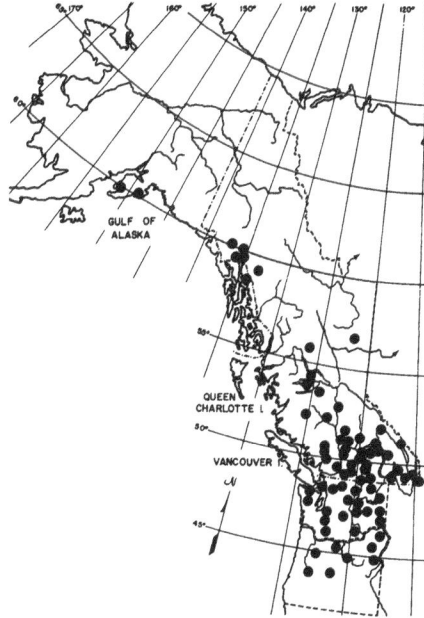

WOODWARDIA J. E. Smith, Mém. Acad. Turin 5:411, 1793

Mostly large, coarse ferns of bogs and wet shady situations. Rhizome very stout; widely creeping or sometimes more or less erect, with brown, ovate, acuminate scales. Fronds stiffly erect or somewhat recurved, several or many in a crown. Stipes usually about equalling the blades. Blades pinnate; pinnae pinnatifid or lobed, segments alternate, glabrous, margins serrulate, sparingly scaly on the rachis and main veins. Sori oblong to linear, in one or more chainlike rows parallel to and near the midrib; indusia persistent, at first completely covering the sporangia, later reflexed, opening on the side next to the midrib. (Named in honour of Thomas J. Woodward, 1745–1820, an English botanist.)

A small genus of temperate and tropical regions of the northern hemisphere.

Woodwardia fimbriata J. E. Smith

[Ex] Rees, Cycl. 38, no. 6, 1818; Henry, Fl. S. British Columbia 5, 1915; Piper & Beattie, Fl. Northwest Coast 5, 1915; Maxon [in] Abrams, Ill. Fl. Pac. States 1:17, *Fig.* 33, 1923; Alston, Philip. J. Sci. **50**:181, 1932; Frye, Ferns Northwest 148, *Figs.* L, LI, 1934; Peck, Man. Higher Pl. Ore. 49, 1941.
W. chamissoi Brack. [in] Wilkes, U.S. Expl. Exped. **16**:138, 1854.
W. spinulosa auth. in part, not Mart. & Gal.

50 cm

5 cm

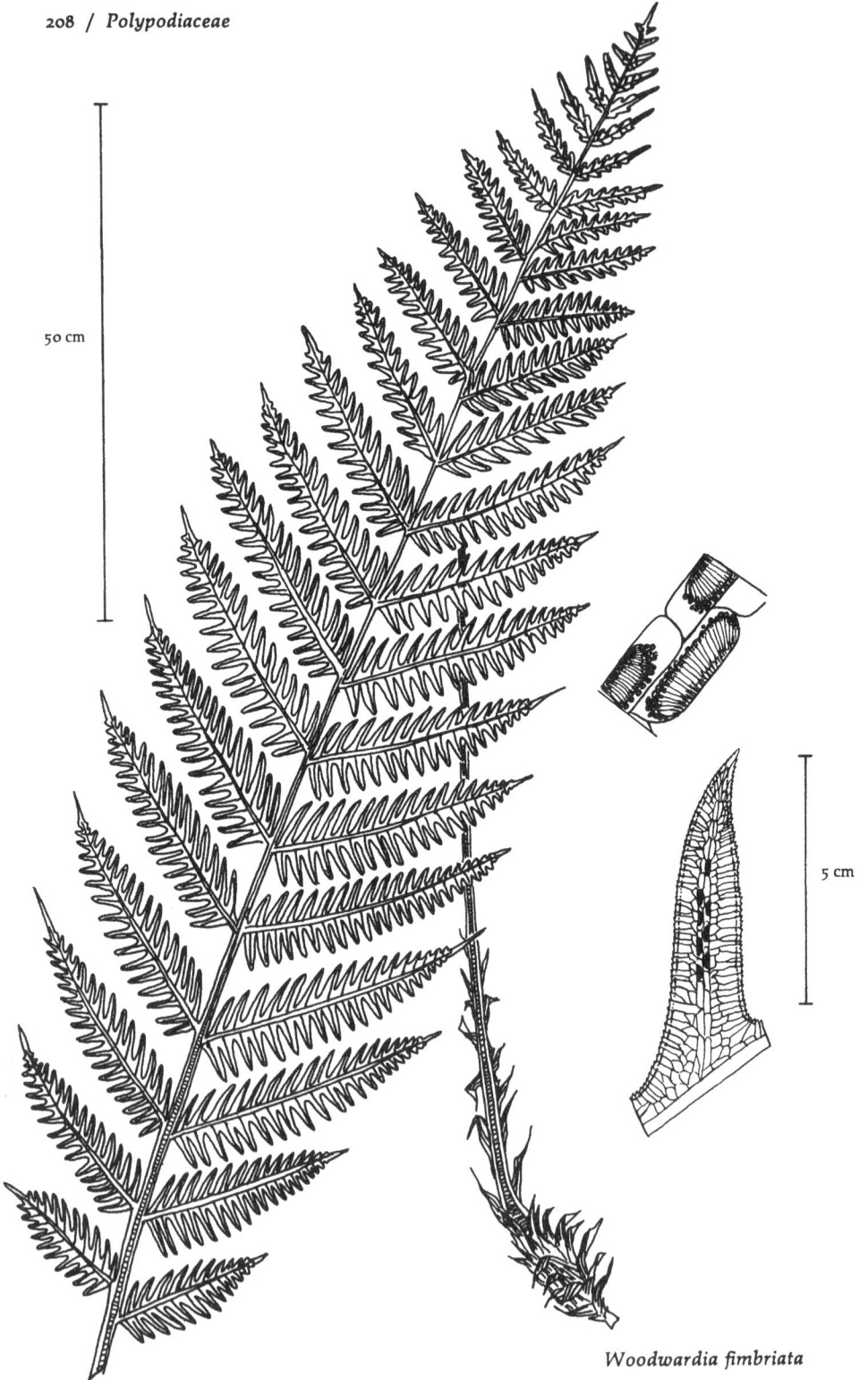

Woodwardia fimbriata

Rhizome very large and stout, woody, ascending, with conspicuous lance-attenuate, glossy, bright chestnut scales. Fronds evergreen, more or less erect, very large and coarse, up to 3 m tall. Stipes short and stout, light straw-coloured from a darker base. Blades broadly linear-oblong to oblong-lanceolate, rather abruptly acuminate, narrowed towards the base, pinnate. Pinnae linear-oblong to ovate, long acuminate, deeply and obliquely pinnatifid. Ultimate divisions narrowly triangular to linear, sub-falcate, attenuate, irregularly and shallowly crenate or lobed, serrate-spinulose, paler on the lower surface. Sori in 2 rows, very large and conspicuous. ($n = 34$)

HABITAT Damp shady woods and banks.

COMMENTS A coarse species of damp forests; ours often coastal. Its rows of elongated sori are unmistakable.

RANGE Southern British Columbia, south to California and Arizona.

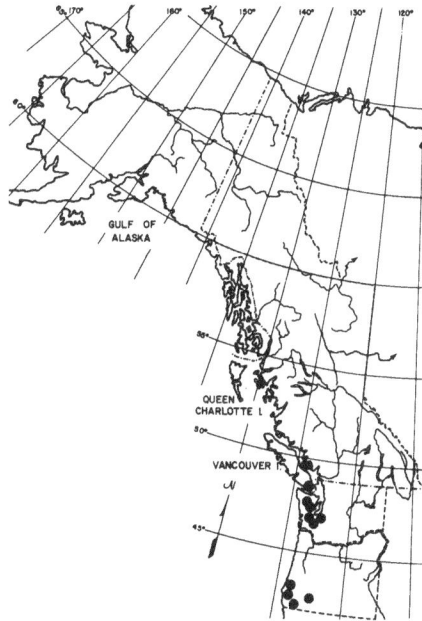

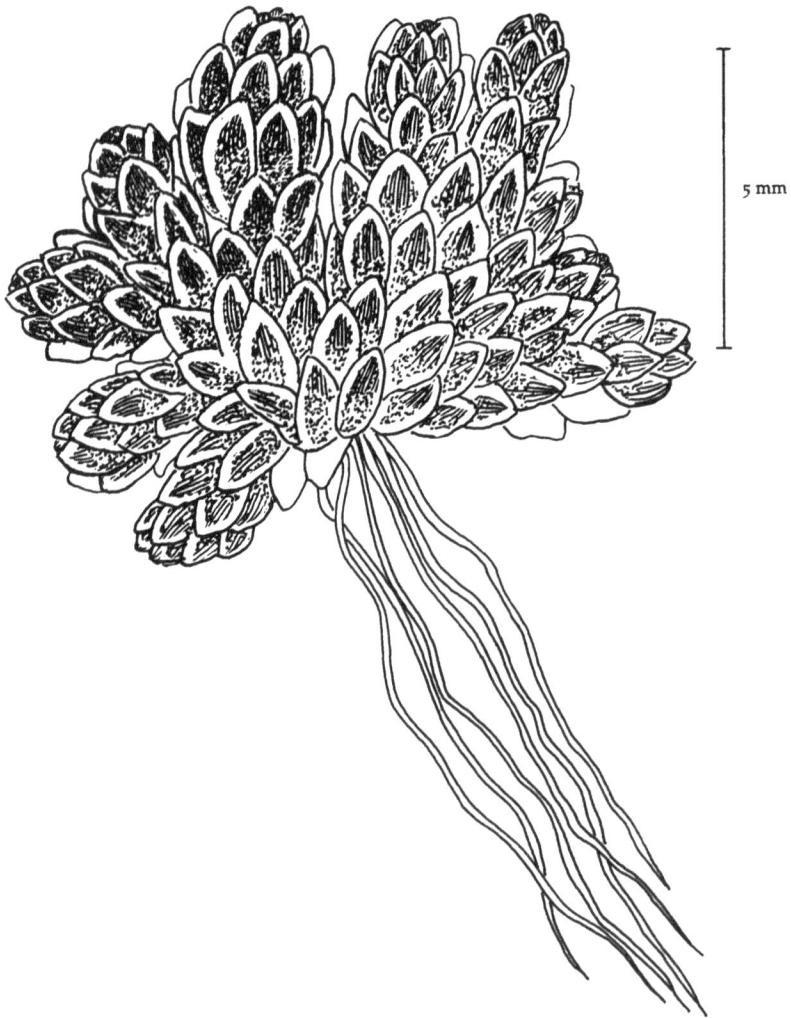

5 mm

Azolla mexicana

Salviniaceae

Small, annual aquatic floating ferns with more or less elongated, and sometimes branching, filiform rhizomes with simple roots, or essentially stemless with some of the leaves modified as roots. Leaves distichous, small, papillose or tuberculate. Sporocarps soft with a thin wall formed by the indusium, borne two or more on a common stalk. Sporangia of two kinds, producing either very numerous microspores or a few, or single, megaspores.

A family of only two genera and a small number of species, widely distributed, mostly in tropical and subtropical regions.

A Z O L L A Lam., Encycl. **1**:343, 1783

Free-floating water plants. Stems pinnately branched from which drop down inconspicuous, unbranched threadlike roots; branches covered with minute, imbricated leaves. Leaves 2-lobed, the lower mostly without chlorophyll and only 1 cell thick; the upper lobe papillose, green or reddish. Sporocarps (sori) in pairs in leaf axils, often unlike, each sorus enclosed by an indusium. Microsporocarps large and globose. Each microsporangium contains 32 or 64 microspores aggregated into 4 to 10 spore masses (massulae) which when released display, in New World species, peculiar barb-tipped hairs (glochidia). Megasporocarp much smaller, acorn-shaped, each containing a single megasporangium with a single spore.

A small genus of less than a dozen species of wide distribution in temperate and tropical regions. The species can be definitely identified only when in the fertile condition and then identification requires the aid of a compound microscope to examine the glochidia.

Azolla mexicana Presl

Abh. Bohm. Ges. Wiss. v **3**:150, 1845; Macoun, Cat. Can. Pl. pt. 5, 294, 1890 (as *A. caroliniana*); Svenson, Am. Fern J. **34**:81, *Plates* 7 & 8, 1944; Mason, Fl. Marshes Calif. 31, *Fig.* 6, 1957; Morton [in] Gleason, Ill. Fl. **1**:22, *Fig.*, 1952; Knobloch & Correll, Ferns Chihuahua 51, *Plate* 7, 1962.

Plants more or less compact, up to 3 cm in diameter, dichotomously branched. Upper leaf lobes imbricated, somewhat variable in shape but mostly broad rhombic, apex broadly rounded, mostly less than 1 mm long, narrow hyaline margin; lower leaf lobe usually much larger than the upper. Microsporangia generally with 4 massulae, glochidia always many septate. Basal portion of megaspore pitted.

HABITAT Ponds and edges of slow streams.

RANGE Southern British Columbia to Bolivia; Nevada, Utah, Missouri, Illinois, and Wisconsin.

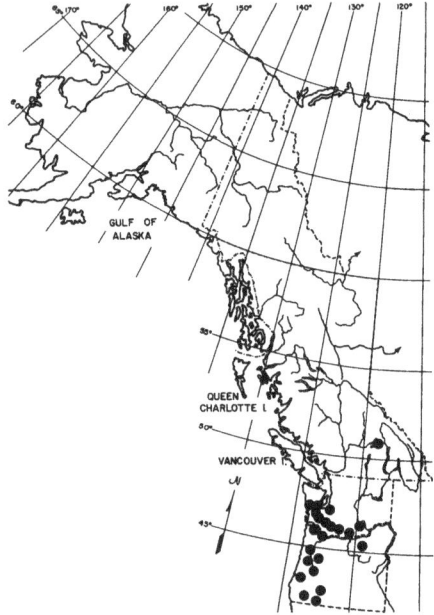

COMMENTS According to Svenson (1944) this is the species of *Azolla* found in our area. The separation from *A. filiculoides* is based on technical details that are only evident in fertile material and unfortunately this is comparatively rare. Very active vegetation reproduction takes place during the growing season. In the autumn the plants turn a bright brick red.

Selaginellaceae BEAUV.

Low, depressed or creeping, mosslike plants (our species). Stems slender, more or less dichotomously branched. Roots few, very fine and thread-like. Leaves small, in 4–6 rows, very numerous, usually imbricate, simple, lanceolate or subulate, all alike and arranged in many ranks, or of two kinds in four ranks. Cones terminal, four-sided or terete in a few species, sporophylls usually similar to the vegetative leaves, each bearing a single axillary sporangium. Sporangia reniform, more or less sessile, of two types, usually the upper ones producing numerous microspores and the lower 1–4 large megaspores. Spores tetrahedral, megaspores yellowish or white with a thick, sculptured wall, microspores reddish or orange, spinulose.

The family contains the single genus *Selaginella* of about 500 species of world-wide, but mostly tropical, distribution. (The name is a diminutive of Selago, a classical name for some species of *Lycopodium* which many of the species resemble superficially.)

SELAGINELLA Beauv., Prodr. Aethéog. 101, 1805

Characters are those of the family.

1	Vegetative leaves clearly in 4 ranks and of 2 sizes, not more than twice as long as broad, rounded-obtuse at apex, or at most somewhat cuspidate. *S. douglasii*
1	Vegetative leaves of one size spirally arranged in many ranks, lanceolate, usually somewhat ligulate and setate.
2	Stems very slender, creeping, not forming mats or festoons; cones upright, conspicuous, thicker than the stem, up to 3 cm long; vegetative leaves spinulose-margined. *S. selaginoides*
2	Stems terrestrial and forming mats or epiphytic and festooned; cones relatively inconspicuous, not appreciably thicker than the stem; vegetative leaves definitely setate.
3	Leaves on stem abruptly adnate and differing in colour from it; ligulate; usually forming somewhat loose mats; leaves rich green and spreading on damp sites, grey-green and appressed on drier sites. *S. wallacei*
3	Leaves decurrent on all sides of the stem.
4	Stems forming flat mats with discrete branches or pendent forming festoons; apex of upper leaves herbaceous to slightly fleshy, in profile,

plane to abruptly bevelled; setae of vegetative leaves nearly 1/2 as long as the blade.

5   Plants usually epiphytic; stems lax and pendent; leaves adnate to stem for nearly 1/2 their length; branches strongly curled in dormant state.   *S. oregana*

5   Plants terrestrial; upper leaves adnate to stem for less than 1/4 of their length; branches not or slightly curled in dormant state.

6   Apex of upper leaves in profile, nearly plane to abruptly bevelled, or, if truncate, the setae lutescent and stem forming compact flat mats with discrete branches.   *S. densa*

6   Stems forming rounded cushion mats with intricate branches; apex forming open spreading mats with intricate branches.   *S. sibirica*

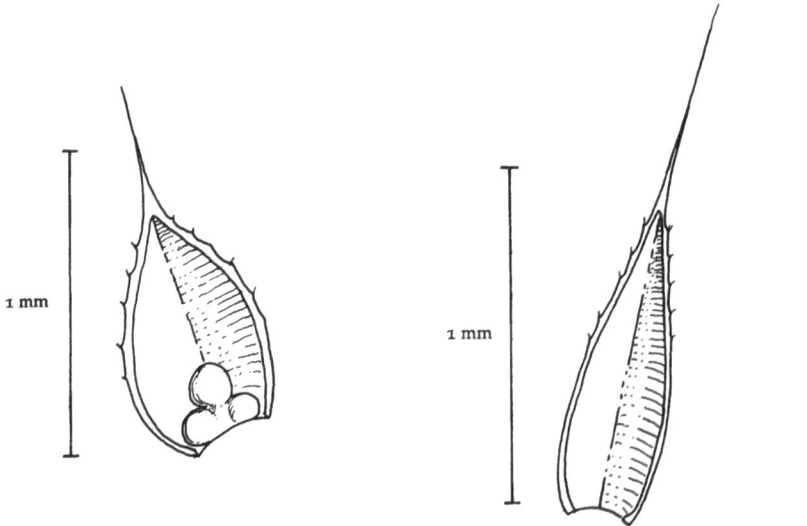

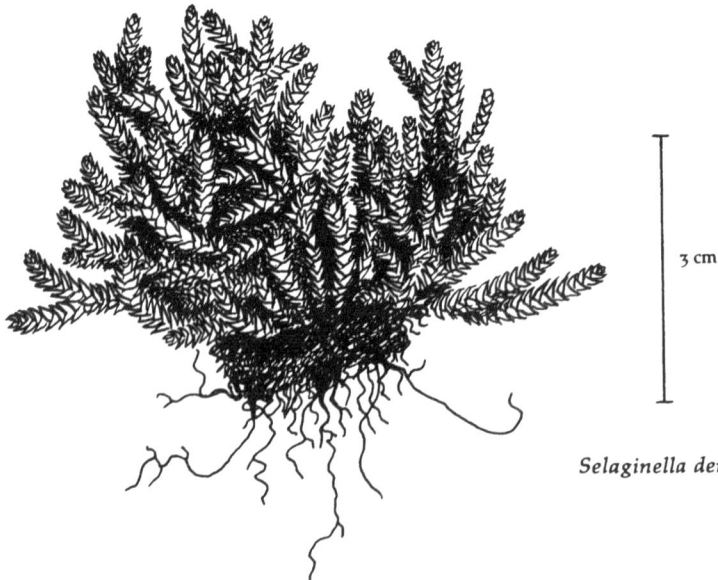

Selaginella densa

4   Apex of upper leaves truncate in profile, setae white to tawny; stems
    of upper leaves fleshy, in profile, truncate or subtruncate; setae of
    vegetative leaves usually less than 1/4 as long as the blade.

S. watsonii

Selaginella densa Rydb.

Mem. N.Y. Bot. Gard. **1**:7, 1900; Macoun, Cat. Can. Pl. pt. 5:292, 1890; Henry,
Fl. S. British Columbia 10, 1915; Piper & Beattie, Fl. Northwest Coast 15, 1915
(as *S. rupestris* (L.) Spring) in part; Maxon [in] Abrams, Ill. Fl. Pac. States **1**:48,
Fig. 105, 1923; Frye, Ferns Northwest 32, 1934; Peck, Man. Higher Pl. Ore. 57,
1941; Tryon, Ann. Mo. Bot. Gard. **42**:66, 1955; Hultén, Fl. Alaska 32, 1968.
S. standleyi Maxon, Smiths. Misc. Coll. **72**:9, 1920.
S. scopulorum Maxon, Am. Fern J. **11**:36, 1921.
S. rupestris of many W. Am. auth. not (L.) Spring.

Densely tufted, main stems creeping to form flat cushion mats with many
short, erect or ascending, crowded, sterile branches. Leaves subglaucous,
densely imbricate and many-ranked, those of the under side longest, all
ligulate to ligulate-lanceolate, grooved on the back, base of upper leaves
adnate to stem for about 1/4 of their length, margins short-ciliate, or eciliate, mostly oblique or incurved; setae whitish, more or less opaque, scabrous or smooth, forming conspicuous tufts at ends of dry branches. Cones
stiffly erect, or the shorter ones curved, sharply 4-angled. Sporophylls
broadly ovate, long acuminate, shorter than the vegetative leaves but
otherwise similar.

HABITAT Alpine meadows, dry rocks and rocky places, exposed hillsides
and dry ground.

COMMENTS This is the common spikemoss found in dry situations.

RANGE Southern Alaska to
southwest Manitoba, south
to northern California,
Arizona and Texas.

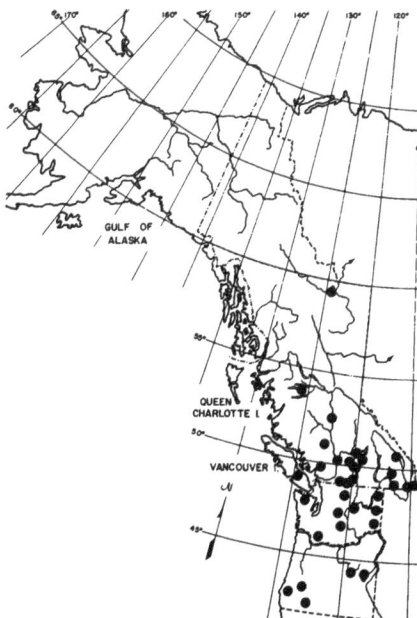

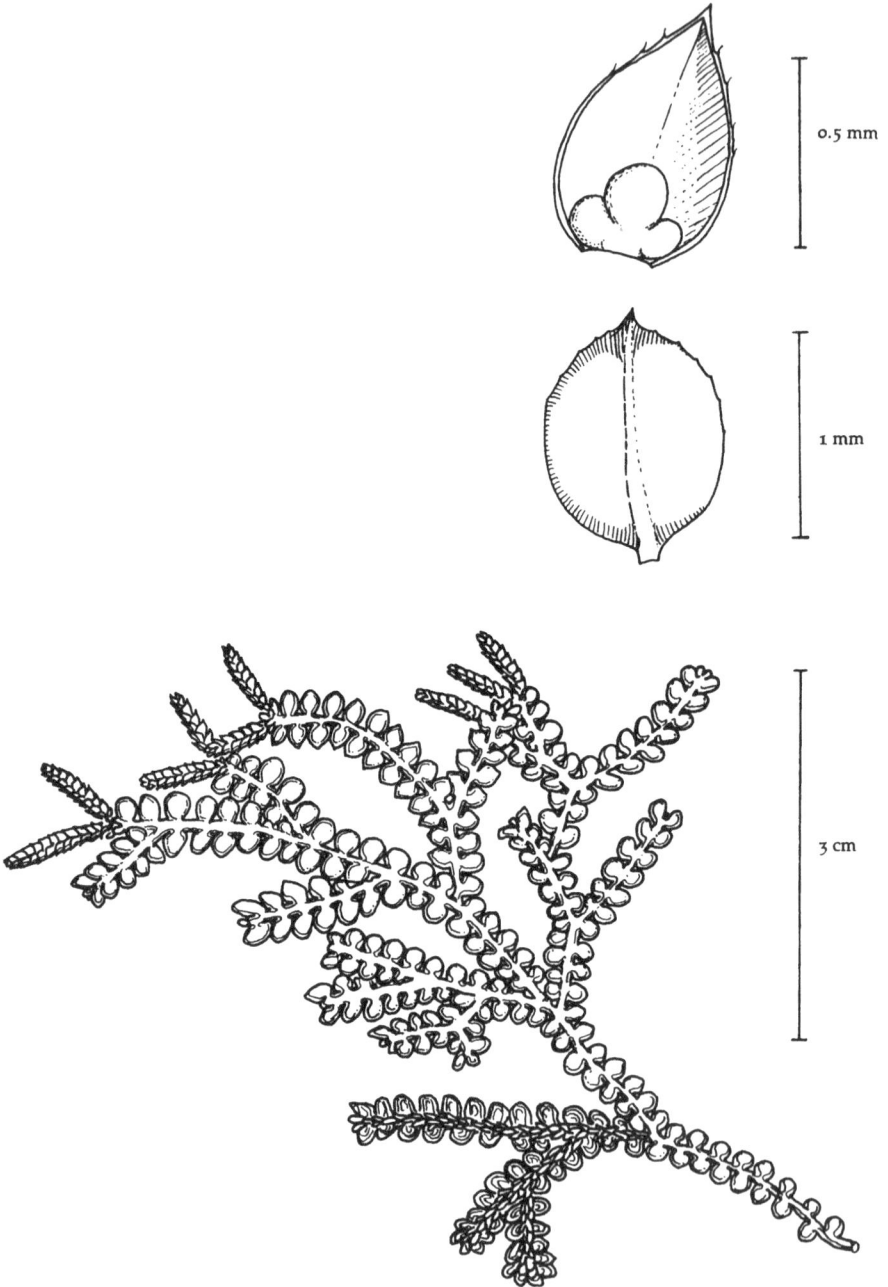

Selaginella douglasii

Selaginella douglasii (Hook. & Grev.) Spring

Mém. Acad. Brux. **24**:92, 1850; Piper & Beattie, Fl. Northwest Coast 14, 1915; Maxon [in] Abrams, Ill. Fl. Pac. States **1**:46, *Fig. 99*, 1923; Frye, Ferns Northwest 29, *Fig.* III, 1934; Peck, Man. Higher Pl. Ore. 57, 1941.

Stems slender, wide-creeping, freely rooting. Branches fairly long and distant, rebranching once or twice, leafy throughout. Leaves of 2 kinds in 4 rows, the lateral larger than those of the two upper rows. Lateral leaves obliquely oval, unequal and auriculate at the base, spreading and usually not overlapping. Upper leaves smaller, about 1/2 as long, rhombic-ovate, short cuspidate, appressed and imbricate. Cones sharply 4-angled; sporophylls acuminate, cordate-ovate, rather prominently keeled.

HABITAT Moist shady places on rocks.

COMMENTS Generally mentioned as occurring in British Columbia but no specimens have been located in support. This is a very distinctive species and unlike any other we have. It closely resembles a number of tropical species (grown as greenhouse plants in our areas).

RANGE Washington, Oregon, and Idaho.

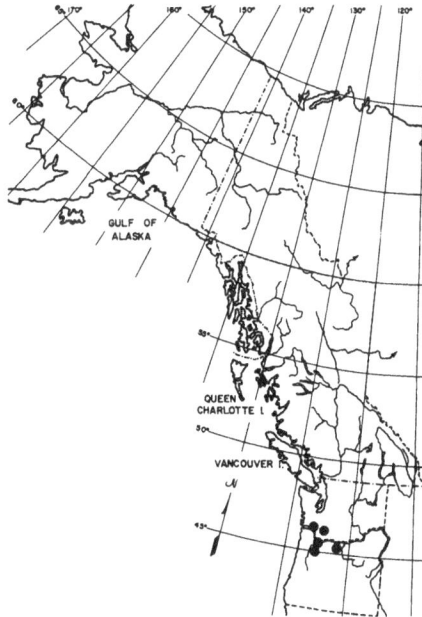

Selaginella oregana D.C. Eaton

[In] S. Wats., Bot. Calif. **2**:350, 1880; Henry, Fl. S. British Columbia 10, 1915; Piper & Beattie, Fl. Northwest Coast 15, 1915; Maxon [in] Abrams, Ill. Fl. Pac. States **1**:48, *Fig. 103*, 1923; Frye, Ferns Northwest 29, *Figs.* III, V, 1934; Peck, Man. Higher Pl. Ore. 57, 1941; Tryon, Ann. Mo. Bot. Gard. **42**:61, 1955. *S. struthioloides* Underw., Bull. Torrey Bot. Club **25**:132, 1898.

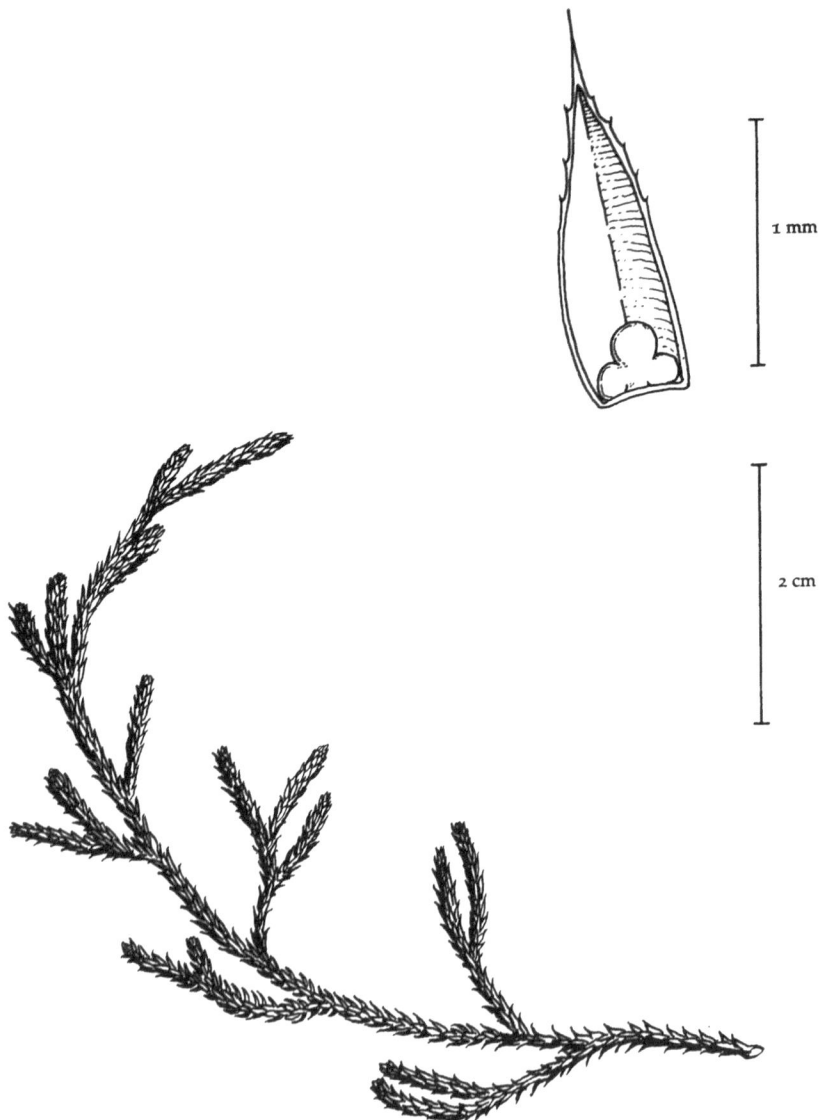

Selaginella oregana

Usually epiphytic, stems long and lax forming pendent festoons, branches long and remote, strongly curled when dormant. Leaves long- to ovate-triangular, bright green, loosely imbricate; bases of the upper leaves long decurrent, margins eciliate or ciliate towards the slightly fleshy, ascending apex; setae short, green to whitish. Cones curved, not conspicuous; sporophylls ovate, long-acuminate, eciliate towards the apex.

HABITAT Mossy trunks or branches of trees, particularly *Acer macrophyllum*, occasionally on soil of shady banks.

RANGE Coastal Washington to northern California.

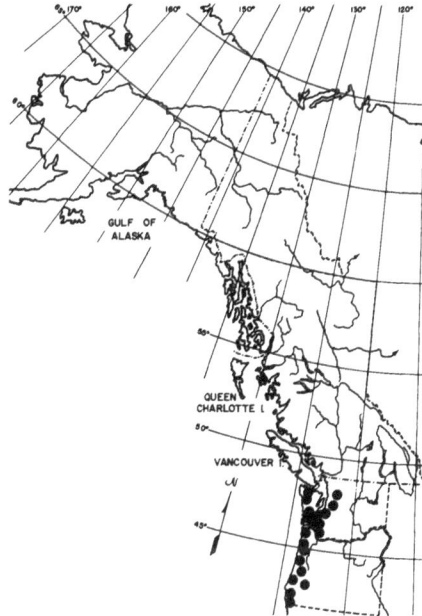

COMMENTS Our only species that is commonly epiphytic. The long decurrent leaf bases and the dried branches curling to form ringlets are two useful distinguishing characters. No specimen can be found to substantiate Henrys' report that the species grows in British Columbia. R. M. Tryon (1955) states (p. 62) that a specimen collected by Scouler and said to have come from Observatory Inlet, British Columbia, must be an error "since this locality is so far north of the otherwise known range."

Selaginella selaginoides (L.) Link

Fil. Hort. Berol. 158, 1841; Macoun, Cat. Can. Pl. pt. 5, 291, 1890; Henry, Fl. S. British Columbia 10, 1915; Morton [in] Gleason, Ill., Fl. **1**:8, *Fig.*, 1952; Anderson, Fl. Alaska 23, 1959; Hultén, Fl. Alaska 31, 1968; Calder & Taylor, Fl. Queen Charlotte Islands, pt. 1, 137, 1968.
S. spinosa Beauv., Prodr. Aethéog. 112, 1805.

2 mm

2 mm

3 cm

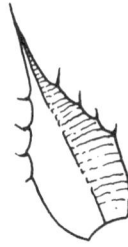

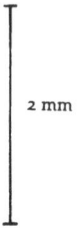

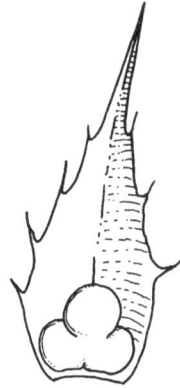

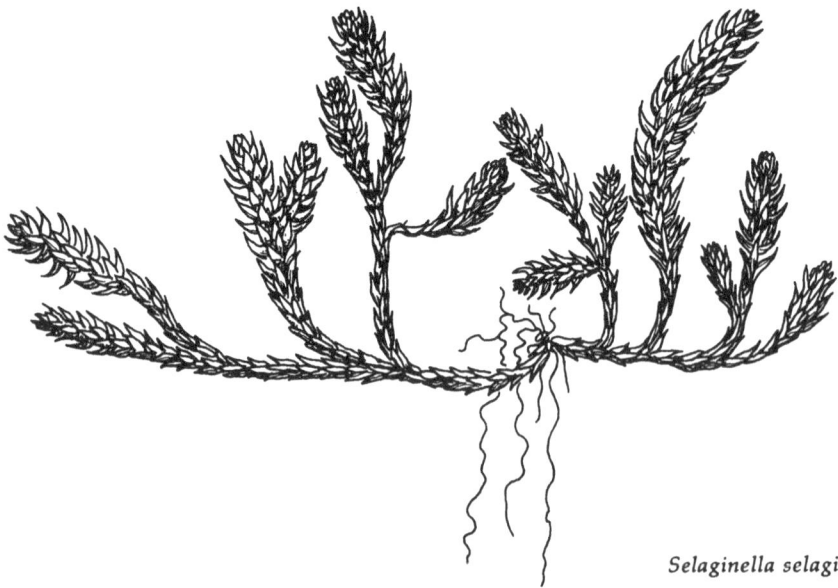

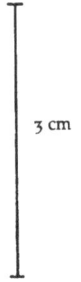

Selaginella selaginoides

Sterile stems weak, prostrate, mostly 2–5 cm long, forming small mats; leaves spirally arranged, uniform, spreading, ascending, lanceolate. Fertile stems erect, up to 15 cm tall, filiform; leaves in many ranks, ascending, acuminate, not bristle-tipped, spinulose-margined (with not more than 3–5 spinules on each side), cuneate at base, inconspicuously 1-nerved. Cones subcylindric, 1–4 cm long and up to 5 mm wide; sporophylls larger than the leaves, more prominently spinulose and nerved, spinules more numerous. ($2n = 18$)

HABITAT Wet banks and rocky ledges, marshy ground, sphagnum bogs.

RANGE Alaska to Newfoundland, south to British Columbia, Idaho, northern Minnesota, northern Michigan, New Hampshire, and northern Maine; Greenland; Iceland; Eurasia.

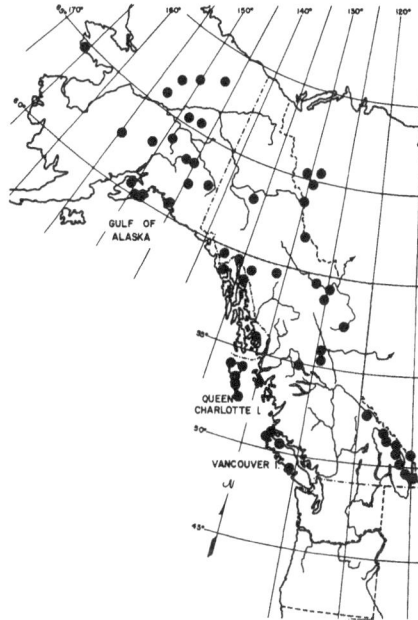

COMMENTS Generally distributed from the Aleutian Islands and coastal Alaska to British Columbia but never very common and apt to be overlooked because of its small size. Its yellowish-green colour and prominent, though slender, cones are distinguishing features.

Selaginella sibirica (Milde) Hieron.

Hedwigia **39**:290, 1900; Tryon, Ann. Mo. Bot. Gard. **42**:71, 1955; Anderson, Fl. Alaska 23, 1959; Wiggins & Thomas, Fl. Alaskan Arctic Slope 35, 1962; Hultén, Fl. Alaska 31, 1968.

Stems forming open, rather spreading mats, branches intricate. Leaves densely imbricated, all more or less equal in length, linear to ligulate to long-triangular; base of the upper leaves decurrent for about 1/6 their length; margins ciliate, mostly spreading; apex of upper leaves fleshy,

4 mm

4 mm

2 cm

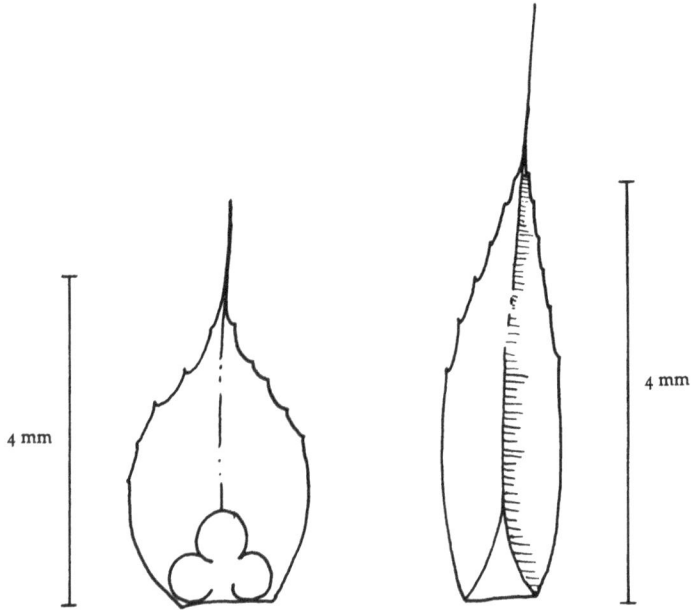

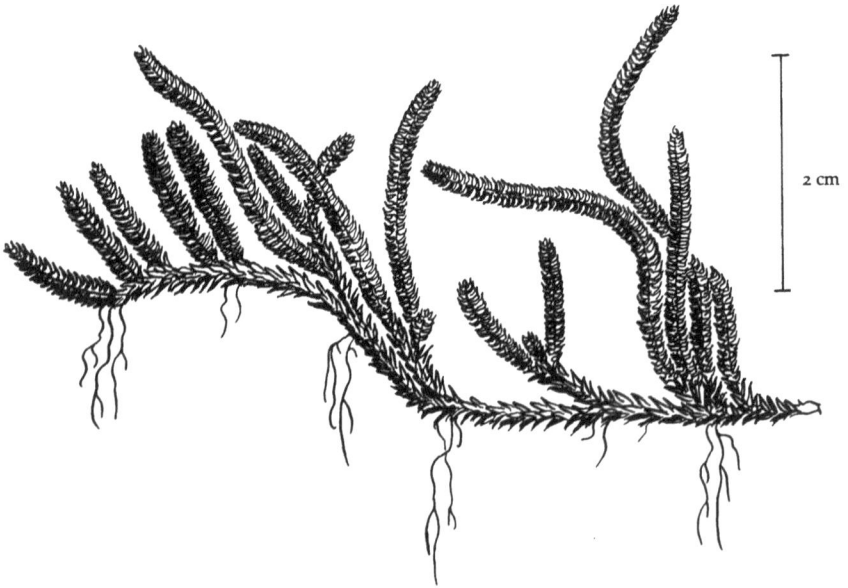

Selaginella sibirica

broadly rounded to keeled, usually truncate in profile; setae about 1/2 as long as the blade, milky white, forming conspicuous tufts at dry branch tips, scabrous, seta base strongly broadened and flattened. Sporophylls eciliate or ciliate, 3–4 times as wide as the leaves.

H A B I T A T Dry, rocky places.

R A N G E Yukon to Manchuria and Japan.

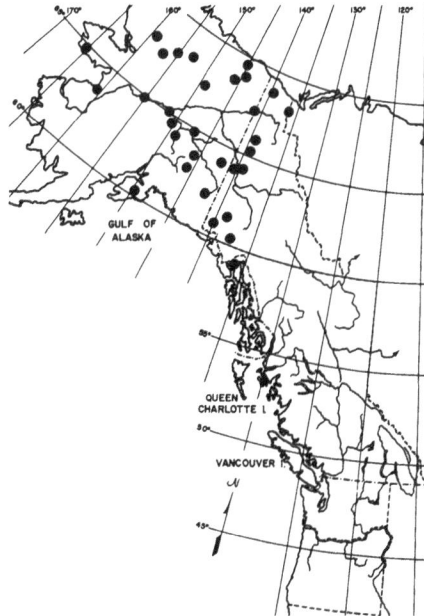

C O M M E N T S So far only known from Alaska and the Yukon in our area. It is similar and probably closely related to *S. densa* var. *standleyi* (Maxon) Tryon from which it may be distinguished by its white to tawny, rather than lutescent setae and by its intricate rather than discrete branches.

Selaginella wallacei Hieron.

Hedwigia **39**:297, 1900; Macoun, Cat. Can. Pl. pt. 5, 292, 1890; Henry, Fl. S. British Columbia 10, 1915; Piper & Beattie, Fl. Northwest Coast 15, 1915; Maxon [in] Abrams, Ill. Fl. Pac. States **1**:49, Fig. 108, 1923; Frye, Ferns Northwest 33, *Fig.* III, 1934; Peck, Man. Higher Pl. Ore. 57, 1941; Tryon, Ann. Mo. Bot. Gard. **42**:43, 1955; St. John, Fl. Southeast. Wash. 2, 1963; Calder & Taylor, Fl. Queen Charlotte Islands pt. 1, 138, 1968.
S. montanensis Hieron., Hedwigia **39**:293, 1900.
S. rupestris (L.) Spring var. *wallacei* (Hieron.) Frye, Ferns Northwest 33, 1934.

Plants forming loose mats. Main stems prostrate, radially symmetrical, rooting sparsely throughout. Branches numerous, ascending, up to 4 cm long, freely but rather remotely branched. Branchlets short, wide-spread-

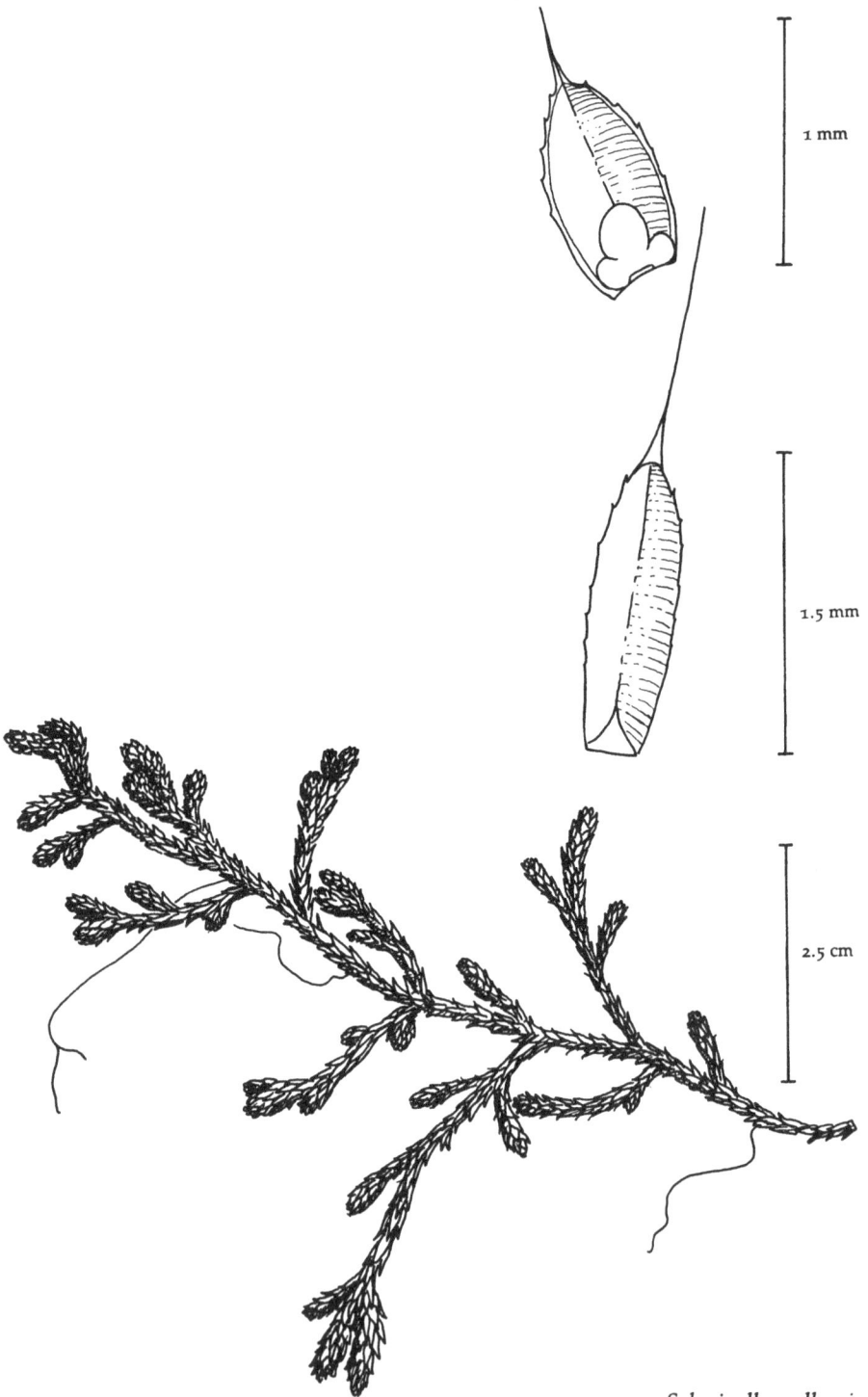

1 mm

1.5 mm

2.5 cm

Selaginella wallacei

ing, rather rigid, cordlike, often intertwined. Leaves papery, rigidly appressed-imbricate on all sides, subglaucous, abruptly adnate to stem at base, mostly oblong-linear with 8–14 oblique cilia on each side, narrowed to a somewhat obtuse apex but prolonged into a stiff, bristlelike seta, leaves about 3 mm long (including the seta); seta about 0.5 mm long, whitish-hyaline, scabrous. Cones numerous, somewhat curved; sporophylls narrowly ovate-deltoid with 10–18 rather variable cilia on each side, shorter than the vegetative leaves, abruptly setigerous with short, stout, whitish-hyaline, nearly smooth setae.

H A B I T A T Rocky bluffs and slopes, both in the open and in light shade.

R A N G E Southern British Columbia to western Montana, south to Oregon, and along the coast to northern California.

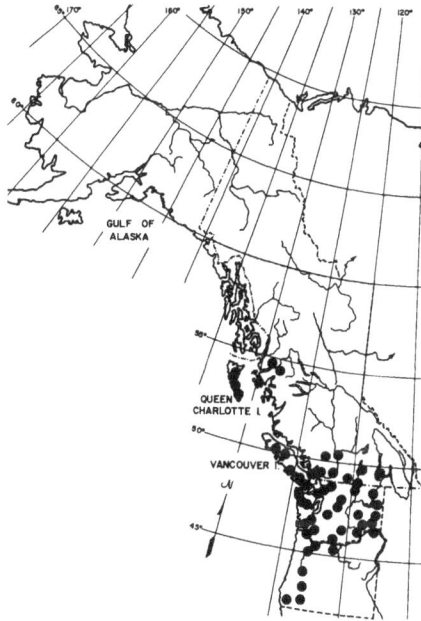

C O M M E N T S This species has a wide range of habitats and to a considerable extent its variability in morphology is correlated with ecological differences. On drier sites it sometimes resembles *S. densa* from which it can be distinguished by its remote branches and its abruptly adnate rather than decurrent leaf bases.

Selaginella watsonii Underw.

Bull. Torrey Bot. Club **25**:127, 1898; Maxon [in] Abrams, Ill. Fl. Pac. States **1**: 49, *Fig.* 106, 1923; Tryon, Ann. Mo. Bot. Gard. **42**:72, 1955.

Usually rather densely cespitose forming rounded cushion mats. Stems short-creeping, branches numerous, short, intricate, ascending, subapproximate. Leaves uniform, densely appressed-imbricate, ligulate to

2 mm

2 mm

2 cm

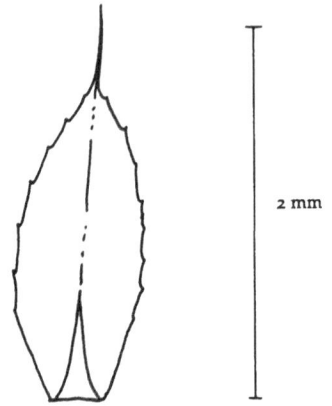

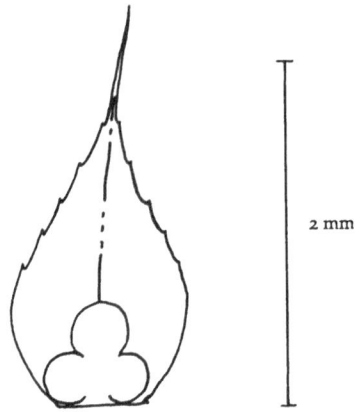

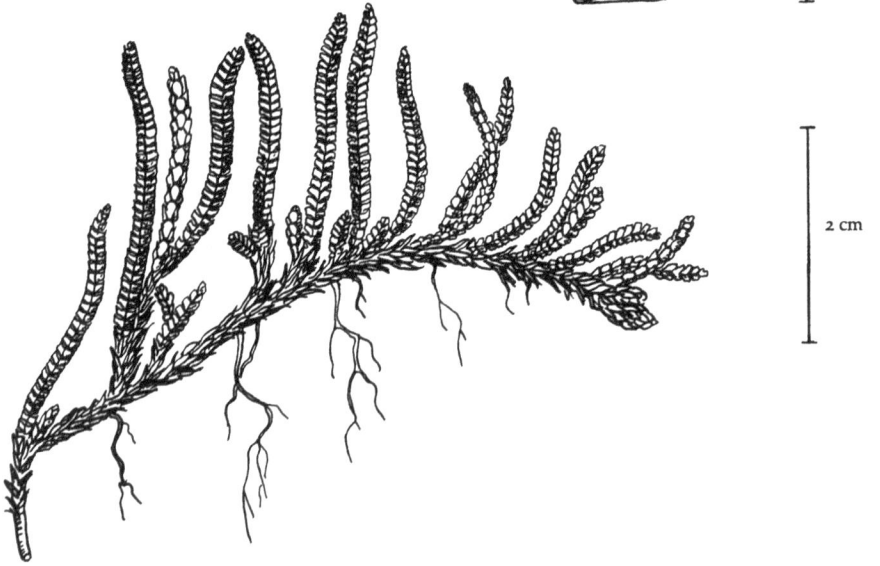

Selaginella watsonii

ligulate-long-triangular; base of upper leaves adnate to stem for about 1/4 of their length; margins usually eciliate; apex of upper leaves fleshy, somewhat keeled, more or less truncate in profile. Setae not forming conspicuous tufts at dry branch tips, usually about 1/4 as long as the blade, essentially smooth, varying from greenish-white to whitish-lutescent, more or less translucent. Sporophylls eciliate towards the apex, the broadest about twice as broad as the leaves.

HABITAT Cliffs, talus slopes, rocky alpine meadows.

RANGE Southwest Montana to northeast Oregon, south to Utah and southern California.

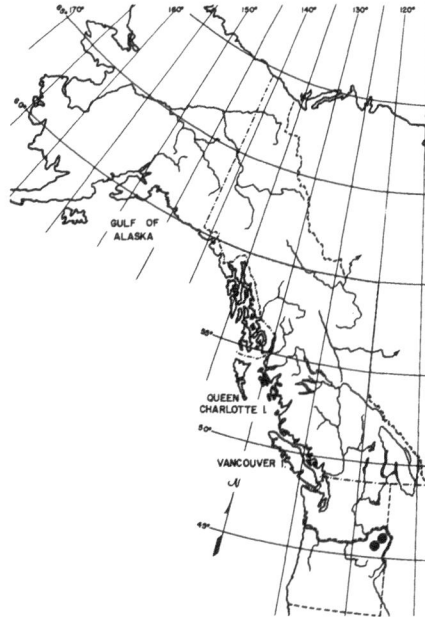

COMMENTS Included on the basis of a single collection made by Wherry in Union County, Oregon.

Excluded Species

Equisetum ramosissimum Desf. Reported from Shawnagin (*sic*) Lake, British Columbia in Macoun, Cat. Can. Pl. pt. 5, p. 252. This is undoubtedly based on a misidentification as this is a Eurasian species that does not occur in North America; probably some form of *E. hyemale* was involved.

Isoetes dodgei A. A. Eaton. Henry, Fl. S. British Columbia, p. 11, includes this species without definite locality. No specimen has been found to substantiate this record. As this is a synonym for *I. riparia* Engelm. var. *canadensis* Engelm., a species that comes no farther west than southern Ontario, its occurrence in British Columbia is extremely unlikely. It is possible that Henry may have misinterpreted Macoun's (Cat. Can. Pl. pt. 5, p. 293) report of its collection in Victoria County, Ontario.

Lycopodium lucidulum Michx. No specimens can be found to substantiate the report of this species by Piper & Beattie, Fl. Northwest Coast, p. 13, or by Macoun, Cat. Can. Pl. pt. 5, p. 288. It is presumed that these records are based on misidentifications of *L. selago*.

Lycopodium sabinaefolium Willd. Reported by Henry, Fl. S. British Columbia, p. 12, from "Rockies and Selkirks, Mt. Arrowsmith, V.I." This is almost certainly based on misidentification with *L. sitchense* as *L. sabinaefolium* apparently does not occur west of Ontario.

Botrychium ramosum (Roth.) Ascher. This name in Henry's Fl. S. British Columbia is a synonym for *B. matricariaefolium* A. Br. No specimens have been found to support this record but it has been found in Kootenai County, Idaho. It is possible that it will be found in our region but should be excluded at the present time.

Woodsia obtusa (Spreng.) Torrey. Included by Frye, Ferns Northwest, p. 111, as a possibility because of a supposed Alaskan record. No specimens are known to support this, however. Morton (1950) says: "all these records go back to a misidentification by W. Trelease as *W. obtusa* of an Alaskan specimen of *Cystopteris fragilis*. The range is correctly stated by Prof. Fernald in the New Gray's Manual."

Azolla filiculoides Lam. An old record from Alaska is almost certainly an aquarium escape.

Selaginella douglasii (Hook. & Grev.) Spring. Generally given as occurring in British Columbia. As no specimens have been found to support this it should be excluded from the flora of this province.

Selaginella rupestris (L.) Spring. R. M. Tryon, Jr. (1955), considers that this species does not occur in our area. All references to it are probably to some form of *S. densa* Rydb. or *S. wallacei* Hieron.

Selaginella sartorii var. *oregonensis* Hieron. Piper & Beattie, Fl. Northwest Coast, p. 15 (under *S. rupestris*), refer to this as being the common rock species within their limits. R. M. Tryon, Jr. (1955), points out that this naming is based on a specimen that incorrectly bears a Lyall label. The true source of the material and who was its collector are unknown. *S. sartorii* Hieron. ranges from Mexico to Colombia and Venezuela.

APPENDIX I

The following is a selected listing of chromosome numbers reported by various authors. More complete summaries can be found in Chiarugi (1960) and Fabri (1963). It will be noted that in some species both diploid and tetraploid races occur; in a few there is a considerable measure of disagreement in the numbers reported. The significance of these different counts has not yet been fully evaluated. As Manton in 1950 provided the first comprehensive list of chromosome numbers in the pteridophytes, this has been taken as the starting date for the list below. Counts based on Pacific Northwest materials are indicated by **, others from North America by *, the remainder are extra-limital, mostly European. The latter are included where data are apparently lacking based on North American materials or to indicate the situation in species whose ranges extend beyond this continent, or both.

	n	$2n$	
Equisetum arvense	c.108		Manton (1950)
	108		Bir (1960)
		216	Löve & Löve (1961a)
	c.108*		Wagner & Wagner (1966)
Equisetum fluviatile	c.108		Manton (1950)
		216	Löve & Löve (1961a)
Equisetum hiemale	c.108		Manton (1950)
		216	Löve & Löve (1961a)
	108*		Wagner & Wagner (1966)
Equisetum laevigatum	meiosis irregular		Hauke (1958)
Equisetum palustre	c.108		Manton (1950)
	108	216	Bir (1960)
Equisetum pratense	c.108		Manton (1950)
		216	Bir (1960)
Equisetum scirpoides	c.108		Manton (1950)
		216	Hagerup & Petersson (1960)
Equisetum sylvaticum	c.108		Manton (1950)
		216	Bir (1960)
Equisetum telmateia	c.108		Manton (1950)
		216	Bir (1960)
Equisetum variegatum	c.108		Manton (1950)
		216	Hagerup & Petersson (1960)
Mecodium wrightii			
Isoetes bolanderi			
Isoetes echinospora		100+	Manton (1950)
		22*	Löve (1962)
Isoetes howellii			
Isoetes nuttallii			
Isoetes occidentalis			
Lycopodium alpinum	24–25	c.48	Manton (1950)

	n	*2n*	
		48*	Löve & Löve (1958)
		46	Löve & Löve (1961b)
		38–40	Hadac & Haskova (1956)
	22–24		Sorsa (1963)
Lycopodium annotinum	c.34	c.68	Manton (1950)
		68*	Löve & Löve (1958)
	34*		Wagner & Wagner (1966)
Lycopodium clavatum	34	68	Manton (1950)
		68*	Löve & Löve (1961b)
	34*		Wagner & Wagner (1966)
Lycopodium complanatum		c.48*	Löve & Löve (1958)
		46	Löve & Löve (1961b)
	22–24		Sorsa (1963)
Lycopodium inundatum	78		Manton (1950)
		156*	Löve & Löve (1958)
	78*		Wagner & Wagner (1966)
Lycopodium obscurum		68*	Löve & Löve (1958)
	34*		Wagner & Wagner (1966)
Lycopodium selago		260	Manton (1950)
		264*	Löve & Löve (1958)
		c.68	Hagerup & Petersson (1960)
		c.88	Löve & Löve (1961b)
	c.45	c.90	Sorsa (1963)
Lycopodium sitchense			
Marsilea vestita	16		Strasburger (1907)
Pilularia americana			
Botrychium boreale		90	Löve & Löve (1961a)
Botrychium lanceolatum	45*		Wagner (1955)
		90	Löve & Löve (1961a)
Botrychium lunaria	45		Manton (1950)
	45*		Wagner & Lord (1956)
Botrychium minganense	90*		Wagner & Lord (1956)
Botrychium multifidum	45*		Wagner (1955)
		90*	Wagner & Chen (1964)
Botrychium pumicola	45**		Wagner (1955)
Botrychium simplex	45*		Wagner (1955)
Botrychium virginianum	91–92*		Britton (1953)
	92*		Wagner (1955)
	90		Gopal-Ayengar (1957)
	92		Niizeki, Nishida & Kurita (1963)
		180	Löve & Löve (1964)
Ophioglossum vulgatum	250–260		Manton (1950)
	410		Verma (1956)
	480		Verma (1956)
	240		Verma (1958)
	c.570		Ninan (1958)
	247–251		Verma (1958)
	150–160	c.300	Roy & Pandey (1963)
		480*	Löve & Löve (1961a)
Adiantum capillus-veneris		60	Manton (1950)
	60*		Wagner (1963)
Adiantum jordanii			

	n	$2n$	
Adiantum pedatum	29*		Britton (1953)
	29**		Manton (1958)
		c.58*	Löve & Löve (1964)
Adiantum X tracyi	59**		Wagner (1962)
Aspidotis densa	30**		Wagner (1963)
Asplenium septentrionale	72	144	Manton (1950)
Asplenium trichomanes	36		Manton (1950)
	36*		Britton (1953)
	72	144	Manton (1950)
	72*		Britton (1953)
	108*		Wagner & Wagner (1966)
Asplenium viride	36	72	Manton (1950)
	36*		Britton (1953)
	36**		Taylor & Lang (1963)
Athyrium distentifolium		80	Manton (1950)
	40**		Taylor & Lang (1963)
Athyrium filix-femina	40		Manton (1950)
	40*		Britton (1953)
Blechnum spicant	34	68	Manton (1950)
Cheilanthes feei		87**	Knobloch (1967)
Cheilanthes gracillima			
Cheilanthes intertexta			
Cryptogramma crispa	60		Manton (1950)
	30**1		Taylor & Lang (1963)
Cryptogramma stelleri	30*		Wagner (1963)
Cystopteris fragilis	84		Manton (1950)
	126		Manton (1950)
	84*		Britton (1953)
		168	Löve & Löve (1961a)
var. *protrusa*	42*		Wagner & Hagenah (1956)
Cystopteris montana	84		Manton (1950)
Dryopteris arguta	41**		Wagner & Chen (1964)
Dryopteris austriaca	41		Sorsa (1958)
"dilatata"		164	Löve & Löve (1961a)
"dilatata"		82**	Walker (1961)
"dilatata"	41*		Wagner (1963)
Dryopteris cristata	82	164	Manton (1950)
	82*		Wagner & Wagner (1966)
Dryopteris filix-mas	82	164	Manton (1950)
	82*		Britton & Soper (1966)
Dryopteris fragrans	41*	82*	Wagner & Hagenah (1962)
	41**		Taylor & Lang (1963)
Gymnocarpium dryopteris	80*		Britton (1953)
		160	Löve & Löve (1961a)
var. *disjunctum*	40**		Wagner (1966b)
Gymnocarpium robertianum	80*	80*	Wagner (1966b)
Matteuccia struthiopteris	40		Sorsa (1962)
	40±1		Britton (1953)
Pellaea andromedaefolia			
Pellaea atropurpurea	87*	87*	Manton (1950)

1 The report of $n = 40$ in this paper is a typographical error and should read $n = 30$.

	n	2n	
Pellaea brachyptera			
Pellaea breweri	29*		A. F. Tryon (1957)
Pellaea glabella	116*	116*	Britton (1953)
	29*		A. F. Tryon (1957)
Pityrogramma triangularis	60**		Manton (1958)
	30**		Alt & Grant (1960)
Polypodium glycyrrhiza		74**	Manton (1951)
	37**		Lang (1969)
Polypodium montense	37**		Lang (1969)
Polypodium hesperium	74**		Lang (1969)
Polypodium scouleri		74**	Manton (1951)
		111**	Manton (1951)
	37**		Taylor & Lang (1963)
Polypodium virginianum	37*		Manton (1950)
	74*		Manton & Shivas (1953)
	111*		Wagner & Wagner (1966)
Polystichum andersonii	82**		Taylor & Lang (1963)
Polystichum braunii	82–84**		Taylor & Lang (1963)
Polystichum kruckebergii	82**		Wagner (1966a)
Polystichum lonchitis	41	82	Manton (1950)
	41*		Wagner & Chen (1964)
Polystichum mohrioides	41**		Taylor & Lang (1963)
Polystichum munitum	41**		Taylor & Lang (1963)
Polystichum scopulinum	82**		Wagner (1966a)
Pteridium aquilinum	52		Manton (1950)
	52*		Britton (1953)
Thelypteris limbosperma	34	68	Manton (1950)
	34**		Taylor & Lang (1963)
Thelypteris nevadensis	26–27**		Taylor & Lang (1963)
Thelypteris phegopteris	90	90	Manton (1950)
	90*	90*	Britton (1953)
Woodsia alpina	82–84		Manton (1950)
		164	Löve & Löve (1961a)
	c.82*		Brown (1964)
Woodsia glabella	39	78	Meyer (1959b)
	39*		Brown (1964)
Woodsia ilvensis	41–42		Manton (1950)
	41±1*		Britton (1953)
		82	Löve & Löve (1961a)
	40–41**		Taylor & Lang (1963)
	41*		Brown (1964)
Woodsia oregana	38*		Brown (1964)
Woodsia scopulina	38*		Wagner (1963)
Woodwardia fimbriata			
Azolla mexicana			
Selaginella densa			
Selaginella douglasii			
Selaginella oregana			
Selaginella selaginoides		18	Reese (1951)
Selaginella sibirica			
Selaginella wallacei			
Selaginella watsonii			

APPENDIX II

Lists of species grouped by distribution patterns

Circumboreal species

Equisetum arvense
Equisetum fluviatile
Equisetum hyemale
Equisetum palustre
Equisetum pratense
Equisetum scirpoides
Equisetum sylvaticum
Fquisetum variegatum
Isoetes echinospora
Lycopodium alpinum
Lycopodium annotinum
Lycopodium clavatum
Lycopodium complanatum
Lycopodium inundatum
Lycopodium selago
Botrychium boreale
Botrychium lanceolatum
Botrychium lunaria
Botrychium multifidum
Botrychium simplex
Botrychium virginianum
Ophioglossum vulgatum

Adiantum capillus-veneris
Asplenium trichomanes
Asplenium viride
Athyrium distentifolium
Cryptogramma crispa
Cryptogramma stelleri
Cystopteris fragilis
Cystopteris montana
Dryopteris cristata
Dryopteris filix-mas
Dryopteris fragrans
Gymnocarpium dryopteris
Gymnocarpium robertianum
Matteuccia struthiopteris
Polystichum braunii
Polystichum lonchitis
Pteridium aquilinum
Thelypteris phegopteris
Woodsia alpina
Woodsia glabella
Woodsia ilvensis
Selaginella selaginoides

Species endemic to North America but widely distributed

Equisetum laevigatum
Marsilea vestita
Pilularia americana
Botrychium minganense
Adiantum pedatum
Aspidotis densa
Cheilanthes feei

Pellaea atropurpurea
Pellaea glabella
Polypodium virginianum
Woodsia oregana
Woodsia scopulina
Azolla mexicana

Species endemic to Western North America

Isoetes bolanderi
Isoetes howellii
Isoetes nuttallii
Isoetes occidentalis
Botrychium pumicola
Adiantum jordanii
Cheilanthes gracillima
Cheilanthes intertexta

Polypodium hesperium
Polypodium montense
Polypodium scouleri
Polystichum andersonii
Polystichum kruckebergii
Polystichum munitum
Thelypteris nevadensis
Woodwardia fimbriata

Pellaea andromedaefolia
Pellaea brachyptera
Pellaea breweri
Pellaea bridgesii
Pityrogramma triangularis
Polypodium glycyrrhiza

Selaginella densa
Selaginella douglasii
Selaginella oregana
Selaginella wallacei
Selaginella watsonii

Western North American and Eurasian species

Equisetum telmateia
Asplenium septentrionale
Athyrium filix-femina
Blechnum spicant

Dryopteris arguta
Dryopteris austriaca
Thelypteris limbosperma

Species with peculiar ranges

Mecodium wrightii	Southern Japan to Korea; extremely localized on west coast of North America.
Lycopodium obscurum	Widespread in North America; Asia but not Europe.
Lycopodium sitchense	Widespread in boreal North America; eastern Asia.
Polystichum mohrioides	Washington south to Tierra del Fuego.
Polystichum scopulinum	Southern British Columbia and western United States; Gaspé Peninsula, Quebec.
Selaginella sibirica	Extreme northwestern North America; extreme northeastern Asia.

Summary

Circumboreal species	44	45.0%
Species endemic to North America and widely distributed	13	13.4
Species endemic to western North America	28	29.0
Western North America and Eurasian species	6	6.2
Species with peculiar ranges	6	6.2
	97	99.8%

GLOSSARY

Acicular Needle-shaped; slender.

Aculeate Armed with prickles; having sharp points.

Acuminate Gradually tapering to a long point.

Acute Ending on a sharp and well-defined angle.

Adnate Having one organ wholly or in part attached to a different organ; fusion of unlike parts.

Allotetraploid Hybrid with two double sets of chromosomes resulting from an interspecific cross.

Amphibious Growing equally well on land and in water.

Anastomose To fuse together, as of veinlets fusing to form a network.

Annulus Any ringlike structure; special elastic ring in fern sporangium by action of which the sporangia open.

Apex The tip of an organ.

Apical At the top or summit of a structure.

Apiculate Ending in a short and sharp, but not stiff, point.

Apogamous Referring to an embryo that develops without prior fertilization.

Appressed Lying flat, pressed against some other structure.

Approximate Very near; close together.

Arcuate Curved to some degree; arching.

Areole Space formed by anastomosing veins.

Aristate Provided with a bristlelike structure, usually at the end.

Articulate Jointed, separating freely by a clean scar.

Articulation Joint between two structures or two parts of the same structure.

Ascending Curving upwards; rising at first obliquely then erect.

Assurgent Ascending; rising upward.

Attenuate Slender tapering; gradually drawn out.

Auricle Small lobe or ear-shaped appendage.

Auriculate With an ear-shaped lobe, usually basal.

Autotetraploid With four sets of chromosomes all derived from a single species.

Axil Upper angle formed between a pinna or pinnule and the rachis, or between an axis and a structure arising from it.

Axis The main or central line around which organs develop.

Biotype Group of individuals of exactly the same genetic constitution.

Bipinnate Twice pinnate; each pinna being itself pinnate.

Bivalvate With two valves.

Blade Expanded portion of a leaf or frond.

Boreal Northern.

Callous Thickened or hardened in texture.

Capillary Slender hairlike structure.

Capitate Shaped like a head or in a headlike arrangement; enlarged or swollen at tip.

Cartilaginous Hard and tough.

Castaneous Chestnut-coloured; dark reddish-brown.

Caudate With a slender, tail-like appendage.

Caudex Woody base of a perennial plant; axis of a plant consisting of stem and root.

Cespitose Pertaining to turf; having low, closely matted stems; growing in bunches or tufts.

Chaffy Covered with small, dry, membranous scales.

Cilia Marginal hairs or hairlike processes.

Ciliate Fringed with hairs, usually along a margin.

Circinate Coiled from the top downward, as the young frond of a fern.

Cline Series of changes in form; a gradient of biotypes along an environmental transition.

Coherent With similar parts united.

Compound Having two or more similar parts aggregated in one organ.

Concolorous Uniform in colour.

Conduplicate Folded together lengthwise.

Confluent Flowing into one another; passing by degrees into one another.

Connate Like parts fused or grown together.

Conspecific One and the same species.

Cordate Heart-shaped with special reference to the base of a structure.

Coriaceous Leathery in texture.

Corm Solid bulblike underground stem or base of stem.

Cortex Tissue lying between the vascular cylinder and the epidermis.

Crenate Margin scalloped with rounded teeth.

Crenulate Finely crenate.

Cuneate Wedge-shaped with the acute angle downwards, applied particularly to leaf bases.

Cuspidate Tipped with a sharp, rigid point.

Deciduous Not persistent; dropping or being shed at the end of a growing season.

Decompound Several times divided, i.e., the main divisions again divided.

Decumbent Prostrate but with the growing tips erect or ascending.

Decurrent Extending downwards on the stem and fused with it.

Dehisce To open spontaneously, usually to release spores or seeds.

Deltoid More or less broadly triangular in shape; shaped like the Greek letter Δ.

Dentate Toothed, usually with sharp teeth directed outwards.

Denticulate Minutely toothed on margin.

Dichotomous Branching regularly into two equal parts.

Dimorphic Occurring in two forms as with differing fertile and vegetative fronds.

Diploid Having a double set of chromosomes, usually expressed as 2*n*.

Discrete Separate, not coalescent.

Distichous Two-ranked; in two vertical rows.

Divergent Inclining away from each other.

Dorsal Upon, or relating to the back or lower surface of a frond; the surface away from the axis.

Dorsiventral Having distinct dorsal and ventral surfaces.

e – Prefix indicating not or without; e.g., eciliate, without cilia.

Echinate Prickly or spiny.

Edaphic Influence of the soil conditions on growth of plants.

Elater One of the four filamentous appendages on the spore of *Equisetum*.

Endemic Confined to a limited region, as an island or a country.

Entire Without teeth or lobing; with a continuous margin.

Epiphyte Plant growing on another plant but not parasitic upon it.

Equilateral Having the sides equal, as in a triangle.

Erose Margin ragged as if gnawed.

Falcate Scythe- or sickle-shaped; flat and curved, tapering gradually to a point.

Filament Threadlike body.

Filiform Threadlike; long and very slender.

Fimbriate Fringed with long slender processes; of hairs longer or coarser than ciliate.

Flabellate Fan-shaped; dilated in a wedge-shape.

Flaccid Flabby, weak and limp.

Flexuous Bent alternately in opposite directions.

Foliate Leaved; having leaves.

Free Not fused or adhering.

Frond Expanded leaflike portion of a fern.

Gametophyte Haploid sexual generation that produces eggs and sperms.

Gemma Vegetative reproductive structure, analogous to a bud, by which a plant propagates itself.

Glabrate Becoming smooth or nearly so.

Glabrous Smooth; devoid of pubescence and hairs of any sort.

Glandular Bearing glands of any sort; of the nature of glands.

Glaucous Covered or whitened with a "bloom"; bluish green.

Globose Nearly spherical, globular.

Glochidium Stout hair or prickle with hooklike barbs.

Hair Epidermal outgrowth consisting of a single elongated cell, or row of cells.

Haploid Having a single set of chromosomes; the number of chromosomes characteristic of the gametophyte generation and usually expressed as *n*.

Herbaceous Somewhat thin in texture like an ordinary leaf or frond; not woody.

Heterosporous Having different sizes of spores; micro- and megaspores in the pteridophytes.

Hirsute Hairy, with stiff or bristly hairs.

Hispid With stiff or bristly hairs, therefore rough in texture.

Hispidulous Somewhat or minutely hispid.

Homosporous Having only one size of spore; opposite to heterosporous.

Hyaline Clear, transparent or translucent.

Imbricate Overlapping, as shingles on a roof.

Incised Sharply and irregularly cut, more or less deeply so.

Indurated Hardened.

Indusium Membrane covering a sorus in ferns.

Inequilateral Of unequal sides.

Inferior Said of one organ when arising below another.

Inflated Blown-up, like a bladder.

Internode Portion of a stem between adjacent nodes.

Intramarginal Within or close to the margin.

Intricate Entangled.

Jointed Separating at a joint; articulated.

Lacerate Irregularly cleft as if torn.

Laciniate Irregularly cut with narrow lobes, as if slashed.

Lanceolate Lance-shaped; several times longer than wide; broadest below the middle and tapering to both ends.

Ligulate Flattened, strap-shaped.

Ligule Minute membranous structure above the sporangium at the base of the leaves in *Isoetes* and *Selaginella*.

Linear Long and narrow with nearly parallel sides.

Lobed Divided into, or bearing, rounded segments which reach less than half-way to the midrib.

Lunate In the shape of a half-moon or crescent.

Lutescent Becoming yellow.

Marcescent Withering without falling off.

Marginal Placed on or attached to the edge.

Margined Furnished with a margin of distinct character, in ferns frequently narrowly winged.

Massula Mass of microspores of *Azolla*.

Medial In the middle, often referring to the position of a sorus midway between the margin and midrib.

Median Pertaining to the middle; lying in the axial plane.

Megaspore Larger of the two sizes of spores in a heterosporous plant.

Meiosis Process of reduction division of chromosome numbers; initiates the gametophyte generation.

Microspore Smaller of the two sizes of spores in a heterosporous plant.

Monomorphic Having only one form, as opposed to dimorphic.

Mucronate Abruptly terminated with a sharp tip or point.

Node Position on a stem at which a leaf or a whorl of leaves normally arises.

Nodose Knobby.

Nodulose With little knobs or knots, chiefly used of roots.

ob – Prefix indicating inversely or oppositely as in "oblanceolate."

Oblique Slanting, unequal sided.

Oblong Two or three times longer than broad, with more or less parallel sides.

Obsolete Not evident; rudimentary.

Obtuse Rounded or blunt at the apex.

Orbicular Circular in outline.

Ovate Shaped like the outline of an egg with the broader end downwards.

Panicle In ferns a loose branched raceme.

Papilla Small pimplelike protuberances.

Papillate With minute conical nipplelike projections.

Papillose Bearing papillae.

Paraphyses Filaments of sterile cells among sporangia.

Pectinate Comb-like.

Pedicel In ferns the stalk supporting a sporangium.

Peduncle In ferns, the stalk supporting a sporocarp as in *Marsilea*, or the cones in *Lycopodium* and *Equisetum*.

Peltate Shield-shaped, with the attachment of a stalk near the centre rather than at the margin, as an indusium.

Pentagonal Five-sided.

Penultimate Last but one.

Perennial Lasting or growing for several years.

Peripheral On or near the margin.

Persistent Long-continuing; evergreen when referring to fronds.

Petiole Stalk of a leaf, in ferns the stipe.

Petiolulate Referring to the petiole of a pinna or pinnule.

Pilose With long, soft hairs; downy.

Pinna Primary division of a pinnately compound blade.

Pinnate Featherlike, with the division extending fully to the rachis.

Pinnatifid Pinnately cut; the divisions extending deeply but not all the way to the rachis or midrib.

Pinnule Primary division of a pinna; the secondary segment of a compound blade.

Proliferous Bearing progeny vegetatively as from buds on fronds.

Pruinose Having a waxy powdery secretion on the surface; a "bloom."

Puberulent Covered with down or fine hair; minutely pubescent.

Pubescent Covered with short, soft hairs or down.

Pungent Ending in a rigid and sharp point.

Raceme In ferns referring to a fertile segment with sporangia on pedicels of equal length.

Rachis Main axis of a compound frond or leaf.

Receptacle Expanded structure that bears other organs.

Reflexed Abruptly bent or turned downward.

Reniform Kidney-shaped.

Reticulate In the form of a network.

Retuse With a shallow notch at a rounded apex.

Revolute Rolled backwards from the margins.

Rhizome An underground stem from which fronds arise.

Rhombic Diamond-shaped.

Rugose Wrinkled.

Salient Projecting forward.

Scabrous With short bristly hairs; rough to the touch.

Scale More or less flattened cellular outgrowth two cells wide or more; a rudimentary or vestigial leaf.

Scarious Thin, dry and membranous, not green.

Sclerotic Hardened.

Septate Divided by cross-walls.

Serrate Having sharp teeth pointing forward; like the edge of a saw.

Serrulate Finely serrate.

Sessile Not stalked.

Seta Bristlelike structure.

Setaceous Bristlelike; set with bristles.

Setate Having setae.

Silex White or colourless, very hard crystalline substance, silica dioxide.

Siliceous Composed of or pertaining to silica.

Simple Of one piece as opposed to compound.

Sinuate With the outline of the margin deeply and strongly wavy.

Sinus Cleft between adjacent lobes of a leaf or other expanded organ.

Sorus Cluster of sporangia in ferns.

Spike Simple axis with sporangia sessile, or nearly so, upon it.

Spinule Small spine.

Spinulose With small spines over the surface.

Sporangium Spore case.

Spore Reproductive structure capable of direct development into a new plant; produced in a sporangium.

Sporocarp Structure containing sporangia, as in *Marsilea* and *Azolla*.

Sporophyll Spore-bearing leaf.

Sporophyte Diploid spore-producing plant that alternates with the gametophyte.

Stellate Star-shaped, applied mostly to hairs and scales.

Stipe Petiole or leaf stalk of a fern.

Strobilus Group of sporophylls forming a conelike structure.

sub- Prefix denoting nearly, almost, or somewhat, as "subopposite," more or less opposite.

Subtend To extend under.

Subulate Awl-shaped; tapering to a fine point.

Sulcate Grooved or furrowed longitudinally.

Superior Growing or arising above another organ.

Synonym Two or more names applied to the same taxon only one of which is correct under the International Rules of Nomenclature.

Taxon General term used when referring to any taxonomic element regardless of its level in classification.

Terete Circular in cross section.

Ternate Arranged in threes or divided into three segments.

Tetrahedral Having, or made up of, four faces.

Tetraploid With four times the normal haploid number of chromosomes, usually expressed as $4n$.

Tomentose Densely pubescent with matted woolly or short hairs.

Trapezoid Pertaining to an irregular four-sided figure.

Trigonous Three-angled.

Tripartite Divided into three parts.

Triploid With three times the normal haploid chromosome number, usually expressed as $3n$.

Truncate Ending abruptly at apex or base, not rounded.

Tubercle Wartlike outgrowth or protuberance.

Ultimate segment Last or smallest division of a frond.

Vallecular Pertaining to the region in the stem of *Equisetum* opposite a groove.

Velum The membranous indusium in *Isoetes*.

Venation Character of the veining.

Ventral Upper or inner face of a leaf; opposite side to dorsal.

Vernation Arrangement of leaves in a bud.

Villous With long, silky, but not matted, hairs.

Viscid Sticky.

Whorl Several structures arising at the same level on an axis.

LITERATURE CITED

Alt, K. S., & V. Grant. 1960. Cytotaxonomic observations on the Goldback fern. Brittonia 12:153–69.

Bir, S. S. 1960. Chromosome numbers of some *Equisetum* species from the Netherlands. Acta Bot. Neerl. 9:224–34.

Britton, D. F. 1953. Chromosome studies on ferns. Am. J. Bot. 40:575–83.

Britton, D. M., & J. H. Soper. 1966. The cytology and distribution of *Dryopteris* species in Ontario. Can. J. Bot. 44:63–78.

Brown, D. F. M. 1964. A monographic study of the fern genus *Woodsia*. Nova Hedwigia 16:1–154.

Chiarugi, A. 1960. Tavole cromosomiche delle Pteridophyta. Caryologia 13:27–150.

Clausen, R. T. 1938. A monograph of the Ophioglossaceae. Mem. Torrey Bot. Club 19(2):1–177.

Ewan, J. 1944. Annotations of West American ferns. III. Am. Fern J. 34:107–20.

Fabri, F. 1963. Primo supplemento alle 'Tavole cromosomiche delle Pteridophyta' di Alberto Chiarugi. Caryologia 16:273–335.

Fernald, M. L. 1924. *Polystichum mohrioides* and some other subantarctic or Andean plants in the northern hemisphere. Rhodora 26:89–95.

Gopal-Ayengar. 1957. Origin and behaviour of chiasmata in *Botrychium*. Proc. Indian Sci. Congr. 44:249.

Gupta, K. M. 1957. Some American species of *Marsilea* with special reference to their epidermal and soral characters. Madroño 14:113–27.

Hadac, O., & V. Haskova. 1956. Taxonomickeposnamby o tatranskych rostlinach ve vztahn cytologigii. Biologia 11:717–23.

Hagerup, O., & V. Petersson. 1960. Botanisk atlas, Bd. II. Kobenhavn.

Hauke, R. L. 1958. Is *Equisetum laevigatum* a hybrid? Am. Fern J. 48:68–72.

— 1963. A taxonomic monograph of the genus *Equisetum* subgenus Hippochaete. Beih. Nova Hedvigia 8:1–123.

Knobloch, I. W. 1967. Chromosome numbers in *Cheilanthes*, *Notholaena*, *Llavea* and *Polypodium*. Am. J. Bot. 54:461–4.

Lang, F. A. 1969. A new name for a species of *Polypodium* from Northwestern North America! Madroño 20:53–60.

Lellinger, D. B. 1968. A note on *Aspidotis*. Am. Fern J. 58:140–1.

Löve, A. 1962. Cytotaxonomy of the *Isoetes echinospora* complex. Am. Fern J. 52:113–23.

Löve, A., & D. Löve. 1958. Cytotaxonomy and classification of the Lycopods. Nucleus 1:1–10.

— 1961a. Some chromosome numbers of Icelandic ferns and fern-allies. Am. Fern J. 51:127–8.

— 1961b. Chromosome numbers of central and northwest European plant species. Opera Bot. (Suppl. Bot. Notiser) 5.

— 1964. In I.O.P.B. chromosome number reports. I. Taxon 13:99–110.

Manton, I. 1950. Problems of cytology and evolution in the Pteridophyta. Cambridge University Press.

— 1951. The cytology of *Polypodium* in America. Nature 167:37.

—— 1958. Chromosomes and fern phylogeny with special reference to the "Pteridaceae." J. Linn. Soc. Bot. **56**:73–92.

Manton, I., & M. G. Shivas. 1953. Two cytological forms of *Polypodium virginianum* in eastern North America. Nature **172**:410.

Meyer, D. E. 1952. Untersuchungen uber Bastardierung in Gattung *Asplenium*. Bibliotheca Bot. Heft **123**. Stuttgart.

—— 1959a. *Polystichum* × *eberlei* (*Polystichum braunii* × *lonchitis*) ein neuer Farnbastard. Naturwiss. **46**:237–8.

—— 1959b. Die chromosomenzahl der *Woodsia glabella* R. Br. Mitteleuropas. Willdenowia **2**:214–17.

Morton, C. V. 1950. Notes on the ferns of Eastern United States. Am. Fern J. **40**:213–25.

Niizeki, S., M. Nishida, & S. Kurita. 1963. Cytotaxonomy of Ophioglossales. I. *Japanobotrychium* in Japan. J. Jap. Bot. **38**:144–8.

Ninan, C. A. 1958. Studies on the cytology and phylogeny of the pteridophytes. VI. Observations of the Ophioglossaceae. Cytologia **23**:291–316.

Reese, G. 1951. Erganzende Mittelungen uber die Chromosomenzahlen mittel-europaishischer Gefässpflanzen. I. Ber. Deut. Bot. Gesell. **64**:240–55.

Root, E. E. 1961. Hybrids in North American Gymnocarpiums. Am. Fern J. **51**:15–22.

Roy, R. P., & S. N. Pandey. 1963. Cytotaxonomic studies of the fern flora of Parasnath Hills. Proc. Indian Sci. Congr. **50**.

Shivas, M. G. 1961. Contributions to the cytology and taxonomy of species of *Polypodium* in Europe and America. J. Linn. Soc. Bot. **58**:13–38.

Sorsa, V. 1958. Chromosome studies on Finnish Pteridophyta. I. Hereditas **44**:541–6.

—— 1962. Chromosomenzahlen finnischer Kormophyten. I. Ann. Acad. Scient. Fennica ser. A. IV. Biologica n. **58**:1–14.

—— 1963. Chromosome studies on Finnish Pteridophyta. III. Hereditas **49**:337–44.

Strasburger, E. 1907. Apogamie bei *Marsilia*. Flora Allg. Bot. Zeit. **97**:123–91.

Svenson, H. K. 1944. The new world species of *Azolla*. Am. Fern J. **34**:69–84.

Taylor, T. M. C. 1953. Further observations on the putative hybrid *Dryopteris filix-mas* × *oreopteris*. Am. Fern J. **43**:69–70.

Taylor, T. M. C., & F. A. Lang. 1963. Chromosome counts in some British Columbia ferns. Am. Fern J. **53**:123–6.

Tryon, A. F. 1957. A revision of the fern genus *Pellaea* section Pellaea. Ann. Mo. Bot. Gard. **44**:125–93.

Tryon, A. F., & D. M. Britton. 1958. Cytotaxonomic studies in the genus *Pellaea*. Evolution **12**:137–45.

Tryon, R. M., Jr. 1942. A new *Dryopteris* hybrid. Am. Fern J. **32**:81.

—— 1948. Some Woodsias from the North Shore of Lake Superior. Am. Fern J. **38**:159–70.

—— 1955. *Selaginella rupestris* and its allies. Ann. Mo. Bot. Gard. **42**:1–99.

Verma, S. C. 1956. Cytology of *Ophioglossum*. Current Sci. **25**:398–9.

—— 1958. Cytology of *Ophioglossum vulgatum*. Acta Bot. Neerl. **7**:629–34.

Wagner, W. H., Jr. 1955. Cytotaxonomic observations on North American ferns. Rhodora **57**:219–40.

—— 1956. A natural hybrid, × *Adiantum tracyi* C. C. Hall. Madroño **13**:195–204.

—— 1962. Cytological observations on *Adiantum* × *tracyi* C. C. Hall. Madroño **16**:158–60.

— 1963. A biosystematic survey of the United States ferns – Preliminary Abstract. Am. Fern J. **53**:1–16.

— 1966a. Two new species of ferns from the United States. Am. Fern J. **56**:3–17.

— 1966b. New data on North American oak ferns, *Gymnocarpium*. Rhodora **68**:121–38.

Wagner, W. H., Jr., & K. L. Chen. 1964. In I.O.P.B. Chromosome number reports I. Taxon **13**:99–110.

Wagner, W. H., Jr., D. R. Farrar, & K. L. Chen. 1965. A new sexual form of *Pellaea glabella* var. *glabella* from Missouri. Am. Fern J. **55**:171–8.

Wagner, W. H., Jr., & D. J. Hagenah. 1956. A diploid variety in *Cystopteris fragilis* complex. Rhodora **58**:79–87.

— 1962. *Dryopteris* in the Huron Mountain area of Michigan. Brittonia **14**:90–100.

Wagner, W. H., Jr., & L. P. Lord. 1956. The morphological and cytological distinctness of *Botrychium minganense* and *B. lunaria* in Michigan. Bull. Torrey Bot. Club **83**:261–80.

Wagner, W. H., Jr., & F. S. Wagner. 1966. Pteridophytes of the Mountain Lake area, Giles Co., Virginia: Biosystematic studies 1964–65. Castanea **31**:121–40.

Walker, S. 1959. Cytotaxonomic studies of some American species of *Dryopteris*. Am. Fern J. **49**:104–12.

— 1961. Cytogenetic studies in the *Dryopteris spinulosa* complex. II. Am. J. Bot. **48**:607–14.

INDEX

This book

was designed by

ANTJE LINGNER

under the direction of

ALLAN FLEMING

and was printed by

University of

Toronto

Press

www.ingramcontent.com/pod-product-compliance
Lightning Source LLC
Chambersburg PA
CBHW080556030426
42336CB00019B/3211